Airborne Circularly Polarized SAR

Airborne Circularly Polarized SAR is a comprehensive resource on airborne synthetic aperture radar (SAR) systems. Readers learn how to build the hardware and software of circularly polarized SAR, the antenna system, and how to generate point target responses and images using the Range-Doppler algorithm (RDA) from raw signal data. The book discusses applications and analyzing techniques using a circularly polarized SAR system and image processing. Images and MATLAB® codes are provided to help professionals and researchers with their applications and future studies.

Features:

1. Provides the theory of circularly polarized wave and polarimetry related to system design, scattering analysis, polarimetric SAR, and applications in microwave remote sensing.
2. Explains the real radio frequency (RF) system and the original antenna, including circuit explanation and know-how of measurement technique to adjust to the required parameter in system design.
3. Discusses the technique of ground test and flight mission to calibrate and validate the performance of airborne circularly polarized SAR.
4. Highlights image signal processing with MATLAB codes and how to obtain a single look complex (SLC) image for further applications.
5. Includes several applications of airborne circularly polarized SAR from international leading experts.
6. Provides exclusive access to support materials curated by the authors, including MATLAB codes, a manual for student exercises, and raw radar data sources that can be downloaded from the website.

This book is beneficial to professionals, researchers, academics, and graduate students from disciplines such as electronic engineering, radar systems, aerospace engineering, signal processing, image processing, and environmental remote sensing.

SAR Remote Sensing

A SERIES
Series Editor
Jong-Sen Lee

Polarimetric SAR Imaging
Theory and Applications
By Yoshio Yamaguchi

Imaging from Spaceborne and Airborne SARs, Calibration and Applications
By Masanobu Shimada

Radar Scattering and Imaging of Rough Surfaces
Modeling and Applications with MATLAB®
By Kun-Shan Chen

Radar Remote Sensing of Tropical Forests
Spatial Analysis Techniques
By Gianfranco D. De Grandi and Elsa Carla De Grandi

Airborne Circularly Polarized SAR
Theory, System Design, Hardware Implementation, and Applications
By Josaphat Tetuko Sri Sumantyo, Ming Yam Chua, Cahya Edi Santosa,
and Yuta Izumi

For more information about this series, please visit: www.routledge.com/SAR-Remote-Sensing/book-series/CRCSRS

Airborne Circularly Polarized SAR

Theory, System Design, Hardware Implementation, and Applications

Josaphat Tetuko Sri Sumantyo,
Ming Yam Chua, Cahya Edi Santosa
and Yuta Izumi

CRC Press
Taylor & Francis Group
Boca Raton London New York

CRC Press is an imprint of the
Taylor & Francis Group, an **informa** business

Designed cover image: Cahya Edi Santosa and Josaphat Tetuko Sri Sumantyo

MATLAB® is a trademark of The MathWorks, Inc. and is used with permission. The MathWorks does not warrant the accuracy of the text or exercises in this book. This book's use or discussion of MATLAB® software or related products does not constitute endorsement or sponsorship by The MathWorks of a particular pedagogical approach or particular use of the MATLAB® software.

First edition published 2023
by CRC Press
6000 Broken Sound Parkway NW, Suite 300, Boca Raton, FL 33487-2742

and by CRC Press
4 Park Square, Milton Park, Abingdon, Oxon, OX14 4RN

CRC Press is an imprint of Taylor & Francis Group, LLC

© 2023 Taylor & Francis Group, LLC

Library of Congress Cataloging-in-Publication Data
Names: Sri Sumantyo, Josaphat Tetuko, author. | Chua, Ming Yam, 1980– author. |
Santosa, Cahya Edi, author. | Izumi, Yuta, author.
Title: Airborne circularly polarized SAR : theory, system design, hardware
implementation, and applications / Josaphat Tetuko Sri Sumantyo, Ming
Yam Chua, Cahya Edi Santosa, Yuta Izumi.
Description: Boca Raton : CRC Press, 2023. |
Series: SAR remote sensing | Includes bibliographical references and index.
Identifiers: LCCN 2022052403 (print) | LCCN 2022052404 (ebook) |
ISBN 9781032250038 (hardback) | ISBN 9781032253206 (paperback) |
ISBN 9781003282693 (ebook)
Subjects: LCSH: Synthetic aperture radar. | Polarimetry. |
Environmental monitoring–Remote sensing.
Classification: LCC TK6592.S95 S86 2023 (print) |
LCC TK6592.S95 (ebook) | DDC 621.3848/5–dc23/eng/20221207
LC record available at https://lccn.loc.gov/2022052403
LC ebook record available at https://lccn.loc.gov/2022052404

ISBN: 978-1-032-25003-8 (hbk)
ISBN: 978-1-032-25320-6 (pbk)
ISBN: 978-1-003-28269-3 (ebk)

DOI: 10.1201/9781003282693

Typeset in Times
by Newgen Publishing UK

Access the support materials: https://www.routledge.com/Sri-Sumantyo-Yam-Chua-Santosa-Izumi/p/book/9781032250038

Contents

Preface

Since the synthetic aperture radar (SAR) was invented in 1951 by Carl Atwood Wiley, various segments of the SAR have continuously improved the performance for various polarization, frequency, platforms, and applications for civilian and military purposes. This technology was implemented for civil application, that is, archeology using real aperture interferometry. Later the SAR sensor was developed for spaceborne, airborne, and UAV onboard purposes. Several SAR onboard spaceborne platforms already have been launched, for example, NASA SEASAT, Russian ALMAZ-1, EU ERS-1, ENVISAT, JAXA JERS-1, ALOS, and ALOS-2, CSA Radarsat, DLR TerraSAR-X, China Yaogang series, and ESA Sentinel-1A/1B, among others.

Many countries already have developed airborne SAR sensors with multi-polarization for various purposes, including military applications as Australian DSTO AuSAR – INGARA, Environmental Canada's SAR580, Denmark's DCRS EMISAR, the French UVSQ/CETP STORM and ONERA RAMSES, Germany's InterMap Technologies AES1, EADS/Donier GmbH, Germany's DLR ESAR, FGAN MEMPHIS/AER II-PAMIR, Japanese JAXA/CRL PiSAR, the Netherlands' TNO-FEL PHARUS, and NASA/JPL AIRSAR.

Most conventional systems were designed and operated in linear and compact SAR and no pure circularly polarized SAR was introduced before. Some books partly discussed the segment of (linear polarized) SAR, that is, system, antenna, signal processing, and so forth. This book focuses on the circularly polarized SAR that was not found in the market before. Therefore, this book introduces concise knowledge on the development of airborne circularly polarized SAR, which is composed of the theory of circularly polarized waves, system design, hardware development, image signal processing, calibration and validation, ground test and flight mission, and applications, based on the authors' experience since 2002 to develop the airborne circularly polarized SAR system. This book provides a single comprehensive resource for engineers, researchers, scientists, professionals, and students in related disciplines to boost the knowledge in the hardware and software, as well as the applications of airborne circularly polarized SAR and future planetary missions.

This book comprises ten chapters. Chapter 1 presents the history of airborne synthetic aperture radar (SAR) in various countries, including the airborne circularly polarized SAR, which is explained using various objectives, platforms, systems, polarization, and applications. Some of our publications and others are reviewed in this chapter.

Chapter 2 begins by reviewing the basic knowledge of SAR polarimetry by explaining the theory of the generation, propagation, and scattering of the circularly polarized wave and polarimetry related to the system design, scattering analysis, polarimetric target decomposition, and several polarimetric observation modes.

Chapter 3 considers the system design of the airborne circularly polarized SAR related to the determination of parameters on airborne circularly polarized SAR (power, altitude, off-nadir angle, beamwidth, resolution, etc.).

Chapter 4 covers the SAR radiofrequency (RF) system related to the transmitter, receiver, and controller, including circuit explanation and know-how of measurement techniques to adjust to the required parameter in system design.

Chapter 5 focuses on the baseband and control unit (BCU) that discusses the know-how to build the chirp generator, modulator and demodulator, solid-state power amplifier (SSPA), low-noise amplifier, ADC/DAC, and control unit to generate local oscillator frequency and timing controllers. The installation of the circularly polarized SAR system in the CN235MPA aircraft to fulfill the requirement of system design, improvement of signal-to-noise ratio (SNR), collecting of full polarimetric raw data, GPS, and inertial measurement unit (IMU) is also discussed.

Chapter 6 introduces the design and manufacturing of the circularly polarized antenna including the target, objective, specification, and characteristics to satisfy the requirements of system design. This chapter explains the curved-truncation and circle-slotted parasitic technique to realize broadband, low-axial ratio, and a narrow beamwidth of the circularly polarized microstrip array antenna for airborne circularly polarized SAR.

Chapter 7 explains the ground test, airborne installation, and flight mission that introduces the technique of ground test to calibrate and validate the performance of a fully polarimetric airborne circularly polarized SAR system, antenna, radome, and image signal processing using several types of corner reflectors (CR) and distance between the SAR system to the CR.

Chapter 8 covers SAR image formation or signal processing by explaining the code of circularly polarized synthetic aperture radar with a Range-Doppler algorithm (RDA) to process the signal (raw data) to obtain single-look complex (SLC) images for further application. This chapter introduces several acquired images from in-flight missions in South Celebes, Indonesia, in March 2021 and the MATLAB® codes, where the reader could learn the code for further study of this system.

Chapter 9 introduces the applications with several examples of indoor polarimetric SAR experiments and airborne circularly polarized SAR analysis using axial ratio images, ellipticity, polarization ratio, and polarimetric SAR (PolSAR) techniques. This chapter also explains the experimental result of the scattering of circularly polarized waves on N219 aircraft models and a rice paddy to compare to conventional linear polarized SAR images.

Chapter 10 concludes the book with a summary of the book related to the history, theory, system, applications, trend issues, future research, and applications.

THE MATLAB® CODE AND RAW DATA

The MATLAB code and raw data can be downloaded from the link www.routledge.com/ Sri-Sumantyo-Yam-Chua-Santosa-Izumi/p/book/9781032250038. This link contains the SAR processor operating manual, the SAR processor user interface in MATLAB code, and the raw data of circularly polarized images. Detailed explanations are included in Chapter 8: SAR Image Formation. The users must prepare the MATLAB software by themselves.

Acknowledgements

Building the airborne radar was a dream of Josaphat Tetuko Sri Sumantyo and a promise to his father at the age of five. He developed the circularly polarized synthetic aperture radar (SAR) for unmanned aerial vehicles (UAV; drone), airborne, stratosphere platform, and spaceborne in Josaphat Microwave Remote Sensing Laboratory (JMRSL), Center for Environmental Remote Sensing, Chiba University, in 2002. This book's authors solved many issues and challenges in hardware, software, calibration techniques, applications, and so forth, of the circularly polarized SAR with our excellent researchers, bachelor's, master's, and doctoral students, friends, and colleagues, and collaborated with companies from several countries. We thank the following people who provided us with useful support, assistance, advice, and information to realize the system of airborne circularly polarized SAR and several flight tests in Indonesia.

Michael Suman Juswaljati, Florentina Srindadi, Wolfgang Martin Boerner, Steven Gao, Yoshio Yamaguchi, Ridha Touzi, Kazuo Ouchi, Kengo Tsushima, Rei Kato, Tsutomu Tokifuji, Good Fried Panggabean, Tomoro Watanabe, Bambang Setiadi, Fransiskus Dwikoco Sri Sumantyo, Karna Sasmita, Agus Mardiyanto, Edy Supartono, Eko Tjipto Rahardjo, Gunawan Wibisono, Muh Aris Marfai, Retnadi Heru Jatmiko, Sudaryatno, Taufik Hery Purwanto, Barandi Sapta, Widartono, Muhammad Kamal, Daniele Perissin, Koichi Ito, Lita Kristiani, Ryutaro Tateishi, Hiroaki Kuze, Ryutaro Tateishi, Fumihiko Nishio, Kim Tu-Hwan, Kim Jae-Hyun, Innes Indreswari Soekanto, Johannes Pandhito Panji Herdento, Kim Hyun-Joo, Agus Supriyatna, Hadi Tjahjanto, Yuyu Sutisna, Supomo, Sri Pulung, Novyan Samyoga, Hilman Lamhot Parasian Ambarita, Shobirin, Ming-Hwang Allen Shie, Chih-Li Chang, Bor-Han Wu, Shiou-Li Chen, Kyohei Suto, Kazuteru Namba, Yohandri, Mohd Zafri Baharuddin, Heein Yang, Joko Widodo, Katia Nagamine Urata, Sevket Demirci, Ayaka Takahashi, Isamu Nagano, Anna Hariyatiningsih, and Coco Innocentia.

This book is dedicated to the Working Group on Remote Sensing Instruments for Unmanned Aerial Vehicles, Technical Committee on Instrumentation and Future Technologies (IFT-TC) of IEEE GRSS to promote future technologies in remote sensing, especially synthetic aperture radar.

We also give great thanks for the continuous support of Jong Sen Lee and Irma Shagla Britton from the proposal stage until publishing, and to Elspeth Mendes, Emily Bonden, Iris Fahrer, and Chelsea Reeves for the kind and detailed proofreading of this book.

For MATLAB® and Simulink® product information, please contact:
The MathWorks, Inc.
3 Apple Hill Drive
Natick, MA, 01760-2098 USA
Tel: 508-647-7000
Fax: 508-647-7001
E-mail: info@mathworks.com
Web: www.mathworks.com

About the Authors

Josaphat Tetuko Sri Sumantyo, PhD, was born in Bandung, Indonesia, in 1970. He earned BEng and MEng degrees in electrical and computer engineering (subsurface radar systems) from Kanazawa University, Japan, in 1995 and 1997, respectively, and his PhD in artificial system sciences (applied radio wave and radar systems) from Chiba University, Japan, in 2002. From 2002 to 2005, he was a lecturer (postdoctoral fellowship researcher) with the Center for Frontier Electronics and Photonics, Chiba University. From 2005 to 2013, he was an associate professor (permanent staff) with the Center for Environmental Remote Sensing, Chiba University, where he is currently a professor (permanent staff). He is also on the academic staff in the Department of Electrical Engineering, Faculty of Engineering, Universitas Sebelas Maret, Indonesia, since 2020. He has been the head of the department of Environmental Remote Sensing, Head Division of Earth and Environmental Sciences, Graduate School of Integrated Science and Technology since 2019, and head of the Division of Disaster Data Analysis, Research Institute of Disaster Medicine, Faculty of Medicine, Chiba University, since 2020. He has authored more than 1,000 papers published in journals and conference proceedings, and 16 books on theoretical wave analysis, synthetic aperture radar (SAR), unmanned aerial vehicles (UAV), microsatellites, and small antennas. His research interests include theoretically scattering microwave analysis and its applications in the microwave (radar) remote sensing, especially SAR and subsurface radar (VLF), including InSAR, DInSAR, and PS-InSAR, analysis, and design of antennas for mobile satellite communications and microwave sensors, development of microwave sensors, including SAR for the UAV, aircraft, stratosphere platform, and microsatellite. Prof. Sri Sumantyo is the General Chair of the 7th and 8th Asia-Pacific Conference on Synthetic Aperture Radar (APSAR 2021 and APSAR 2023) in Bali, Indonesia, and has given more than 300 invited talks and lectures related to SAR and its applications. He is the co-leader of the Working Group on Remote Sensing Instrumentation and Technologies for Unmanned Aerial Vehicles of the IEEE Geoscience and Remote Sensing Society (GRSS) Technical Committee on Instrumentation and Future Technologies (IFT-TC), and associate editor of the *IEEE Geoscience and Remote Sensing Letters* (GRSL), and *IEEE Antennas and Wireless Propagation Letters* (AWPL).

Ming Yam Chua, PhD, was born in Malacca, Malaysia, in 1980. He earned his BEng (Hons.) in electronics and MEngSc and PhD in engineering, focusing on radar system design and radar waveform synthesis techniques from Multimedia University, Malaysia, in 2003, 2007, and 2016, respectively. He was with Multimedia University as a Research Associate from 2003 to 2009, lecturer from 2009 to 2012, senior lecturer from 2012 to 2022, and is currently an assistant professor at the School of Electrical Engineering and Artificial Intelligence, Xiamen University, Malaysia. Dr. Chua completed a 23-month postdoctoral research fellowship from November 2016 to September 2017 at the Center for Environmental Remote Sensing, Chiba University, Japan, working on developing synthetic aperture radar systems and conducting flight missions in Japan and Indonesia. Dr. Chua has delivered numerous grants and consultancy projects in developing SAR systems, developing FPGA applications for radar systems, and human sensing techniques.

Cahya Edi Santosa, PhD, earned a bachelor's degree in electrical engineering from Gadjah Mada University, Yogyakarta, Indonesia, in 2003, a master's degree in electrical engineering from the Bandung Institute of Technology, Bandung, Indonesia, in 2013, and a PhD in information processing and computer science from Chiba University, Japan, in 2019. From 2003 to 2021, he was a researcher with the National Institute of Aeronautics and Space of Indonesia (LAPAN), Bogor, Indonesia. He was a research associate (postdoctoral research fellowship) with the Center of Environmental and Remote Sensing, Chiba University, from 2019 to 2021. Currently, he is a senior researcher at the Research Center for Aeronautics Technology, National Research and Innovation Agency BRIN, Jakarta. His primary interest includes analysis of broadband microstrip array antenna, conformal antenna for mobile communication, circularly polarized horn antenna, and research on synthetic aperture radar onboard UAVs, airborne, and satellite. Dr. Santosa is a member of the IEEE Transaction Antenna and Propagation Society and the Institute of Electronics, Information, and Communication Engineers of Japan. He was a recipient of the National Institute of Aeronautics and Space of Indonesia (LAPAN) award for Best Young Researcher in 2008 for his outstanding PhD, the Excellence Academic Award from Chiba University in 2018, and several awards related to his research.

Yuta Izumi, PhD, earned his BEng and MEng from Chiba University, Japan, in 2016 and 2018, respectively, and his PhD in radar remote sensing from Tohoku University, Japan, in 2021. From 2018 to 2020, he was a JSPS research fellow DC1. From 2019 to 2020, he worked at the Swiss Federal Institute of Technology (ETH) Zurich, Switzerland. From 2021 to 2022, he was a JSPS research fellow (postdoctoral) at the Institute of Industrial Science, the University of Tokyo. He is currently an assistant professor at Muroran Institute of Technology. His research interests include SAR interferometry and polarimetry and their applications for disaster mitigation. Dr. Izumi is a recipient of the Student Paper Award at the ISRS in 2017, President Award from Chiba University in 2018, Dean Award from the Graduate School of Advanced Integration in 2018, Science Young Researcher Award from IEEE Geoscience and Remote Sensing Society All Japan Joint Chapter in 2019, and the Student Poster Award at the IEICE-AP in 2020.

1 Introduction of Synthetic Aperture Radar

Josaphat Tetuko Sri Sumantyo

1.1 INTRODUCTION

Synthetic aperture radar (SAR) is well-known as a multi-purpose sensor that can be operated in all-weather conditions, day and night [1]–[3], which has more benefits than optical remote sensing in cloudy and rainy areas. Previously, various linearly polarized SARs have been developed for satellites as well as airborne and unmanned aerial vehicles (UAV) for different applications, including ground-based radar [4]–[8].

The general operation modes of SAR are known as the traditional single-polarized (SP), dual-polarized (DP), and full-polarized (FP) SAR. SP transmits a single polarized wave, then receives it. FP mode transmits two orthogonally polarized waves at a separate time and simultaneously receives each echo wave in two orthogonal polarizations. DP transmits on one polarization and receives on two polarizations that are the same and orthogonal to it at the same time. FP mode obtains the complete scattered mode to construct the scattering matrix compared to SP and DP modes. However, the FP mode needs two linearly or circularly polarized transmitters, horizontal (H) and vertical (V), or right-handed (RH) and left-handed (LF) polarization, and also needs a complex system, including an accurate time controller for pulse generation and receiving, power consumption, and system power consumption. The pulse repetition frequency (PRF) needs two systems and the coverage is half of the DP mode.

Nowadays, compact polarimetric SAR offers an option for polarimetric SAR (PolSAR) applications by considering the need for a compromise among swath coverage, hardware requirements, and scattering information of measured scenes [9]. There are essentially two ways to investigate the polarization information available from compact polarimetric SAR images using information resulting from HH, VV, VH, and HV channels that cannot be obtained directly from compact polarimetric SAR [10]. The first approach is the reconstruction of a pseudo-quadrature polarization (QP) signature from compact polarimetric SAR using the QP data in existing algorithms. The second approach is focused on using the backscatter parameters to obtain polarimetric information directly from the hybrid polarization scattering vectors.

As a special DP configuration, compact polarimetric SAR transmits signals in a special polarization, such as a 45° linear polarization or a circular polarization, and receives echoes simultaneously in two orthogonal polarizations, such as horizontal

DOI: 10.1201/9781003282693-1

1

polarization (HP) and vertical polarizations (VP), or right-handed circular polarization (RHCP) and left-handed circular polarization (LHCP). Such systems can provide a greater amount of polarization information than the standard DP linear systems while covering twice the swath width of conventional QP systems and achieving the benefits (e.g., large ranges of feasible incidence angles and less energy budget) that QP systems cannot provide [9][11].

In the application assessment, the land-use and land-cover classifications from compact polarimetric SAR achieve accuracy levels comparable to those from a QP SAR to within a few percent, but compact polarimetric SAR has the advantages of large-scale coverage and compact data volume [12]–[14]. compact polarimetric SAR has been considered a suitable option for spaceborne radar, such as RISAT-1, ALOS-2, and the future RADARSAT constellation. This significantly increases the interest of the remote sensing community in the compact polarimetric SAR domain and poses new scientific and technological problems involving radar systems, data interpretation, performance assessment, and practical applications.

To get undistorted polarimetric information, the calibration technique of compact polarimetric measurements must be proposed by several researchers [15]–[19]. First, Freeman and Chen proposed methods for compact polarimetric SAR calibration merely using calibrators, which require at least three calibrators at a range line and would require many more calibrators when monitoring the change of the system across the swath is necessary [15]–[16]. It has been proved that the fully polarimetric (FP) SAR calibration methods exploring scattering characteristics of natural distributed targets (DTs) can reduce the requirement of the number of deployed calibrators. However, it is much more complicated in the case of compact polarimetric (CP) SAR calibration, as only two channels of CP measurements are available. Second, a numerical compact polarimetric calibration method using natural DTs and only one corner reflector (CR) for a reciprocal compact polarimetric SAR system was proposed by Tan et. al [17]. The validity of the method for three typical compact polarimetric modes is confirmed by simulations with ALOS-2 PALSAR-2 data, where sensitivity analysis, a dihedral at 0°, is selected as the best CR for the calibration algorithm. Third, the compact polarimetric calibration of the RISAT-1 dataset using a combination of trihedral and dihedral CRs with the Cloude compact polarimetric decomposition technique was employed to assess the ground target characterization quality of the dataset before and after polarimetric calibration to improve the system-induced polarimetric distortions, which include the channel imbalances, phase bias, and crosstalk between the channels, and the ionospheric distortion [18]. Fourth, transmit wave polarization (i.e., ellipticity) deviation from the ideal or intended to transmit mode affects compact polarimetric products and associated data exploitation. Sabry [19] evaluated this effect on the compact polarimetric model and accounted for the associated variations through data exploitations by proposing an approach to model and estimate ellipticity variations for calibration by taking advantage of a variational model and adjoint polarimetric SAR data using Radarsat-2 data.

As introduced above, several existing compact polarimetric systems have been developed to generate compact polarimetric SAR data, and calibration techniques were also proposed and developed by using linear polarization and combinations [15]–[21]; therefore in this book, we introduce a broadband (maximum 400 MHz bandwidth) C-band circularly polarized SAR system to improve the resolution and enrich the information (spatial, frequency, and time domains) of existing SAR system, especially circularly polarized SAR systems and preliminary investigation of the calibration technique of circularly polarized SAR using conventional corner reflector for further investigation of calibration techniques of circularly polarized SAR. We selected the "Compromise" C-band (5.3 GHz) in this book for land deformation and environmental remote sensing to collect information on the land surface and vegetation simultaneously.

1.2 HISTORY OF SYNTHETIC APERTURE RADAR

Referring to [22]–[24], radar had invented and proved as a defensive tool in World War II to identify ships and planes in darkness and through cloud cover, haze, and fog. Therefore, it promised to be a tool for all-weather aerial reconnaissance, supplementing traditional photography. However, radar technology in the 1940s and 1950s had an issue with providing the high resolution required, so that the aircraft as the flying platform had to carry a large antenna almost as big as a football field. In 1950, Carl Wiley started to work at Goodyear Aircraft (now Lockheed Martin) to challenge this problem for airborne radars. At the same time, Motorola developed the Boeing 737 onboard Side Looking Airborne Modular Multi-Mission Radar (SLAMMR) AN/APS-94 for ship and flying targets in the 1950s as the real aperture radar (RAR). Figure 1.1 shows the SLAMMR installed Boeing 737–200 and an example of a picture (bottom) captured by SLAMMR. The author improved the SLAMMR and the same picture (center above) shows the result.

 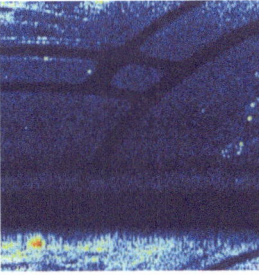 

Boeing 737-200 with SLAMMR   Improved SLAMMR's image   Optic image for comparison

SLAMMR's images of Java island in 1980s

FIGURE 1.1   An example of a picture captured by RAR of SLAMMR.

Wiley's breakthrough came in 1951 with the realization that radar reflections from discrete objects in a passing radar beam field each had a minute Doppler, or speed, shift relative to the antenna. By precisely analyzing the frequency of the return signal, Wiley found he could create a detailed radar image using an antenna that was 1/100th the size of the football field-long antenna traditional radar required. Wiley's patent called "Simultaneous Buildup Doppler" in 1954 is considered the first SAR patent. Thereafter, the teams at the University of Illinois and the University of Michigan developed the same concept of processing signals to create a "synthetic aperture," which sharpened and focused the returning radar signal.

Lockheed Martin continued to lead SAR development at the end of the 1950s when the system called Douser was deployed on a C-47 as the first SAR to create an image with only 500 feet in swath width. In 1960, the next version of SAR was improved with a resolution of 50 feet, and the SAR system was developed for SR-71 reconnaissance aircraft with a resolution of 30 feet. By the latter part of the 1960s, Goodyear SAR systems were much improved and operating with a resolution of one foot. Another innovation developed in the 1960s was a foliage-penetrating SAR named FOPEN to distinguish the object structures from tree trunks and it still plays a role in today's advanced airborne weaponry.

The development of SAR technology for non-military purposes began in the 1970s. Lockheed Martin developed the SAR for mapping systems to map areas all over the world, especially areas along the equator that was often covered by clouds. In 1974, the National Oceanic and Atmospheric Administration and the Jet Propulsion Laboratories developed the spaceborne SAR for oceanic observations, called the Seasat satellite, which was launched in 1978 and started the competition in the world for remote sensing from space. It improved the optical cameras onboard earlier Landsat satellites that were not able to see through cloud cover and had resolutions in the tens of meters, compared to less than a meter for Seasat. Further, many SAR onboard satellites were launched for various missions.

1.3   SPACEBORNE SYNTHETIC APERTURE RADAR

Several SAR onboard spaceborne platforms already have been launched as shown in Figure 1.2 and Table 1.1, for example, the US government with NASA considered the benefit of SAR sensors that could penetrate clouds and observe the sea surface, and the US developed and launched the first SAR onboard satellite called Seasat [25] in 1978 that works on L-band frequency (1.27 GHz) with a resolution of 25m.

The Cosmos-1870 satellite mission [26] was the first USSR radar demonstration mission that was launched on 25 July 1987. The spacecraft had a mass of 18,550kg, payload mass of 1,950kg, mean altitude of 275km, an inclination of 73°, and an orbital period of 92 minutes. Ekor-A (Sword), the sidelooking SAR instrument named Ekor-A in S-band has a center frequency of 3.125 GHz and HH polarization. The observation data provided a spatial resolution of about 25–30m on a swath of 20km. The SAR data was recorded by video tape recorder and real-time data transmission with a transmission rate of 90Mbit/s.

The Almaz program (Russian for "diamond") [27], a military space program of the former Soviet Union, had its origin in 1964. The Almaz-1 spacecraft was launched

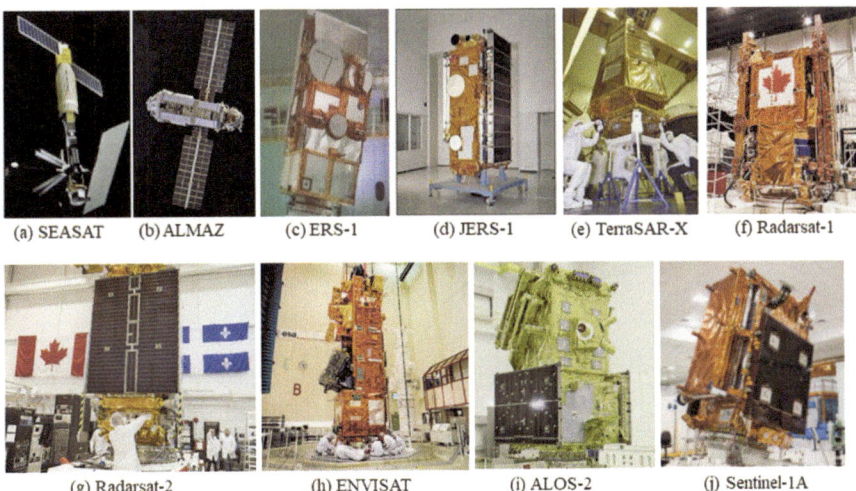

(a) SEASAT   (b) ALMAZ   (c) ERS-1   (d) JERS-1   (e) TerraSAR-X   (f) Radarsat-1

(g) Radarsat-2   (h) ENVISAT   (i) ALOS-2   (j) Sentinel-1A

FIGURE 1.2   Spaceborne synthetic aperture radar.

on 31 March 1991 with Ekor-A (Sword), an SAR instrument in S-band (center frequency = 3.125 GHz). The satellite has a total mass of 18,550kg and a payload mass of 3,420kg. The SAR sensor has an output with an average power of 2.4kW, a peak power of 7.5–10kW, and two three-panel SAR antennas (1.5m x 15m). The Almaz-1 orbit is non-sun-synchronous circular orbit, reaching an altitude of 270–380km, with an inclination of 72.7°, and orbital period of 92 minutes. In 1991, Russia or the former Soviets launched Almaz-1. During the same period, the European Union launched the ERS-1 (C-band, 5.3 GHz) and ENVISAT (C-band, 5.331 GHz) in 1991 for Earth resources monitoring, with Japan following with Japanese Earth Resources observation satellite (JERS-1) in 1992 [28]–[30].

ALOS is a Japanese Earth-observation satellite [31], developed by JAXA (Japan Aerospace Exploration Agency) that was launched on 24 January 2006. The orbit is a sun-synchronous near-recursive circular orbit, reaching an altitude of 691.65km, inclination of 98.16°, repeat cycle of 46 days, local time at descending node of 10.30am (±15 min), period of 98.51 minutes, and the orbits/day of 14 27/46. Phased Array L-band Synthetic Aperture Radar (PALSAR) was the heritage of SAR on JERS-1. PALSAR is a sidelooking phased array L-band instrument with a pointing capability from 8° to 60° of incidence angle. The SAR antenna is 8.9m (length) and 3.1m (width). An array of 80 transmitting/receiving modules (T/R) is mounted behind the antenna panels to provide the electrical beam steering in elevation. PALSAR has main operational modes – fine resolution beam (FB), polarimetry, ScanSAR, and direct transmission modes. FB mode comprises 18 selections in the off-nadir angle range between 9.9° and 50.8°, each with four alternative polarizations: single polarization HH or VV, and dual polarization HH+HV or VV+VH. The bandwidth is 28MHz in single polarization and 14MHz in the dual-polarization modes. Out of the 72 possible FB modes, two have been selected for operational use. The polarimetric mode employed 14MHz

TABLE 1.1

Specifications of Spaceborne Synthetic Aperture Radar

Name	Country	Launched Year	Inclination Angle (degree)	Repetition Days	Weight (kg)	Height (km)	Frequency (GHz)
Seasat	USA	1978	108.0	17	2,300	760	1.275
Cosmos-1870	USSR	1987	73.0	-	1,950	275	3.125
Almaz-1	USSR	1991	72.7	-	6,500	270-380	3.125
ERS-1	EU	1991	98.5	3, 35, 176	2,157	785	5.300
JERS-1	Japan	1992	97.7	44	1,340	568	1.275
Radarsat	Canada	1995	98.6	24	1,540	793-821	5.300
Envisat	EU	2002	98.55	35	8,211	799.8	5.331
ALOS	Japan	2006	98.16	46	4,000	691.65	1.270
Terra SAR-X	Germany	2007	97.44	11	1,250	475-525	9.650
COSMO-SkyMed-1	Italy	2007	97.86	16	1,700	619.6	9.6
COSMO-SkyMed-2	Italy	2007	97.86	16	1,700	619.6	9.6
COSMO-SkyMed-3	Italy	2008	97.86	16	1,700	619.6	9.6

Polarization	Swath Width (km)	Resolution		PRF (Hz)	Band width (MHz)	Peak Power (W)	Antenna Size (m)
		Azimuth (m)	Range (m)				
HH	100	25	25	1,463-1,640	19	1,000	2.16x10.74
HH	20	25-30	25-30	-	-	-	-
HH	35-55	15	15-30	-	-	7,500-10,000	15x15
VV	100	30	30	1,640-1,720	19	4,800	1x10
HH	75	18	18	1,505.8, 1,530.1, 1,555.2, 1,581.1, 1,606.0	15	325 (spec 1,300)	2.2x11.9
HH	50-500	9-147	6-147	1,361	11.6/17.3/ 30.0	5,000	1.5x15
HH,VV, HH+VV, VV+VH, HH+HV	56.5-104.8	30-1,000	30-1,000	1,650-2,100	8.48-16	1,400	1.3-8.9
HH,VV, HH+HV, VV+VH, HH+VV +HV+VH	20-350	10-100 7-100	10-100 7-100	2,776	14/28	2,300	3.1x8.9
HH+ VV, HH+ HV, VV+HV	Along 5km Across 15,30, 100	1,2,3, 15	1.2, 1.2, 3, 16	3,000-6,500	5-300	2,260	4.78x0.70
HH, VV, HV, VH	10-200	1 3 5 15 30 100	1 3 5 15 30 100	-	400	14,000	5.7x1.4
HH, VV, HV, VH	10-200	1 3 5 15 30 100	1 3 5 15 30 100	-	400	14,000	5.7x1.4
HH, VV, HV, VH	10-200	1 3 5 15 30 100	1 3 5 15 30 100	-	400	14,000	5.7x1.4

(continued)

TABLE 1.1 (Continued)
Specifications of Spaceborne Synthetic Aperture Radar

Name	Country	Launched Year	Inclination Angle (degree)	Repetition Days	Weight (kg)	Height (km)	Frequency (GHz)
COSMO-SkyMed-4	Italy	2010	97.86	16	1,700	619.6	9.6
ALOS-2	Japan	2014	97.9	14	656.8	628	1257.5, 1236.5 1278.5
Sentinel-1A	EU	2014	98.2	12	2,300	693	5.405
Sentinel-1B	EU	2016	98.2	12	2,300	693	5.405
SAOCOM-1A	Argentina	2018	97.86	16	3,050	619.6	1,275
SAOCOM-1B	Argentina	2020	97.86	16	3,050	619.6	1,275

Polarization	Swath Width (km)	Resolution		PRF (Hz)	Band width (MHz)	Peak Power (W)	Antenna Size (m)
		Azimuth (m)	Range (m)				
HH, VV, HV, VH	10-200	1 3 5 15 30 100	1 3 5 15 30 100	-	400	14,000	5.7x1.4
HH/VV/ HV, HH + HV,VV + VH, HH + HV + VV + VH	25-350	1,3, 6, 10,100	3,6,10, 100	1,500- 3,000	84/42/28/ 14	3,300 -6,100	
HH+HV, VV+VH, HH	20 80 240 400	20 5 5 25	5 5 20 80	1,000-3,000	0-100 MHz	4,368-4,075	12.3x 0.821
HH+HV, VV+VH, HH	20 80 240 400	20 5 5 25	5 5 20 80	1,000-3,000	0-100 MHz	4,368-4,075	12.3x 0.821
HH, VV, HH+HV, VV+VH	65 170 320	10 25 50 100	10 25 50 100	-	<45 MHz	3,100	10x3.5
HH, VV, HH+HV, VV+VH	65 170 320	10 25 50 100	10 25 50 100	-	<45 MHz	3,100	10x3.5

of bandwidth for the full quad-polarization (HH+HV+VH+VV) scattering matrix with 12 alternative off-nadir angles between 9.7° and 26.2°. The ScanSAR mode is available at a single polarization only (HH or VV) and can be operated with three, four, or five sub-beams transmitted in short (14MHz) or long bursts (28MHz). Out of the 12 ScanSAR modes available, the sort-burst, HH polarization, and five-beam modes have been selected for operational support. It features a 350km swath width with an incidence angle range of 18–43°. Finally, the direct transmission (or downlink) mode is a contingency backup mode that allows the downlink of the (degraded, 14MHz) FB mode data to local ground stations in case the high-speed DRTS (Data Relay and Test Satellite) becomes unavailable.

ALOS-2 [31]–[32] was launched on 24 May 2014, with a sun-synchronous near-circular sub-recurrent orbit, altitude of 628km, inclination of 97.9°, period of 97.4 minutes, revisit time of 14 days, the number of orbits per day at 15 per 3/14, and the LSDN (Local Sun Time on Descending Node) of 12pm ± 15 min. ALOS-2 PALSAR-2 operated the center frequencies of either 1257.5, 1236.5, or 1278.5MHz (L-band) with the selected bandwidth at either 84MHz, 42MHz, 28MHz, or 14MHz. The antenna of ALOS-2 is an active phased array antenna that steers the beam in both elevation and azimuth direction (±30° in elevation and ±3.5° in azimuth). It has a size of 9.9m (azimuth) x 2.9m (elevation) and is composed of five electrical panels. The antenna consists of 1,080 radiation elements that are driven by 180 TRMs (Transmit and Receive Modules). The design enables to steer and form the beam in elevation and azimuth directions for several imaging modes: Stripmap, Spotlight, and ScanSAR. The full aperture (five panels) or partial aperture (three of five panels, No 2, 3, and 4) of the antenna aperture may be used for signal transmission (Tx). The peak radiation power is 3,300W with three panels for spotlight mode and ultra-fine mode, or 5,100W with the full aperture for high sensitivity mode, Fine mode, and ScanSAR mode. The transmitter-receiver module (TRM) enables selection of the polarization of single (HH/VV/HV), dual (HH + HV=VV + VH), quad (HH + HV + VV + VH), and compact polarimetry (Tx: oriented 45° or circular, Rx: H or V) by transmitting H and V polarization simultaneously. The SAR instrument features a compact polarimetry mode as an experimental mode that can transmit the H and V polarization simultaneously resulting in a linear polarization oriented at 45° or circular (LHCP or RHCP), selectable by command.

This led to Radarsat (C-band 5.3GHz) for land surface and vegetation monitoring [33]–[34], TerraSAR-X (X-band, 9.65 GHz) [35] for multi-purpose monitoring, and in China with the Yaogang series of SAR onboard satellites for military purposes, among others.

1.4 AIRBORNE SYNTHETIC APERTURE RADAR

As shown in Figure 1.3 and Table 1.2, many countries already have developed aircraft onboard SAR sensors with multi-polarization for various purposes, including military applications.

The Australian DSTO developed AuSAR – INGARA using X-band (quadrature/full polarization) onboard DC3 (97) KingAir 350 (00) Beach 1900C. Then Environmental Canada developed SAR580 onboard Convair CV-580 using C and

EMISAR; L and C bands
(DCRS)

DOSAR; S,C,X and K bands
(EADS / Donier GmbH)

DLR; P. L. S. C and X bands
(DLR)

MEMPHIS/AERII-PAMIR
Ka, W and X bands
(UVSQ / CETP)

NASA; P,L and C bands
(JPL)

AuSAR-INGARA; X band
(DSTO)

AES1; X and P bands
(InterMap Technologies)

RAMSES; P, L, S, X, Ku, Ka,
and W bands (ONERA)

PHARUS; C band
(TNO-FEL)

PISAR; L and X bands
(Mitsubishi Electric)

SAR580; C and X bands
(Environmental Canada)

CN235MPA; L and C bands
(JMRSL Chiba University)

FIGURE 1.3   Synthetic aperture radar onboard aircraft.

X-bands (quadrature/full polarization), and Denmark's DCRS that operates the EMISAR using L and C-bands (quadrature/full polarization) using G3 Aircraft. France UVSQ/CETP developed STORM onboard Merlin IV that works on C-band (quadrature/full polarization). ONERA also developed RAMSES onboard Transal C160 that works on P, L, S, C, X, Ku, Ka, and W-bands (quadrature/full polarization). Germany InterMap Technologies developed AES1 using Gulfstream Commander aircraft for X-band (Horizontal-Horizontal, HH) and P-band (quadrature/full polarization); EADS/Donier GmbH developed S, C, X-bands (quadrature) and Ka-band (vertical-vertical, VV) using DO228(89), C160 (98) and G222 aircraft; DLR developed ESAR using DO 228 aircraft for P, L, S-bands (quadrature) and C, X-bands (single polarization) SAR sensors; and FGAN using Transal C160 for MEMPHIS/AER II-PAMIR for Ka, W, and X-band (quadrature polarization).

The Japanese NASDA (now JAXA) and CRL (now NICT) developed PiSAR onboard Gulfstream with L and X-bands (quadrature/full polarization) SAR. Josaphat Microwave Remote Sensing Laboratory (JMRSL) in the Center for Environmental Remote Sensing, Chiba University, developed the airborne circularly polarized SAR with L, C, and X-bands. The Netherlands' TNO-FEL developed PHARUS onboard CESSNA – Citation II using C-band (Quadrature/Polarization), and the United States, NASA, and JPL developed AIRSAR onboard DC 8 for P, L, and C-bands (quadrature/full polarization).

1.5   UNMANNED AERIAL VEHICLE ONBOARD SYNTHETIC APERTURE RADAR

As shown in Figure 1.4, low-cost and light-weighted SAR onboard UAVs for various purposes has also been developed by several countries, especially for multi-purposes as listed below. Nowadays, many new SARs onboard UAVs were proposed by many institutions and companies, where the user could get them at a low cost.

TABLE 1.2
Specification of Airborne Synthetic Aperture Radar

Name	Institution	Country	Platform	Frequency Band	Polarization *)
AuSAR - INGARA	DSTO	Australia	DC3 (97) KingAir 350 (00) Beach 1900C	X	FP
SAR580	Environmental Canada	Canada	Convair CV-580	C, X	FP
EMISAR	DCRS	Denmark	G3	L, C	FP
STORM	UVSQ / CETP	France	Merlin IV	C	FP
RAMSES	ONERA	France	Transal C160	P, L, S, C, X, Ku, Ka, and W	FP
AES1	InterMap Technologies	Germany	Gulfstream Commander	X P	SP (HH) FP
EADS	Donier GmbH	Germany	DO228(89) C160 (98) G222	S, C, X Ka	FP SP (VV)
ESAR	DLR	Germany	DO 228	P, L, S C, X	FP SP
MEMPHIS/AER II-PAMIR	FGAN	Germany	Transal C160	Ka, W, X	FP
PiSAR	JAXA	Japan	Gulfstream	L, X	FP
PHARUS	TNO-FEL	Netherland	CESSNA – Citation II	C	FP
Hinotori (Firebird)	JMRSL Chiba University	Japan	Cessna 172 CN235MPA Boeing 737-200	L, X C	SP FP
AIRSAR	NASA/JPL	United States	DC 8	P, L, C	FP
MITL SAR	Chinese Academy of Sciences	China	-	P	FP

*) SP – SP – Single Polarization, FP – Full Polarization, DP – Dual Polarization, CP – Compact Polarization

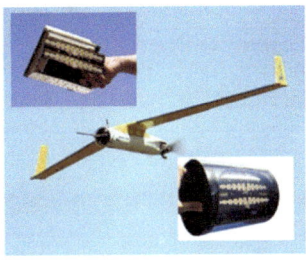

NanoSAR for ScanEagle UAV; Ku, X, UWB, and UHF bands

MicroSAR (C and L bands) and NuSAR (L and X bands)

ARBRES (C and X bands)

JMRSL JX Series UAV (L and C bands)

FIGURE 1.4 SAR onboard Unmanned Aerial Vehicle (UAV).

IMSAR developed NanoSAR C with ground resolution of 0.2, 0.5, 1, 2, 5 and 10m, multi-modes (stripmap, spotlight, circular and MTI, serial and ethernet communication, power of 25–70W, and weight from 0.45kg to 1.4kg [36]. The Space Dynamics Laboratory (SDL) of Utah State University joined the research undertaken by Naval Research Laboratory (NRL), Brigham Young University (BYU), and Artemis Inc to develop MicroSAR and NuSAR [37]. The specification of MicroSAR and NuSAR as the MicroSAR are: this system works in the C-band (5.4GHz) and L-band (1.75GHz), with output power of 1W, 1m resolution, multi-modes (stripmap, bistatic, continuous-wave design), less than 2 kg, and raw data stored to flash drive or output via ethernet. The NuSAR is a high-performance UAV SAR with L (1.75GHz) and X-band (9.75GHz), 25W transmitter, 33cm ground resolution, full real-time SAR image formation, ethernet output in NITF image format/raw I and Q data, and is available to operate in multi-modes (Stripmap, monostatic, and pulsed design modes). The Remote Sensing Laboratory, Department of Signal Theory and Communications of Universitat Politecnica de Catalunya developed lightweight CW/FM SAR sensors for small UAVs called ARBRES [38]. This system works in C (5.3GHz, 50MHz bandwidth) for <3m resolution and X-bands (9.65GHz, 100MHz bandwidth) for <1.5m resolution with a mass of less than 5kg. JMRSL CEReS developed the Circularly Polarized Synthetic Aperture Radar onboard the Josaphat Laboratory Ground Experimental Unmanned Aerial Vehicle (JX series) [39] using L-band (1.27GHz, <150 MHz bandwidth), altitude 2,000m, output power 50W (47 dBm), with 1m resolution. JMRSL also developed a C-band SAR system with 200–400 MHz and output peak power of about 400W with a resolution of 37.5cm.

1.6 FOCUS DISCUSSIONS IN THIS BOOK

This book will focus on the theory, design, system building, system testing, ground test, flight mission, and applications of airborne circularly polarized SAR. First, we discuss the history of airborne SAR in various countries, including the airborne and UAV circularly polarized SAR, which was explained in this chapter with various objectives, platforms, systems, polarization, and applications. Then, we introduce the basic knowledge of SAR Polarimetry by explaining the theory of the generation, propagation, and scattering of the circularly polarized wave and polarimetry related to the system design, scattering analysis, polarimetric SAR, and the applications of airborne circularly polarized SAR. The system design of the airborne circularly polarized SAR is related to determining parameters of airborne circularly polarized SAR (e.g., power, altitude, off-nadir angle, beamwidth, etc.). Further, the Radiofrequency (RF) system related to the transmitter, receiver, and controller is discussed including circuit explanation and know-how of measurement techniques to adjust to the required parameter in system design. The chapter related to the Baseband and Control Unit will explain the know-how to build the chirp generator, modulator and demodulator, solid-state power amplifier (SSPA), low noise amplifier, ADC/DAC, and control unit to generate local oscillator frequency and timing controller; how to install the circularly polarized SAR system in the CN235MPA aircraft to fulfill the system design requirement, improve the signal-to-noise ratio (SNR), and collect full polarimetric raw data; GPS and inertial measurement unit (IMU) are also discussed in this chapter. We learn about the design and manufacturing of the circularly polarized antenna in this book, including the target, objective, specification, and characteristics to satisfy the system design requirement. Here, we propose the curved-truncation and circle-slotted parasitic technique to realize broadband, a low-axial ratio, and a narrow beamwidth of the circularly polarized microstrip array antenna for airborne circularly polarized SAR.

After the integration of the RF system, antenna, GPS, IMU, and so forth, we will explain the ground test, airborne installation, and flight mission that introduces the technique of ground test to calibrate and validate the performance of a full polarimetric airborne circularly polarized SAR system, antenna, radome, and image signal processing using several types of the corner reflector (CR) and distance between the SAR system to the CR.

Another important part of SAR is image processing. We will share the raw images and the code of circularly polarized SAR with a Range-Doppler algorithm (RDA) to process the signal (raw data) to obtain single-look complex (SLC) images for further application. This book introduces several acquired images from in-flight missions in South Celebes, Indonesia, in March 2018, as well as the MATLAB codes, where readers, especially students and young scientists, can learn the code and study this system further.

Finally, we also introduce applications with several examples of airborne circularly polarized SAR analysis using axial ratio images, ellipticity, polarization ratio, and polarimetric SAR (PolSAR) techniques. We also explain the experimental result of the scattering of circularly polarized waves on N219 aircraft models and a rice paddy compared to conventional linear polarized SAR images. We hope the readers will continue the research development and study of circularly polarized SAR regarding the system, image processing, and applications.

REFERENCES

[1] I.G. Cumming and F.H. Wong, *Digital Processing of Synthetic Aperture Radar*, Artech House, Boston, 2005.

[2] M. Soumekh, *Synthetic Aperture Radar Signal Processing*, Wiley Interscience, New York, 1999.

[3] H. Maitre, *Processing of Synthetic Aperture Radar Images*, Wiley, London, 2008.

[4] A. Gustavsson, P.O. Frolind, H. Hellsten, T. Jonsson, B. Larsson, and G. Stenstrom, "The Airborne VHF SAR System CARABAS," Proc. IEEE Intern. Geosci. and Rem. Sens. Symp. (IGARSS '93), 18–21 Aug. 1993.

[5] A. Reigber, R. Scheiber, M. Jager, P.P. Iraola, I. Hajnsek, T. Jagdhuber, K. P. Papathanassiou, M. Nannini, E. Aguilera, S. Baumgartner, R. Horn, A. Nottensteiner, and A. Moreira, "Very-High-Resolution Airborne Synthetic Aperture Radar Imaging: Signal Processing and Applications, Proc. the IEEE, Vol.101, No.3, Mar. 2013.

[6] E. Michael and G. Weiss, "Second-Generation Ka-Band UAV SAR System," The 38th Eur. Microwave Conf. (EuMC 2008), Dec 2008.

[7] H. Essen, W. Johannes, S. Stanko, R. Sommer, A. Wahlen, and J. Wilcke, "High Resolution W-Band UAV SAR," IEEE Geosci. and Rem. Sens. Symp. (IGARSS 2012), Jul 2012.

[8] P. Rosen, S. Hensley, K. Wheeler, G. Sadowy, T. Miller, S. Shaffer, R.J. Muellerschoen, C. Jones, S. N. Madsen, and H. Zebker, "UAVSAR: New NASA Airborne SAR System for Research," IEEE Aerosp. and Elect. Syst. Mag., Vol.22, No.11, pp.21–28, Dec 2007.

[9] Editors, "Foreword to the Special Issue on Compact Polarimetric SAR," IEEE Journ. Selected Topics in Appl. Earth Observ.and Rem. Sens., Vol.12, No.10, pp.3708–3711, Oct. 2019.

[10] H. Li, W. Perrie, Q. Li, and Y. Hou, "Estimation of Melt Pond Fractions on First Year Sea Ice Using Compact Polarization SAR," Journ. Geophy. Res.: Oceans, Vol.122, No.10, pp.8145–8166, Oct. 2017.

[11] P.C.D. Fernandez, J.C. Souyris, S. Angelliaume, and F. Garestier, "The Compact Polarimetry Alternative for Spaceborne SAR at Low Frequency," IEEE Trans. Geosci. Remot. Sens., Vol.46, No.10, pp.3208–3222, Oct. 2008.

[12] R.K. Rancy, "Comparing Compact and Quadrature Polarimetric SAR Performance," IEEE Geosci. Remot. Sens. Lett., Vol.13, No.6, pp.861–864, Jun. 2016.

[13] M. Ohki and M. Shimada, "large-Area Land Use and Land Cover Classification with Quad, Compact, and Dual Polarization SAR Data by PALSAR-2," IEEE Trans. Geosci. Remot. Sens., Vol.56, No.9, pp.5550–5557, Sep. 2018.

[14] K. Dasari and A. Lokam, "Exploring the Capability of Compact Polarimetry (Hybrid Pol) C Band RISAT-1 Data for Land Cover Classification," IEEE Access, Vol.6, pp.57981–57993, Oct. 2018.

[15] M.-L. Truong-Loï, P. Dubois-Fernandez, E. Pottier, A. Freeman, and J.C. Souyris, "Potentials of a Compact Polarimetric SAR System," in Proc. IEEE IGARSS, Honolulu, HI, USA, 2010, pp. 742–745.

[16] J. Chen and S. Quegan, "Calibration of Spaceborne CTLR Compact Polarimetric Low-frequency SAR Using Mixed Radar Calibrators," IEEE Trans. Geosci. Remote Sens., vol. 49, no. 7, pp. 2712–2723, Jul. 2011.

[17] H. Tan and J. Hong, "Calibration of Compact Polarimetric SAR Images Using Distributed Targets and One Corner Reflector," IEEE Trans. Geosci. Remote Sens., Vol.54, No.8, pp.4433–4444, Aug. 2016.

[18] A. Babu, S. Kumar, and S. Agrawal, "Polarimetric Calibration of RISAT-1 Compact-Pol Data," IEEE Journ. Selected Topics in Appl. Earth Observ. and Remote Sens., Vol.12, No.10, pp.3731–3736, Oct. 2019.

[19] R. Sabry, "A Tool for Analysis and Calibration of Compact Polarimetry SAR Mode Anomaly," IEEE Geosci. Remote Sens. Lett., Vol.15, No.3, pp.424–428, Mar. 2018.

[20] J.T.S. Sumantyo, C.M. Yam, C.E. Santosa, G.F. Panggabean, T. Watanabe, B. Setiadi, F.D.S. Sumantyo, K. Tsushima, K. Sasmita, A. Mardiyanto, E. Supartono, E.T. Rahardjo, G. Wibisono, M.A. Marfai, R.H. Jatmiko, Sudaryatno, T.H. Purwanto, B.S. Widartono, M. Kamal, D.l Perissin, S. Gao, and K. Ito, "Airborne Circularly Polarized Synthetic Aperture Radar," IEEE Selected Topics in Applied Earth Observations and Remote Sensing (JSTARS), Vol.14, pp.1676–1692, January 2021.

[21] C.M. Yam, J.T.S. Sumantyo, C.E Santosa, G.F. Panggabean, F.D. Sri Sumantyo, T. Watanabe, Y.Q. Ji, P.P Sitompul, M. Nasucha, F. Kurniawan, B. Purbantoro, A. Awaludin, K. Sasmita, E.T. Rahardjo, G. Wibisono, R.H. Jatmiko, Sudaryatno, T.H. Purwanto, B.S. Widartono, and M. Kamal, "The Maiden Flight of Hinotori-C: The First C Band Full Polarimetric Circularly Polarized Synthetic Aperture Radar in the World," IEEE Aerospace and Electronic Systems Magazine, Vol.34, No.2, pp.24–35, February 2019.

[22] www.lockheedmartin.com/en-us/news/features/history/sar.html (accessed on 27 August 2022)

[23] S. Lasswell, "History of SAR at Lockheed Martin (formerly Goodyear Aerospace)," Society of Photo-Optical Instrumentation Engineers, Proceedings Paper, 16 May 2005.

[24] R. Colburn, "Synthetic Aperture Radar," IEEE-USA Today's Engineer Online, March 2009.

[25] SEASAT: https://theconversation.com/comment-les-satellites-permettent-de-surveiller-letat-des-infrastructures-118468

[26] ALMAZ: https://space.skyrocket.de/doc_sdat/almaz-1v.htm

[27] ALMAZ: https://eoportal.org/web/eoportal/satellite-missions/a/almaz

[28] ERS-1: https://earth.esa.int/eogateway/missions/ers?text=ERS-1

[29] ENVISAT: www.esa.int/ESA_Multimedia/Images/2002/02/ENVISAT_satellite_integration2

[30] JERS-1: https://directory.eoportal.org/web/eoportal/satellite-missions/j/jers-1

[31] ALOS-2: https://spaceflight101.com/spacecraft/alos-2/

[32] ALOS-2: https://global.jaxa.jp/projects/sat/alos2/

[33] Radarsat-1: www.un-spider.org/news-and-events/news/canadas-radarsat-1-towards-retirement-after-28-successful-years

[34] Radarsat-2: www.thestar.com/business/2021/02/02/mda-starts-working-on-radarsat-2-surveillance-satellite-replacement.html

[35] TerraSAR-X: https://fineartamerica.com/art/terrasar-x

[36] www.imsar.com/uploads/files/59_IMSAR_NanoDS_Jul2014.pdf

[37] www.sdl.usu.edu/downloads/nusar.pdf

[38] A. Aguasca, R. Acevo-Herrera, A. Broquetas., J.J. Mallorqui, and X. Fabregas, "ARBRES: Light-weight CW/FM SAR Sensors for Small UAVs," Sensors, Vol.13, No.3, pp.3204–3216, March 2013. DOI: 10.3390/s130303204

[39] J.T.S. Sumantyo, K.V. Chet, L.T. Sze, T. Kawai, T. Ebinuma, Y. Izumi, M. Z. Baharuddin, S. Gao and K. Ito, "Development of Circularly Polarized Synthetic Aperture Radar Onboard UAV JX-1," International Journal of Remote Sensing, Special Issue Papers on Drones, UAVs, RPASs for Environmental Research, Vol.38, No.8–10, pp.2745–2756, Online: December 2016, Printed: July 2017, DOI:10.1080/01431161.2016.1275057.

2 Basics of SAR Polarimetry

Yuta Izumi and Josaphat Tetuko Sri Sumantyo

Polarization is a unique characteristic of electromagnetic (EM) waves that are defined as the electric field vector trajectory as a function of time at a fixed space reference plane. In addition to the major EM characteristics of frequency, propagation direction, and wave intensity, polarization plays an important role in SAR application.

Polarimetric radar imaging gives us additional target physical information realized by multi-polarimetric observations. At the beginning of this research field, the description of the scattering process over the target was investigated, as represented by the Sinclair scattering matrix for coherent scattering [1]. Later, practical descriptions for naturally distributed scatter based on second-order statistics extracted from polarimetric covariance matrix (or coherency matrix) has generally been exploited. With these polarimetric descriptions, current research focuses mainly on the target's physical parameter derivation based on the polarimetric target decomposition theories [2].

With the increasing demand for SAR polarimetry, the Jet Propulsion Laboratory (JPL) first conducted airborne fully polarimetric (FP) observation at L-band called the Airborne SAR (AIRSAR) campaign in 1988. Subsequently, the first spaceborne polarimetric SAR mission, SIR-C/X-SAR, was conducted onboard the space shuttles at C-band and L-band. Nowadays, we can access a wide variety of spaceborne polarimetric SAR data acquired by ALOS (PALSAR), TerraSAR-X, RADARSAT-2, TanDEM-X, RISAT, ALOS-2 (PALSAR2), Gaofen-3, SAOCOM, NovaSAR, and Kompsat-6 [3]. Quad polarimetric data by ALOS-4 (PALSAR3) and NISAR, as well as P-band polarimetric data by Biomass mission, will be available.

This chapter will demonstrate the fundamental concept of SAR polarimetry, beginning with a description of the polarization state, EM reflection and scattering, matrix representations of the scattering process for coherent and distributed targets, and polarimetric target decomposition for subsequent chapters, especially for Chapter 9. This chapter also provides polarimetric SAR theories for various polarimetric modes expanded from the original theory based on linear polarization.

2.1 POLARIZATION STATE REPRESENTATION

2.1.1 MAXWELL EQUATION

Maxwell's equations that govern the electromagnetic fields are given as [4]

$$\nabla \times E(r,t) = -\frac{\partial B(r,t)}{\partial t},$$

DOI: 10.1201/9781003282693-2

$$\nabla \times \boldsymbol{H}(r,t) = \boldsymbol{J}(r,t) + \frac{\partial \boldsymbol{D}(r,t)}{\partial t},$$

$$\nabla \cdot \boldsymbol{B}(r,t) = 0,$$

$$\nabla \cdot \boldsymbol{D}(r,t) = \rho \qquad \text{(Equation 2.1)}$$

where $\boldsymbol{E}(r,t)$, $\boldsymbol{H}(r,t)$, $\boldsymbol{D}(r,t)$, $\boldsymbol{B}(r,t)$, and $\boldsymbol{J}(r,t)$ are the electric field, magnetic field, electric flux density, magnetic flux density, and electric current density, as well as ρ, the electric charge density. By using free-space constitutive relations without any conduction current density (current source) as

$$\boldsymbol{D}(r,t) = \varepsilon_0 \boldsymbol{E},$$

$$\boldsymbol{B}(r,t) = \mu_0 \boldsymbol{H},$$

$$\boldsymbol{J}(r,t) = \sigma \boldsymbol{E}(r,t) \qquad \text{(Equation 2.2)}$$

where ε_0, μ_0, and σ are the free space permittivity, permeability, and conductivity, respectively. For vacuum (non-dissipative media $\sigma = 0$) with source-free ($\rho = 0$) assumption, Maxwell's equations in Equation 2.1 can be written as

$$\nabla \times \boldsymbol{E}(r,t) = -\mu_0 \frac{\partial \boldsymbol{H}(r,t)}{\partial t},$$

$$\nabla \times \boldsymbol{H}(r,t) = \varepsilon_0 \frac{\partial \boldsymbol{E}(r,t)}{\partial t},$$

$$\nabla \cdot \boldsymbol{E}(r,t) = 0,$$

$$\nabla \cdot \boldsymbol{H}(r,t) = 0 \qquad \text{(Equation 2.3)}$$

2.1.2   WAVE EQUATION AND PROPAGATION

According to (Equation 2.3), for non-dissipative media, the vector wave equation for the electric field can be obtained as

$$\nabla^2 \boldsymbol{E}(r,t) - \mu_0 \varepsilon_0 \frac{\partial^2 \boldsymbol{E}(r,t)}{\partial t^2} = 0 \qquad \text{(Equation 2.4)}$$

For the monochromatic electromagnetic wave, the vector wave equation in (Equation 2.4) can be simplified by the phasor representation $\boldsymbol{E}(r)$ under the assumption of time harmonics $e^{j\omega t}$. Phasor representation is the complex quantity independent of time, expressed by the relationship of $E(r,t)= Re\{E(r)e^{j\omega t}\}$ where $E(r,t)$ is the instantaneous electric field vector. When we express the vector wave equation in (Equation 2.4) as the phasor representation, the time derivative by $\dfrac{\partial}{\partial t} = j\omega$, and (Equation 2.4) is rewritten as

$$\nabla^2 E(r)+k^2 E(r)= 0 \qquad \text{(Equation 2.5)}$$

where $k = \omega\sqrt{\varepsilon_0\mu_0}$ is the wavenumber and ω is the angular frequency. A solution of the wave equation in (Equation 2.5) for $E(r)$ can be obtained

$$E(r)= E_+ e^{-jkr} + E_- e^{+jkr} \qquad \text{(Equation 2.6)}$$

where E_+ and E_- are the amplitude vectors of wave propagations toward the $+r$ and $-r$ direction, respectively. Now the electric field is represented on an orthogonal basis (x,y,z). Assuming the propagation in the z-direction, i.e., $E_z = 0$, the electric field can be defined as

$$E(z)= \hat{x}\left|E_x\right|e^{-j(kz+\varphi_x)} + \hat{y}\left|E_y\right|e^{-\left(jkz+\varphi_y\right)} \qquad \text{(Equation 2.7)}$$

where $\left|E_x\right|$ and $\left|E_y\right|$ are the amplitudes and φ_x and φ_y are the absolute phase. Then, the electric field $E(z)$ of x and y orthogonal components as a function of space and time can be obtained by using $E(z,t)= \operatorname{Re}\{E(z)e^{j\omega t}\}$ as

$$E_x(z,t)=\left|E_x\right|\cos\left(\omega t - kz + \varphi_x\right),$$

$$E_y(z,t)=\left|E_y\right|\cos\left(\omega t - kz + \varphi_y\right) \qquad \text{(Equation 2.8)}$$

where E_x and E_y have maximum amplitude displacement in x and y directions in any z positions, respectively. Hence both electric field E_x and E_y are defined as linear polarization.

2.1.3 POLARIZATION ELLIPSE

Polarization is defined as the time evolution of the electric field vector trajectory observed at a fixed position of the space, seen back from the propagation direction, as shown in Figure 2.1 (a) [5]. According to this definition, (Equation 2.8) gives the temporal wave trajectory at a fixed z_0 position by removing the ωt using both $E_x(z,t)$ and $E_y(z,t)$,

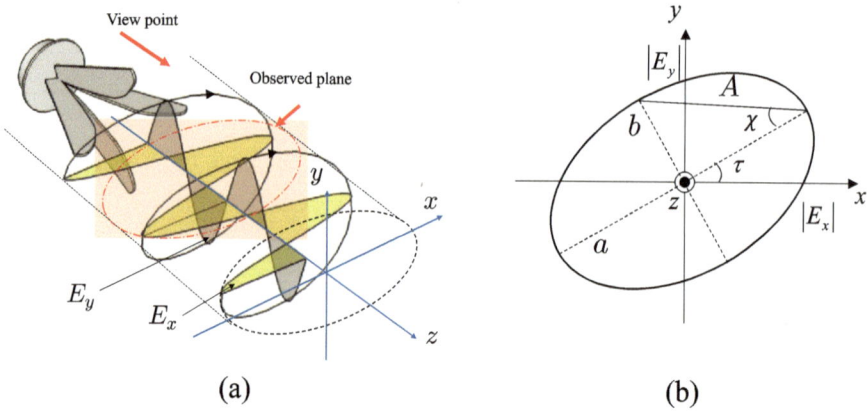

(a)   (b)

FIGURE 2.1 The elliptically polarized wave. (a) Vector trajectory of the elliptically polarized wave. (b) The polarization ellipse characterized by three geometrical parameters.

$$\frac{E_x^{\,2}\left(z_0,t\right)}{\left|E_x\right|^2}-\frac{2E_x\left(z_0,t\right)E_y\left(z_0,t\right)}{\left|E_x\right|\left|E_y\right|}\cos\delta+\frac{E_y^{\,2}\left(z_0,t\right)}{\left|E_y\right|^2}=\sin\delta, \qquad\text{(Equation 2.9)}$$

where $\delta = \varphi_y - \varphi_x$. The expression in (Equation 2.9) explicitly shows the equation of a tilted ellipse, called the polarization ellipse, describing the wave polarization [6]. The shape of the polarization ellipse is thus determined by the relative phase $\delta = \varphi_y - \varphi_x$ and relative strength of the amplitudes $\left|E_x\right|$ and $\left|E_y\right|$. The polarization ellipse with $\delta = 0°$ corresponds to the linear polarization, while perfect circular polarization is defined when $\delta = \pm90°$ and $\left|E_x\right| = \left|E_y\right|$. On the other hand, if E_x and E_y have different amplitudes and exhibit $\delta = \pm90°$, the electric field becomes an elliptical polarization. The positive and negative signs of δ determine the sense of the field rotation. For circularly polarized waves, anti-clockwise and clockwise rotation corresponds to left-handed circular polarization (LHCP) and right-handed circular polarization (RHCP), respectively.

In general, the polarization ellipse shape is characterized by geometrical parameters, which are ellipticity angle χ, tilt angle τ, and size $A = \sqrt{a^2 + b^2}$, as shown in Figure 2.1 (b). The ellipse aperture and the sense of rotation are given by χ and its sign, respectively. The phase and amplitude of E_x and E_y determine χ and τ by following relationships:

$$\sin2\chi = \frac{2\left|E_x\right|\left|E_y\right|\sin\delta}{\left|E_x\right|^2+\left|E_y\right|^2} \qquad\text{(Equation 2.10)}$$

$$\tan2\tau = \frac{2\left|E_x\right|\left|E_y\right|\cos\delta}{\left|E_x\right|^2-\left|E_y\right|^2} \qquad\text{(Equation 2.11)}$$

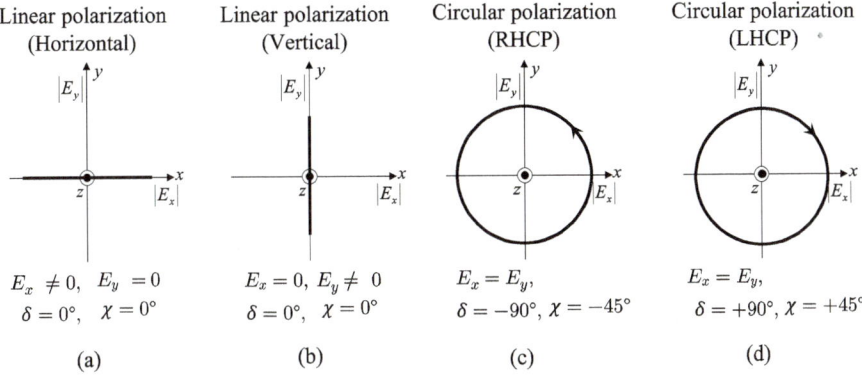

Linear polarization (Horizontal)

$E_x \neq 0, \; E_y = 0$
$\delta = 0°, \; \chi = 0°$

(a)

Linear polarization (Vertical)

$E_x = 0, \; E_y \neq 0$
$\delta = 0°, \; \chi = 0°$

(b)

Circular polarization (RHCP)

$E_x = E_y,$
$\delta = -90°, \; \chi = -45°$

(c)

Circular polarization (LHCP)

$E_x = E_y,$
$\delta = +90°, \; \chi = +45°$

(d)

FIGURE 2.2 Linear polarization and circular polarization. (a) Horizontal polarization, (b) Vertical polarization, (c) RHCP, and (d) LHCP.

Figure 2.2 visualizes polarization ellipses of linear polarization and circular polarization. Note that here we define that E_x and E_y are horizontal (H) and vertical (V) polarization, respectively (x and y directions correspond to horizontal and vertical directions, respectively).

2.1.4 JONES VECTOR

When we represent the 2D vector of (Equation 2.7) as the phasor representation at $z = 0$, the corresponding vector in linear polarization basis (H-V) becomes

$$E_0(HV) = E(z=0) = \begin{bmatrix} |E_H| e^{j\varphi_H} \\ |E_V| e^{j\varphi_V} \end{bmatrix} \qquad \text{(Equation 2.12)}$$

This vector is called the *Jones vector* that can be utilized to describe field components for wave polarization description with a simple form proposed by R. Clark Jones. The Jones vector can describe the polarization state with phase information; therefore, this expression is only applied to the coherent electric field. We can express the Jones vector by two parameters, namely α_w and relative phase δ by

$$E_0(HV) = \sqrt{E_H^2 + E_V^2} \begin{bmatrix} \cos\alpha_w \, e^{j\varphi_H} \\ \sin\alpha_w \, e^{j\varphi_V} \end{bmatrix}$$

$$= \sqrt{E_H^2 + E_V^2} \begin{bmatrix} \cos\alpha_w \\ \sin\alpha_w \, e^{j\delta} \end{bmatrix} \qquad \text{(Equation 2.13)}$$

where α_w is related to geometrical parameters of polarization ellipse [7]

$$\cos 2\alpha_w = \cos 2\tau \cos 2\xi \qquad \text{(Equation 2.14)}$$

The Jones vector can also be expressed by *polarization ratio* ρ which is defined on (H-V) basis as

$$\rho_{HV} = \frac{E_V}{E_H} = \frac{|E_V|}{|E_H|} e^{j(\varphi_V - \varphi_H)} = \tan \alpha_w e^{j\delta} \qquad \text{(Equation 2.15)}$$

Employing ρ, the Jones vector representation in (Equation 2.12) becomes

$$E_0(HV) = \frac{\sqrt{E_H E_H^* + E_V E_V^*} \, e^{j\varphi_H}}{\sqrt{1 + \rho_{HV} \rho_{HV}^*}} \begin{bmatrix} 1 \\ \rho_{HV} \end{bmatrix} \qquad \text{(Equation 2.16)}$$

Now let us derive the LHCP and RHCP from the Jones vector representation on (H-V) basis. According to the definition of LHCP, i.e., $\delta = +90°$ and $|E_x| = |E_y|$, ρ_{HV} becomes j. Therefore, LHCP on (H-V) basis becomes

$$LHCP(HV) = \frac{1}{\sqrt{2}} \begin{bmatrix} 1 \\ j \end{bmatrix} \qquad \text{(Equation 2.17)}$$

Similarly, RHCP can be derived as

$$RHCP(HV) = \frac{1}{\sqrt{2}} \begin{bmatrix} 1 \\ -j \end{bmatrix} \qquad \text{(Equation 2.18)}$$

The expression of RHCP in (Equation 2.18), however, does not agree with the condition of unitary transformation for vector transformation demonstrated in the subsequent subsection. Instead, the following representation of RHCP is usually employed in radar polarimetry [3],

$$RHCP(HV) = \frac{1}{\sqrt{2}} \begin{bmatrix} j \\ 1 \end{bmatrix} \qquad \text{(Equation 2.19)}$$

where the orthogonality between RHCP and LHCP is still valid under this representation [6].

2.1.5 Vector Transformation

The Jones vector can be mathematically transformed into any basis. Let us consider the basis vector transformation from (H-V) basis into arbitrary (A-B) basis. The following expression allows this mathematical transformation

$$\begin{bmatrix} E_A \\ E_B \end{bmatrix} = [U] \begin{bmatrix} E_H \\ E_V \end{bmatrix},$$

$$E_0(AB) = [U]E_0(HV) \qquad \text{(Equation 2.20)}$$

where $[U]$ is the vector transformation matrix [8]. This vector transformation should be a unitary transformation to preserve amplitude and phase information. Hence, $[U]$ should follow the unitary matrix conditions,

$$[U]^{-1} = [U]^{*T},$$

$$|\det[U]| = 1 \qquad \text{(Equation 2.21)}$$

Polarization ratio and geometrical parameters of polarization ellipse determine the vector transformation matrix $[U]$,

$$[U] = \frac{1}{\sqrt{1+\rho\rho^*}} \begin{bmatrix} e^{-j\xi} & 0 \\ 0 & e^{j\xi} \end{bmatrix} \begin{bmatrix} 1 & \rho^* \\ -\rho & 1 \end{bmatrix} \qquad \text{(Equation 2.22)}$$

where

$$\xi = \tan^{-1}(\tan\tau\tan\chi) \qquad \text{(Equation 2.23)}$$

Let us now consider the vector basis transformation from linear polarization (H-V) basis to circular polarization (L-R) basis via defined $[U]$. In this case, due to the conditions of $\rho_{HV} = j$ and $\xi = 0$ according to the LHCP definition in (H-V) basis, vector transformation matrix becomes

$$[U]_{HV \to LR} = \frac{1}{\sqrt{2}} \begin{bmatrix} 1 & -j \\ -j & 1 \end{bmatrix} \qquad \text{(Equation 2.24)}$$

Therefore, transformation is mathematically described by

$$\begin{bmatrix} E_L \\ E_R \end{bmatrix} = \frac{1}{\sqrt{2}} \begin{bmatrix} 1 & -j \\ -j & 1 \end{bmatrix} \begin{bmatrix} E_H \\ E_V \end{bmatrix}$$

$$= \frac{1}{\sqrt{2}} \begin{bmatrix} E_H - jE_V \\ -jE_H + E_V \end{bmatrix} \qquad \text{(Equation 2.25)}$$

where E_l and E_R are orthogonal electric fields, indicating LHCP and RHCP, respectively.

2.1.6 STOKES VECTOR

In the SAR observation scenario, the received wave is the sum of the scattered waves interacted with numerous scattered points over observed media. When the portion of the scattered waves shows temporally random phase fluctuation, the summed received wave becomes a partially incoherent wave. This phenomenon is called depolarization (unpolarized wave) that is the result of the coherent transmitted wave into an incoherent received wave, while a coherent wave is called a completely polarized wave. The Jones vector is not able to treat such depolarized waves because it accounts for phase information of completely polarized waves. Stokes vector $\mathbf{g}$ can represent the depolarized wave with four Stokes parameters g_0, g_1, g_2, and g_3 which are real numbers and described on (H-V) basis as

$$
\mathbf{g} = \begin{bmatrix} \langle g_0 \rangle \\ \langle g_1 \rangle \\ \langle g_2 \rangle \\ \langle g_3 \rangle \end{bmatrix} = \begin{bmatrix} \langle |E_H|^2 \rangle + \langle |E_V|^2 \rangle \\ \langle |E_H|^2 \rangle - \langle |E_V|^2 \rangle \\ 2\mathrm{Re}\langle E_V E_H^* \rangle \\ 2\mathrm{Im}\langle E_V E_H^* \rangle \end{bmatrix} = \begin{bmatrix} \langle |E_H|^2 \rangle + \langle |E_V|^2 \rangle \\ \langle |E_H|^2 \rangle - \langle |E_V|^2 \rangle \\ \langle 2|E_H||E_V|\cos\delta \rangle \\ \langle 2|E_H||E_V|\sin\delta \rangle \end{bmatrix}
$$

$$
= A^2 \begin{bmatrix} 1 \\ \cos 2\tau \cos 2\zeta \\ \sin 2\tau \cos 2\zeta \\ \sin 2\zeta \end{bmatrix}
$$

(Equation 2.26)

where $\langle \cdot \rangle$ indicates the ensemble averaging with the assumptions of both the stationarity and the ergodicity, practically realized by spatial or temporal averaging. Note that the completely polarized wave does not necessarily equal the ensemble averaging, and $g_0^2 = g_1^2 + g_2^2 + g_3^2$ are valid. Once the Stokes parameters are measured, polarization ellipse geometrical parameters can be determined according to (Equation 2.26), and the following relationships are derived

$$
A^2 = g_0,
$$

$$
\chi = \frac{1}{2}\sin^{-1}\frac{g_3}{g_0},
$$

$$
\tau = \frac{1}{2}\tan^{-1}\frac{g_2}{g_1}
$$

(Equation 2.27)

Similarly, relative phase δ can be expressed as

$$
\delta = \tan^{-1}\frac{g_3}{g_2}
$$

(Equation 2.28)

Employing Stokes parameters, the degree of polarization m is derived as

$$m = \frac{\sqrt{\langle g_1 \rangle^2 + \langle g_2 \rangle^2 + \langle g_3 \rangle^2}}{\langle g_0 \rangle} \qquad \text{(Equation 2.29)}$$

which measures the polarized wave ratio, while $m = 1$ and $m = 0$ correspond to completely polarized and completely unpolarized waves, respectively.

The vector basis transformation in (Equation 2.25) leads to the Stokes vector in (L-R) basis as

$$\begin{bmatrix} \langle g_0 \rangle \\ \langle g_1 \rangle \\ \langle g_2 \rangle \\ \langle g_3 \rangle \end{bmatrix} = \begin{bmatrix} \langle |E_H|^2 \rangle + \langle |E_V|^2 \rangle \\ \langle |E_H|^2 \rangle - \langle |E_V|^2 \rangle \\ 2\operatorname{Re}\langle E_V E_H^* \rangle \\ 2\operatorname{Im}\langle E_V E_H^* \rangle \end{bmatrix} = \begin{bmatrix} \langle |E_L|^2 \rangle + \langle |E_R|^2 \rangle \\ 2\operatorname{Im}\langle E_L E_R^* \rangle \\ 2\operatorname{Re}\langle E_L E_R^* \rangle \\ \langle |E_L|^2 \rangle - \langle |E_R|^2 \rangle \end{bmatrix} \qquad \text{(Equation 2.30)}$$

On a circular polarization basis, the LHCP and RHCP power components give us the two Stokes parameters g_0 and g_3. Accordingly, the χ is given by circular polarization power components as

$$\chi = \frac{1}{2}\sin^{-1}\left\{ \frac{\left[\langle |E_L|^2 \rangle - \langle |E_R|^2 \rangle \right]}{\left[\langle |E_L|^2 \rangle + \langle |E_R|^2 \rangle \right]} \right\} \qquad \text{(Equation 2.31)}$$

2.1.7 AXIAL RATIO

In antenna engineering, the axial ratio measures the purity of circular polarization. It is defined by a ratio of the major axis a to the minor axis b of the polarization ellipse [9]

$$AR = 20\log\left|\frac{a}{b}\right| \qquad \text{(Equation 2.32)}$$
$$= 20\log\left(\cot(\chi)\right)$$

which is the function of ellipticity angle χ. Since the χ can be obtained by two Stokes parameters as shown in (Equation 2.27), the AR can be represented by two Stokes parameters as well. In particular, (Equation 2.31) demonstrates the possibility of AR measurement by the received power of both LHCP and RHCP.

In principle, 0dB of *AR* represents an ideal circular polarization, while an infinite value corresponds to an ideal linear polarization. However, it is usually difficult to obtain an ideal 0dB of *AR* over the entire antenna bandwidth; thus, a value of less than 3dB is considered to be sufficient for most practical cases [10].

2.2 EM REFLECTION, TRANSMISSION, AND SCATTERING IN CIRCULAR POLARIZATION

This section discusses the scattering characteristics of circular polarization for determining the parameter of the airborne circularly polarized SAR systems [41]. Figure 2.3 shows the scattering model using a plane wave that illustrates the scattering and transverse waves on the dielectric discontinuity surface

$$\theta_r = \theta_i \qquad \text{(Equation 2.33)}$$

$$sin\theta_t = \frac{\eta_1}{\eta_2} sin\theta_i \qquad \text{(Equation 2.34)}$$

where η_n ($n = 1, 2$) is the characteristic impedance of each medium as shown by

$$\eta_n = \sqrt{\frac{j\omega\mu_n}{\sigma_n + j\omega\varepsilon_o \varepsilon_{rn}}} \qquad \text{(Equation 2.35)}$$

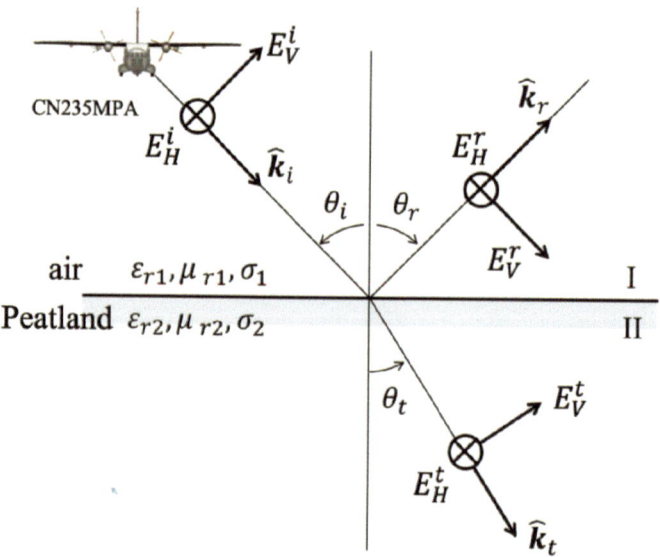

FIGURE 2.3   Scattering model of circular polarization.

where $n=1, 2$. Fresnel reflection (R_V and R_H) and transmission (T_V and T_H) coefficients are

$$R_V = \frac{\eta_2 cos\theta_t - \eta_1 cos\theta_i}{\eta_2 cos\theta_t + \eta_1 cos\theta_i} \qquad \text{(Equation 2.36)}$$

$$T_V = \frac{2\eta_2 cos\theta_i}{\eta_2 cos\theta_t + \eta_1 cos\theta_i} \qquad \text{(Equation 2.37)}$$

$$R_H = \frac{\eta_2 cos\theta_i - \eta_1 cos\theta_t}{\eta_2 cos\theta_i + \eta_1 cos\theta_t} \qquad \text{(Equation 2.38)}$$

$$T_H = \frac{2\eta_2 cos\theta_i}{\eta_2 cos\theta_i + \eta_1 cos\theta_t} \qquad \text{(Equation 2.39)}$$

For the following discussion, we consider the dielectric constant or permittivity of reflection surface as 3.1 (measured value of peat soil sample collected in Indonesia with a frequency of 5.3GHz (C-band)) and lossless medium for simplicity ($\sigma = 0$). Hence, we consider ε_{r1} and ε_{r2} are air and soil permittivity with values 1 and 3.1, respectively.

If the incident angle θ_i increases from 0, R_V will be 0 at an angle θ_B (Brewster's angle) as shown in Figure 2.4. When $\theta_i = \theta_B$, transmitted wave and reflected wave are perpendicular or $\hat{\boldsymbol{k}}_r \cdot \hat{\boldsymbol{k}}_t = 0$. If $R_V = 0, \mu_{r1} = \mu_{r2} = 1, \sigma_1 = \sigma_2 = 0$, Brewster's angle θ_B is to be

$$\theta_B = tan^{-1} \sqrt{\frac{\varepsilon_{r2}}{\varepsilon_{r1}}} \qquad \text{(Equation 2.40)}$$

Assuming incident wave $\boldsymbol{E}^i = \begin{bmatrix} E_H^i & E_V^i \end{bmatrix}^T$, reflected wave $\boldsymbol{E}^r = \begin{bmatrix} E_H^r & E_V^r \end{bmatrix}^T$, and transmitted wave $\boldsymbol{E}^t = \begin{bmatrix} E_H^t & E_V^t \end{bmatrix}^T$, the relationship below could be found

$$\begin{bmatrix} E_H^r \\ E_V^r \end{bmatrix} = \begin{bmatrix} R_H & 0 \\ 0 & R_V \end{bmatrix} \begin{bmatrix} E_H^i \\ E_V^i \end{bmatrix} \qquad \text{(Equation 2.41)}$$

$$\begin{bmatrix} E_H^t \\ E_V^t \end{bmatrix} = \begin{bmatrix} T_H & 0 \\ 0 & T_V \end{bmatrix} \begin{bmatrix} E_H^i \\ E_V^i \end{bmatrix} \qquad \text{(Equation 2.42)}$$

where E_H and E_V are two perpendicular components of the circularly polarized wave. Reflection of circular polarization on boundary conditions is given as

$$\begin{bmatrix} E_L^r \\ E_R^r \end{bmatrix} = \begin{bmatrix} R_c & R_x \\ R_x & R_c \end{bmatrix} \begin{bmatrix} E_L^i \\ E_R^i \end{bmatrix} \qquad \text{(Equation 2.43)}$$

where

$$R_c = \frac{1}{2}\left(R_H - R_V\right)$$
(Equation 2.44)

$$R_x = \frac{1}{2}\left(R_H + R_V\right)$$
(Equation 2.45)

where R_c is the reflection coefficient of main polarization (co-polarization) of circular polarization, and R_x is the reflection coefficient of cross-polarization.

Transmission of circular polarization on boundary conditions could be derived as

$$\begin{bmatrix} E_L^t \\ E_R^t \end{bmatrix} = \begin{bmatrix} T_c & T_x \\ T_x & T_c \end{bmatrix}\begin{bmatrix} E_L^i \\ E_R^i \end{bmatrix}$$
(Equation 2.46)

where

$$T_c = \frac{1}{2}\left(T_H + T_V\right)$$
(Equation 2.47)

$$T_x = \frac{1}{2}\left(T_H + T_V\right)$$
(Equation 2.48)

where T_c is the transmission coefficient of main polarization (co-polarization) of circular polarization, and T_x is the transmission coefficient of cross-polarization.

Figure 2.4 shows that R_V is zero at θ_B (Brewster's angle). R_V and R_H are negative at $\theta_i < \theta_B$ that shows the reflected wave changed the phase 180°. R_V changed to be

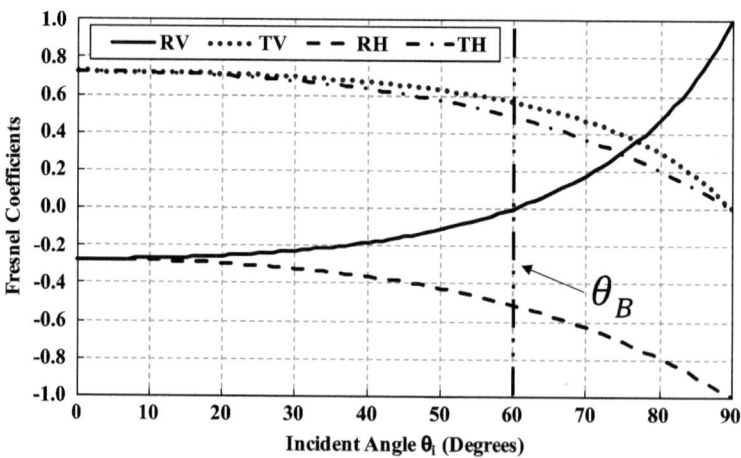

FIGURE 2.4   Fresnel coefficient of C-band circular polarization.

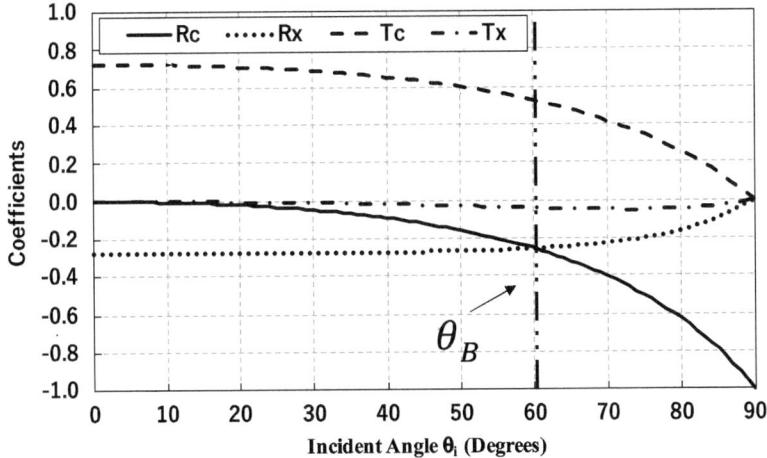

FIGURE 2.5   Reflection and transmission coefficient of C-band circular polarization.

positive, and R_H remained negative. T_V and T_H are positive, that it means the phase does not change.

In the case where $\theta_i = \theta_B$, as shown in Figure 2.5, R_c and R_x of Equations (2.44)-(2.45) have the same value, where the scattered wave's intensity of the main polarized (co-polarization) wave and the cross-polarized wave is the same. This combination generates a linearly polarized wave, which is shown by the infinite axial ratio in Figure 2.6. In this case, the scattered wave only has R_H, which is shown by R_V with zero value in Figure 2.4. It means only E_H^r scatters on the surface and E_V^r does not scatter.

In the case where $\theta_i < \theta_B$, as shown in Figure 2.5, R_c and R_x have a negative value, but the absolute value of R_x is larger than R_c. This means the cross-polarized wave is dominant after being scattered. It shows electric fields perpendicular and parallel components phases are different, and then switch of 180° after the waves are scattered, as shown in Figure 2.4 where R_H and R_V are negative. If we consider the propagation direction after scattering, it has the opposite direction, while the circularly polarized wave after scattering has the opposite sense of polarization.

If T_c and T_x respectively have positive and negative values, as shown in Figure 2.5, which means the main polarized (co-polarization) wave penetrated in the medium without a phase change, but the cross-polarized wave penetrated in the medium with a 180° phase change.

The absolute value of T_c is larger than T_x, which means the main polarized wave (co-polarization) is predominant in penetrating the medium.

In the case where $\theta_i > \theta_B$, as shown in Figure 2.5, the absolute value shows R_c and R_x have negative values, but the absolute value of R_c is larger than that of R_x. It means the main polarized (co-polarization) wave is predominant after scattering. E_H^r component has the opposite direction after scattering, and the E_V^r component keeps the same direction, as shown by negative R_H and positive R_V in Figure 2.4. This means the sense of scattered circular polarization is the same. The absolute value of R_c is larger than the one of R_x, which means the main polarized (co-polarization) wave of the incident wave is predominant.

As shown in Figure 2.5, the relationship between T_c and T_x under the conditions of $\theta_i > \theta_B$ has the same result as the conditions of $\theta_i < \theta_B$; it also means the main polarized (co-polarization) wave penetrated the medium without a phase change, but the cross-polarized wave penetrated in the medium with a 180° phase change. The absolute value of T_c is larger than T_x, which means the main polarized (co-polarization) wave is predominant in penetrating the medium.

We assume the axial ratio of the incident wave is 0dB, or perfectly, circular polarization where the axial ratio is given by $AR = 20\log\left|\dfrac{E_V}{E_H}\right|$. As the incident angle θ_i gets closer to 0°, the incident wave sense changes to the opposite sense, or the incident wave becomes cross-polarized. The axial ratio of the scattered wave gets closer to 0dB, and the polarization tends to be circular, as shown in Figure 2.6. But for incident angles θ_i larger than 30°, the axial ratio is larger than 3dB, and the polarization of scattered waves becomes elliptically polarized at an incident angle of $30° < \theta_i < \theta_B$. The axial ratio reaches its largest value at $\theta_i = \theta_B$, which indicates linear polarization, where only perpendicular waves scatter, and linear polarization (LP) with a perpendicular direction remains after scattering. At $\theta_i > \theta_B$, the scattering wave closes parallel to the scattered plane, and the axial ratio decreases to be circular polarization. In this case, the rotation direction of polarization of a scattered wave is the same as the polarization of the incident wave or main wave.

As the summary of Figures 2.6 and 2.7, the sense of wave changes to the opposite sense or changes to be a cross-polarized wave at an incident angle θ_i closes to 0°, and the axial ratio of the scattered wave closes to 0dB or closes to circular polarization. But at an incident angle θ_i, larger than 30°, as shown in Figure 2.6, the axial ratio

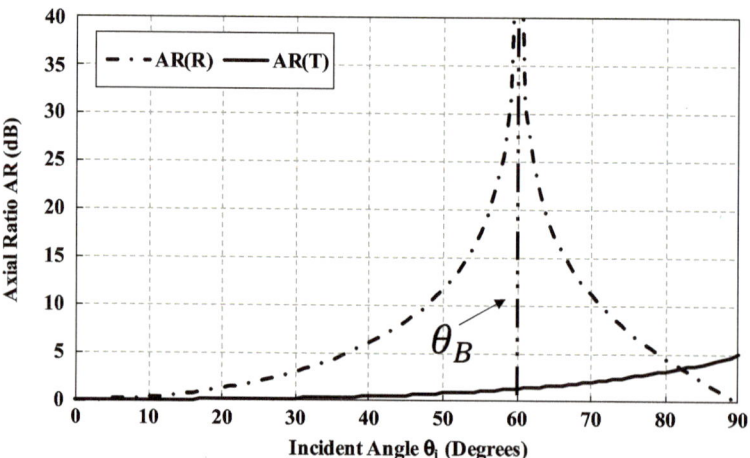

FIGURE 2.6   The axial ratio of circular polarization.

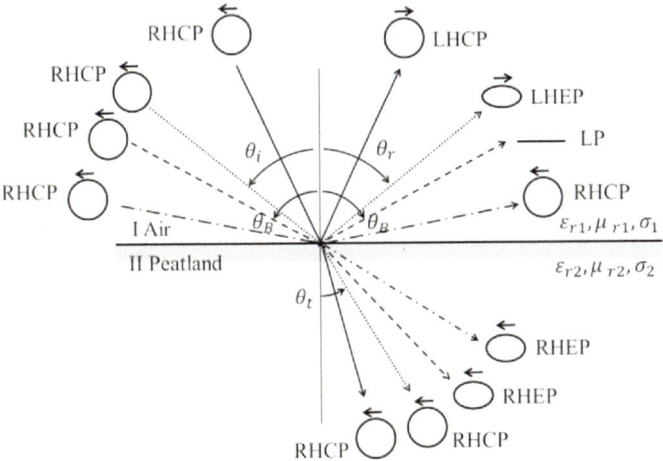

FIGURE 2.7 Scattering polarization of the circularly polarized wave.

is larger than 3 dB, and the polarization of a scattered wave is elliptical polarization (EP) at an incident angle $30°<\theta_i<\theta_B$, where θ_B is Brewster's angle. Axial ratio is largest at $\theta_i=\theta_B$, which means elliptical and linear polarization (LP), where only horizontal polarization scatters and linear polarization with a perpendicular direction remains after scattering. At $\theta_i>\theta_B$, scattering wave close parallel to scattered plane, the axial ratio decreases to be circular polarization. In this case, the rotation direction of polarization of a scattered wave is the same as the polarization of the incident wave, or main wave.

The transmission or penetrated wave has the same polarization as an incident wave with an axial ratio of less than 3dB at an incident angle of less than 30°, as shown in Figure 2.6. Based on these results, the design of the Circularly Polarized SAR must be efficient on power consumption and keep a stable axial ratio for circular polarization, where we must employ a low-nadir angle or incidence angle (less than 30°) to reduce the transmitting power, particularly on the mission of C-band.

2.3 MATRIX REPRESENTATION OF THE SCATTERING PROCESS

2.3.1 Scattering Matrix in Linear Polarization Basis

The monostatic radar transmits and receives the EM wave at the same position with the same coordinates. In this configuration, the radar receives the backscattering EM wave interacting with the target. Defining the transmitting field by the Jones vectors E_t in the (H-V) polarization basis as

$$E_t(HV) = \begin{bmatrix} E_{tH} \\ E_{tV} \end{bmatrix} = \begin{bmatrix} |E_{tH}|e^{j\varphi_{tH}} \\ |E_{tV}|e^{j\varphi_{tV}} \end{bmatrix}$$

(Equation 2.49)

the received field is expressed by

$$E_r = \begin{bmatrix} E_{rH} \\ E_{rV} \end{bmatrix} = \frac{e^{-jkr}}{r} \begin{bmatrix} S_{HH} & S_{HV} \\ S_{VH} & S_{VV} \end{bmatrix} \begin{bmatrix} E_{tH} \\ E_{tV} \end{bmatrix} \qquad \text{(Equation 2.50)}$$

where S_{ij} are called complex scattering coefficients with i and j denoting the received and transmitted waves, respectively, while $[S]$ is the *scattering matrix* (i.e., Sinclair scattering matrix). The monostatic radar configuration makes $S_{HV} = S_{VH}$ due to the reciprocity condition valid under the far-field planner wave assumption without any Faraday rotation effect [11]. Diagonal channels of $[S]$ are called co-polarization

---

TABLE 2.1

Scattering Matrix in (H-V) Basis and (L-R) Basis of Some Canonical Targets

Target	Scattering matrix in (H-V) basis	Scattering matrix in (L-R) basis
Sphere Trihedral Plate	$\begin{bmatrix} 1 & 0 \\ 0 & 1 \end{bmatrix}$	$\begin{bmatrix} 0 & j \\ j & 0 \end{bmatrix}$
Horizontal dihedral	$\begin{bmatrix} 1 & 0 \\ 0 & -1 \end{bmatrix}$	$\begin{bmatrix} 1 & 0 \\ 0 & -1 \end{bmatrix}$
Tilted dihedral	$\begin{bmatrix} \cos 2\theta & \sin 2\theta \\ \sin 2\theta & -\cos 2\theta \end{bmatrix}$	$\begin{bmatrix} e^{j2\theta} & 0 \\ 0 & e^{-j2\theta} \end{bmatrix}$
Horizontal wire	$\begin{bmatrix} 1 & 0 \\ 0 & 0 \end{bmatrix}$	$\frac{1}{2}\begin{bmatrix} 1 & j \\ j & -1 \end{bmatrix}$
Tilted wire	$\begin{bmatrix} \cos^2 \theta & \frac{1}{2}\sin 2\theta \\ \frac{1}{2}\sin 2\theta & \sin^2 \theta \end{bmatrix}$	$\frac{1}{2}\begin{bmatrix} e^{j2\theta} & -j \\ -j & e^{-j2\theta} \end{bmatrix}$

(co-pol) where the Tx polarization state is the same as Rx. On the other hand, off-diagonal channels of $[S]$ are called cross-polarization (cross-pol) where the Tx polarization state is different from Rx.

The scattering matrix is employed to characterize the target and represent the interaction between the transmitted wave and a given target [6]. The complex scattering coefficients are the function of parameters regarding the target geometrical structure and dielectric properties as well as the transmitted wave frequency and imaging geometry. Therefore, the scattering matrix varies pixel-to-pixel. In general, the relative scattering matrix is employed to represent the polarimetric characteristic [3]

$$[S(HV)]_{relative} = \begin{bmatrix} |S_{HH}| & |S_{HV}|e^{j(\phi_{HV}-\phi_{HH})} \\ |S_{VH}|e^{j(\phi_{VH}-\phi_{HH})} & |S_{VV}|e^{j(\phi_{VV}-\phi_{HH})} \end{bmatrix} \qquad \text{(Equation 2.51)}$$

For examples, the scattering matrices in the (H-V) basis of some canonical targets are displayed in Table 2.1. Odd-bounce targets such as the sphere, trihedral corner reflector, and plate give only a co-pol response with the same response between HH and VV. The horizontal dihedral corner reflector (even-bounce target) also gives only a co-pol response with a 180° phase difference between HH and VV. On the other hand, when a horizontal dihedral is rotated around the radar line-of-sight (LOS), the cross-pol response is increased, and consequently, the co-pol response approaches 0 when the rotation angle reaches 45°. The horizontal wire target shows only HH response, while tilted wire target gives both co-pol and cross-pol responses.

2.3.2 SCATTERING MATRIX IN CIRCULAR POLARIZATION BASIS

The scattering matrix can be mathematically transformed into any basis via unitary transformation. The basis transformation matrix $[T_b]$ transforms the scattering matrix in (H-V) basis into arbitrary (A-B) basis, shown as

$$[S(AB)] = [T_b]^T [S(HV)][T_b] \qquad \text{(Equation 2.52)}$$

where $[T_b]$ is defined as an inverse of vector transformation matrix $[U]$ in (Equation 2.22) as,

$$[T_b] = [U]^{-1}$$
$$= \frac{1}{\sqrt{1+\rho\rho^*}}\begin{bmatrix} 1 & -\rho^* \\ \rho & 1 \end{bmatrix}\begin{bmatrix} e^{j\xi} & 0 \\ 0 & e^{-j\xi} \end{bmatrix} \qquad \text{(Equation 2.53)}$$

It is important to note that the information contents of the scattering matrix in the arbitrary basis (A-B) are mathematically equivalent to that of the original basis since the basis transformation is performed under unitary transformation. Therefore, the polarimetric analysis, such as polarimetric decomposition, can be tested on any basis by this "emulation" without any additional measurements.

Employing the relationship in (Equation 2.52), the scattering matrix on an (L-R) basis can be obtained from the scattering matrix on an (H-V) basis. With $\rho = j$ and $\xi = 0$ for LHCP, basis transformation matrix $\left[T_b \right]_{HV \to LR}$ is given as

$$\left[T_b \right]_{HV \to LR} = \frac{1}{\sqrt{2}} \begin{bmatrix} 1 & j \\ j & 1 \end{bmatrix} \qquad \text{(Equation 2.54)}$$

As a result, the scattering matrix on an (L-R) basis is given by the following transformation

$$\begin{bmatrix} S_{LL} & S_{LR} \\ S_{RL} & S_{RR} \end{bmatrix} = \frac{1}{2} \begin{bmatrix} 1 & j \\ j & 1 \end{bmatrix} \begin{bmatrix} S_{HH} & S_{HV} \\ S_{VH} & S_{VV} \end{bmatrix} \begin{bmatrix} 1 & j \\ j & 1 \end{bmatrix}$$
$$= \frac{1}{2} \begin{bmatrix} S_{HH} - S_{VV} + 2jS_{HV} & j\left(S_{HH} + S_{VV} \right) \\ j\left(S_{HH} + S_{VV} \right) & S_{VV} - S_{HH} + 2jS_{HV} \end{bmatrix} \qquad \text{(Equation 2.55)}$$

where we use reciprocity condition $S_{HV} = S_{VH}$.

The scattering matrices on (L-R) basis are given in Table 2.1. In circular polarization, the rotation around radar LOS does not alter the amplitude but only the phase information. Odd-bounce targets provide only cross-pol response while even-bounce targets provide only co-pol response regardless of the rotation angle. On the other hand, wire targets provide both co-pol and cross-pol responses.

2.3.3   SECOND-ORDER POLARIMETRIC MATRICES IN LINEAR POLARIZATION BASIS

One of the advantages of radar polarimetry is the extraction of the target physical information from polarimetric data. The scattering matrix can only deal with the coherent scattering wave or coherent sum of some point targets within the single resolution cell (resulting in a so-called compound scattering matrix under specific conditions [12]). On the other hand, coherent interference of scattered waves from many elementary scatterers results in a granular noise pattern in radar images, called *speckle noise* (multiplicative noise), generated over the natural distributed area [13]. In this case, each elementary scatter (distributed scatterers) is assumed to be randomly located, and the reflected wave intensity and phase from each scatterer are also random, resulting in an incoherent received wave. Hence the target interpretation by the scattering matrix is not straightforward under the speckle. In practice, speckle filtering has to be applied before the physical interpretation of the polarimetric data. The speckle filtering applies the ensemble averaging to each pixel typically realized by spatial multi-looking averaging. The resulted polarimetric data of the speckle filtering shows second-order statistics that should be exploited for distributed scatterers.

The two well-known 3×3 monostatic covariance matrices $\left[C_3 \right]$ and the coherency matrix $\left[T_3 \right]$ are defined by second-order statistics, defined as

$$
\begin{aligned}
\left[C_3 (HV) \right] &= \left\langle k_L (HV) k_L (HV)^{*T} \right\rangle \\
&= \begin{bmatrix}
\left\langle S_{HH} S_{HH}^* \right\rangle & \sqrt{2} \left\langle S_{HH} S_{HV}^* \right\rangle & \left\langle S_{HH} S_{VV}^* \right\rangle \\
\sqrt{2} \left\langle S_{HV} S_{HH}^* \right\rangle & 2 \left\langle S_{HV} S_{HV}^* \right\rangle & \sqrt{2} \left\langle S_{HV} S_{VV}^* \right\rangle \\
\left\langle S_{VV} S_{HH}^* \right\rangle & \sqrt{2} \left\langle S_{VV} S_{HV}^* \right\rangle & \left\langle S_{VV} S_{VV}^* \right\rangle
\end{bmatrix}
\end{aligned}
$$

(Equation 2.56)

$$
\begin{aligned}
\left[T_3 (HV) \right] &= \left\langle k_p (HV) k_p (HV)^{*T} \right\rangle \\
&= \frac{1}{2} \begin{bmatrix}
\left\langle \left| S_{HH} + S_{VV} \right|^2 \right\rangle & \left\langle \left(S_{HH} + S_{VV} \right)\left(S_{HH} - S_{VV} \right)^* \right\rangle & \left\langle 2 S_{HV}^* \left(S_{HH} + S_{VV} \right) \right\rangle \\
\left\langle \left(S_{HH} - S_{VV} \right)\left(S_{HH} + S_{VV} \right)^* \right\rangle & \left\langle \left| S_{HH} - S_{VV} \right|^2 \right\rangle & \left\langle 2 S_{HV}^* \left(S_{HH} - S_{VV} \right) \right\rangle \\
\left\langle 2 S_{HV} \left(S_{HH} + S_{VV} \right)^* \right\rangle & \left\langle 2 S_{HV} \left(S_{HH} - S_{VV} \right)^* \right\rangle & \left\langle 4 \left| S_{HV} \right|^2 \right\rangle
\end{bmatrix}
\end{aligned}
$$

(Equation 2.57)

where $\langle \cdot \rangle$ denotes ensemble averaging, assuming homogeneity of the random medium. The k_L and k_p are the lexicographic and Pauli scattering vectors in (H-V) basis and are the vector notations of the scattering matrix, defined as

$$
k_L (HV) = \begin{bmatrix} S_{HH} \\ \sqrt{2} S_{HV} \\ S_{VV} \end{bmatrix}
$$

(Equation 2.58)

$$
k_p (HV) = \frac{1}{\sqrt{2}} \begin{bmatrix} S_{HH} + S_{VV} \\ S_{HH} - S_{VV} \\ 2 S_{HV} \end{bmatrix}
$$

(Equation 2.59)

One important aspect is that both $\left[C_3 \right]$ and $\left[T_3 \right]$ are positive semi-definite Hermitian matrices. Therefore, all eigenvalues are real and positive values, accepting zero. Note that the off-diagonal of $\left[C_3 \right]$ show correlation coefficients between complex scattering coefficients, so-called *polarimetric correlation coefficients* [14]. Such correlation information is not taken into account in the scattering matrix. Therefore, more polarimetric target information can be extracted from the second-order polarimetric matrices than the scattering matrix. More specifically, polarimetric matrices show nine totally independent polarimetric parameters, while the scattering matrix has five independent parameters for monostatic radar configuration. When polarimetric matrices are formed as the single-look, the number of independent polarimetric parameters of such matrices is equal to the scattering matrix because the modulus of off-diagonal elements of polarimetric matrices becomes one.

2.3.4　SECOND-ORDER POLARIMETRIC MATRICES IN CIRCULAR POLARIZATION BASIS

Like the scattering matrix, covariance and coherency matrices defined in circular polarization basis can be derived from those of the linear polarization basis via unitary transformation.

The 3D monostatic k_L and k_p in (L-R) basis can be defined as the vector notation of the scattering matrix,

$$k_L(LR) = \begin{bmatrix} S_{LL} \\ \sqrt{2}S_{LR} \\ S_{RR} \end{bmatrix} \qquad \text{(Equation 2.60)}$$

$$k_P(LR) = \frac{1}{\sqrt{2}} \begin{bmatrix} S_{LL} + S_{RR} \\ S_{LL} - S_{RR} \\ 2S_{LR} \end{bmatrix} \qquad \text{(Equation 2.61)}$$

Employing the relationship between (L-R) basis and (H-V) basis in (Equation 2.55), the following expressions of the basis transformation are derived

$$\begin{aligned} k_L(LR) &= \frac{1}{2} \begin{bmatrix} S_{HH} - S_{VV} + 2jS_{HV} \\ +\sqrt{2}j(S_{HH} + S_{VV}) \\ -(S_{HH} - S_{VV}) + 2jS_{HV} \end{bmatrix} \\ &= \frac{1}{2} \begin{bmatrix} 1 & j\sqrt{2} & 1 \\ j\sqrt{2} & 0 & j\sqrt{2} \\ -1 & j\sqrt{2} & 1 \end{bmatrix} \begin{bmatrix} S_{HH} \\ \sqrt{2}S_{HV} \\ S_{VV} \end{bmatrix} \\ &= \left[U_{L(HV \to LR)} \right] k_L(HV) \end{aligned} \qquad \text{(Equation 2.62)}$$

$$\begin{aligned} k_P(LR) &= \frac{1}{\sqrt{2}} \begin{bmatrix} 2jS_{HV} \\ S_{HH} - S_{VV} \\ j(S_{HH} + S_{VV}) \end{bmatrix} \\ &= \frac{1}{\sqrt{2}} \begin{bmatrix} 0 & 0 & j \\ 0 & 1 & 0 \\ j & 0 & 0 \end{bmatrix} \begin{bmatrix} S_{HH} + S_{VV} \\ S_{HH} - S_{VV} \\ 2S_{HV} \end{bmatrix} \\ &= \left[U_{P(HV \to LR)} \right] k_P(HV) \end{aligned} \qquad \text{(Equation 2.63)}$$

where $\left[U_{L(HV \to LR)} \right]$ and $\left[U_{P(HV \to LR)} \right]$ are unitary matrix. The covariance and coherency matrices on an (L-R) basis are then defined by $k_L(LR)$ and $k_P(LR)$ and linked with (H-V) basis

$$
\begin{aligned}
\left[C_3(LR) \right] &= \left\langle k_L(LR) k_L(LR)^{*T} \right\rangle \\
&= \begin{bmatrix}
\left\langle S_{LL} S_{LL}^* \right\rangle & \sqrt{2}\left\langle S_{LL} S_{LR}^* \right\rangle & \left\langle S_{LL} S_{RR}^* \right\rangle \\
\sqrt{2}\left\langle S_{LR} S_{LL}^* \right\rangle & 2\left\langle S_{LR} S_{LR}^* \right\rangle & \sqrt{2}\left\langle S_{LR} S_{RR}^* \right\rangle \\
\left\langle S_{RR} S_{LL}^* \right\rangle & \sqrt{2}\left\langle S_{RR} S_{LR}^* \right\rangle & \left\langle S_{RR} S_{RR}^* \right\rangle
\end{bmatrix} \\
&= \left[U_{L(HV \to LR)} \right] \left\langle k_L(HV) k_L(HV)^{*T} \right\rangle \left[U_{L(HV \to LR)} \right]^{*T} \\
&= \left[U_{L(HV \to LR)} \right] \left[C(HV) \right] \left[U_{L(HV \to LR)} \right]^{*T}
\end{aligned}
$$

(Equation 2.64)

$$
\begin{aligned}
\left[T_3(LR) \right] &= \left\langle k_P(LR) k_P(LR)^{*T} \right\rangle \\
&= \frac{1}{2} \begin{bmatrix}
\left\langle |S_{LL} + S_{RR}|^2 \right\rangle & \left\langle (S_{LL} + S_{RR})(S_{LL} - S_{RR})^* \right\rangle & \left\langle 2 S_{LR}^* (S_{LL} + S_{RR}) \right\rangle \\
\left\langle (S_{LL} - S_{RR})(S_{LL} + S_{RR})^* \right\rangle & \left\langle |S_{LL} - S_{RR}|^2 \right\rangle & \left\langle 2 S_{LR}^* (S_{LL} - S_{RR}) \right\rangle \\
\left\langle 2 S_{LR} (S_{LL} + S_{RR})^* \right\rangle & \left\langle 2 S_{LR} (S_{LL} - S_{RR})^* \right\rangle & \left\langle 4 |S_{LR}|^2 \right\rangle
\end{bmatrix} \\
&= \left[U_{P(HV \to LR)} \right] \left\langle k_P(HV) k_P(HV)^{*T} \right\rangle \left[U_{P(HV \to LR)} \right]^{*T} \\
&= \left[U_{P(HV \to LR)} \right] \left[C(HV) \right] \left[U_{P(HV \to LR)} \right]^{*T} .
\end{aligned}
$$

(Equation 2.65)

Therefore, the basis transformation of covariance and coherency matrices from (H-V) basis to (L-R) basis is realized by unitary matrices $\left[U_{L(HV \to LR)} \right]$ and $\left[U_{P(HV \to LR)} \right]$, respectively. Note that the information content is mathematically preserved under that unitary basis transformation, that is, full linear polarimetric measurement data and full circular polarimetric measurement data are mathematically equivalent to each other.

2.4 POLARIMETRIC SAR TARGET DECOMPOSITION THEOREM

2.4.1 INTRODUCTION OF POLARIMETRIC SAR TARGET DECOMPOSITION

The polarimetric target decomposition has been widely studied and applied for target classification and scattering data inversion, and was first formalized by Huynen [15]. The coherent decomposition aims at decomposing the scattering matrix into a canonical scattering mechanism. However, such decomposition can only be applied to images with fewer speckle. For natural distributed terrain surface observation monitoring, incoherent decomposition concepts that decompose the polarimetric matrices (the covariance or the coherency matrices) are more appropriate, as these aim at providing the dominant or averaged scattering mechanism.

Among the number of incoherent target decomposition theorems, the *eigenvector-based polarimetric decomposition theorem* is herein demonstrated. The *model-based decomposition* is another incoherent polarimetric decomposition class that fits the pre-defined physical scattering models to polarimetric matrices to decompose the scattering power [16], [17]. This power decomposition has been utilized in a wide application range thanks to its easy understanding of the scattering mechanism interpretation.

2.4.2 TARGET DECOMPOSITION FOR FULLY POLARIMETRIC SYSTEMS

Cloude first introduced eigenvector-based decomposition to derive the dominant scattering mechanism based on eigenvalues and eigenvectors of the Coherency matrix [18], [19]. Subsequently, Cloude and Pottier introduced a scheme for extracting average polarimetric parameters, called $H/\bar{\alpha}$ polarimetric decomposition theorem [2], [20]. In eigenvector-based decomposition, the scattering mechanisms and their relative magnitudes correspond to the eigenvectors and eigenvalues, respectively, obtained by diagonalization of the averaged coherency matrix as

$$\left[\boldsymbol{T}_3\right] = \left[\boldsymbol{U}_3\right] \begin{bmatrix} \lambda_1 & 0 & 0 \\ 0 & \lambda_2 & 0 \\ 0 & 0 & \lambda_3 \end{bmatrix} \left[\boldsymbol{U}_3\right]^{-1} = \sum_{i=1}^{3} \lambda_i \boldsymbol{u}_i \boldsymbol{u}_i^{*T} \,,$$

$$\lambda_1 \geq \lambda_2 \geq \lambda_3 \geq 0 \qquad\qquad \text{(Equation 2.66)}$$

where $\left[\boldsymbol{U}_3\right] = \left[u_1, u_2, u_3\right]$. Since the $\boldsymbol{T}_3$ matrix is defined as the positive semi-definite Hermitian matrix, derived eigenvalues are real and positive and may have zero values. In this framework, the eigenvectors are parameterized as

$$\boldsymbol{u}_i = \left[\cos\alpha_i \quad \sin\alpha_i \cos\beta_i e^{j\phi_i} \quad \sin\alpha_i \cos\beta_i e^{j\gamma_i}\right]^T \qquad \text{(Equation 2.67)}$$

where phase ϕ_i expresses the relative phase between $S_{HH} - S_{VV}$ and $S_{HH} + S_{VV}$, similarly, phase γ_i is the relative phase between S_{HV} and $S_{HH} + S_{VV}$, α_i represents the intrinsic scattering type, and β_i is related to the polarimetric orientation angle about the radar line-of-sight. The expression in (Equation 2.66) explicitly shows mathematical decomposition as the sum of three independent scattering targets, and the scattering matrix can represent each. To represent the averaged scattering mechanism of three independent scatterings, mean α and β are employed, realized by the statistically weighted average

$$\bar{\alpha} = \sum_{i=1}^{3} p_i \alpha_i \qquad\qquad \text{(Equation 2.68)}$$

$$\bar{\beta} = \sum_{i=1}^{3} p_i \beta_i \qquad \text{(Equation 2.69)}$$

where scattering probabilities are

$$p_i = \frac{\lambda_i}{\sum_{k=1}^{3} \lambda_k} \text{ with } \sum_{k=1}^{3} p_i = 1 \qquad \text{(Equation 2.70)}$$

Moreover, the degree of the scattering randomness is defined by employing eigenvalues, called *scattering entropy H* shown as

$$H = -\sum_{i=1}^{3} p_i \log_3 (p_i) \qquad \text{(Equation 2.71)}$$

where the higher H indicates high scattering randomness. Because the entropy H is not a unique function, the polarimetric anisotropy A can be used as the complemental parameter of H, defined as

$$A = \frac{\lambda_2 - \lambda_3}{\lambda_2 + \lambda_3} \qquad \text{(Equation 2.72)}$$

The anisotropy A becomes important for the targets showing higher λ_2 and λ_3 equivalent to lower H.

To evaluate the α parameter for canonical targets, we here consider the scattering from a single canonical target, that is, $\lambda_1 \neq 0, \lambda_2 = \lambda_3 = 0$. Under such a condition, $[T_3] = \lambda_1 u_1 \cdot u_1^{*T} = k_p \cdot k_p^{T*}$; hence, $k_p = \sqrt{\lambda_1} u_1$.

Because u_1 is the unit vector, normalized Pauli vector $\left(1/\|k_p\|\right) k_p$ corresponds to the eigenvector u_1. Based on this fact, we can relate α to the components of k_p

$$\cos^2 \alpha = \frac{|S_{HH} + S_{VV}|^2}{|S_{HH} + S_{VV}|^2 + |S_{HH} - S_{VV}|^2 + 4|S_{HV}|^2} \qquad \text{(Equation 2.73)}$$

Hence, α of the three basic scatterers, that are, odd-bounce scattering (e.g., plate, sphere, and trihedral), dipole scattering (e.g., wire target), and even-bounce scattering (e.g., dihedral) tabulated in Table 2.1, is derived as

$$\text{odd}: \alpha = 0°, \text{ dipole}: \alpha = 45°, \text{ even}: \alpha = 90° \qquad \text{(Equation 2.74)}$$

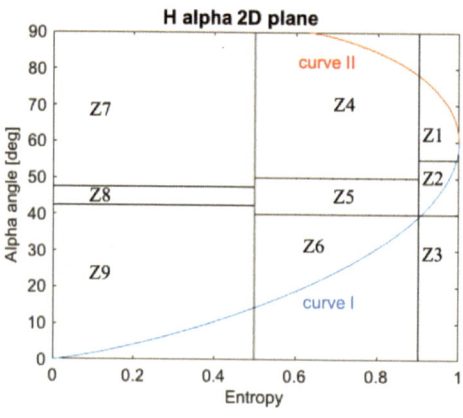

FIGURE 2.8 $H/\bar{\alpha}$ 2D plane.

It is worth noting that the three eigenvalues and the α_i are the roll-invariant parameters, while β_i, ϕ_i, and γ_i are rotational variant parameters. Therefore, frequently employed H, A, and $\bar{\alpha}$ remain rotational invariant parameters, which is one of the important aspects of $H/\bar{\alpha}$ decomposition.

The obtained pairs of H and $\bar{\alpha}$ are then plotted on the $H/\bar{\alpha}$ 2D plane with some boundaries to clarify the feasible zones, as shown in Figure 2.8 [20]. This plane will help to classify the target based on the scattering mechanism and randomness according to defined zones. Because we apply averaging operation, the distribution of $\bar{\alpha}$ in the 2D plane is bounded as a function of H[20]. We can deduce two boundary curves I (lower bound) and II (upper bound) and determine the feasible zone. We define curve I by setting $\bar{\alpha} = 0$ (odd-bounce or surface scattering) for dominant eigenvector with depolarizing parameter m_d. As m_d increases, $\bar{\alpha}$ moves toward 60° and H tends to 1. On the other hand, we define curve II by setting $\bar{\alpha} = \pi/2$ for dominant eigenvector at $0 \leq m_d \leq 0.5$, but for $0.5 \leq m_d \leq 1$, curve II toward 60° ($H = 1$). Corresponding diagonal coherency matrices are given in (Equation 2.75) as a function of m_d. As a result, $\bar{\alpha}$ takes the full range of values from 0 to 90° for $H = 0$, while $\bar{\alpha}$ can take only 60° for $H = 1$.

$$\text{Curve I}\left[\boldsymbol{T}_3\right]_I = \begin{bmatrix} 1 & 0 & 0 \\ 0 & m_d & 0 \\ 0 & 0 & m_d \end{bmatrix} \quad 0 \leq m_d \leq 1$$

$$\text{Curve II}\begin{cases} \left[\boldsymbol{T}_3\right]_{II} = \begin{bmatrix} 0 & 0 & 0 \\ 0 & 1 & 0 \\ 0 & 0 & 2m_d \end{bmatrix} \quad 0 \leq m_d \leq 0.5 \\[20pt] \left[\boldsymbol{T}_3\right]_{II} = \begin{bmatrix} 2m_d-1 & 0 & 0 \\ 0 & 1' & 0 \\ 0 & 0 & 1 \end{bmatrix} \quad 0.5 \leq m_d \leq 1 \end{cases} \quad \text{(Equation 2.75)}$$

Furthermore, the 2D plane can be extended to the 3D plane by adding the anisotropy A for better classification [6].

In the FP SAR system, the scattering matrix observed on an (L-R) basis can be transformed into an (H-V) basis. Consequently, we can follow the same decomposition procedure on an (H-V) basis. Instead, diagonalization of the averaged coherency matrix on an (L-R) basis also leads to H and $\bar{\alpha}$ parameters estimations,

$$
\begin{aligned}
\left[T_3(LR) \right] &= \left[U_{P(HV \to LR)} \right] \left[C(HV) \right] \left[U_{P(HV \to LR)} \right]^{*T} \\
&= \left[U_{P(HV \to LR)} \right] \left[U_3 \right] \begin{bmatrix} \lambda_1 & 0 & 0 \\ 0 & \lambda_2 & 0 \\ 0 & 0 & \lambda_3 \end{bmatrix} \left[U_3 \right]^{-1} \left[U_{P(HV \to LR)} \right]^{*T} \\
&= \left[U_{LR} \right] \begin{bmatrix} \lambda_1 & 0 & 0 \\ 0 & \lambda_2 & 0 \\ 0 & 0 & \lambda_3 \end{bmatrix} \left[U_{LR} \right]^{*T} = \sum_{i=1}^{3} \lambda_i l_i l_i^{*T}
\end{aligned}
\qquad \text{(Equation 2.76)}
$$

where $\left[U_{LR} \right] = \left[l_1, l_2, l_3 \right]$ and the eigenvectors are parameterized as

$$
l_i = \left[j \sin\alpha_i \sin\beta_i e^{j\gamma_i} \quad \sin\alpha_i \cos\beta_i e^{j\phi_i} \quad j\cos\alpha_i \right]^T . \qquad \text{(Equation 2.77)}
$$

2.4.3 Eigenvector-Based Decomposition for Dual Polarimetric Systems

Cloude investigated $H / \bar{\alpha}$ decomposition scheme for dual linear polarimetric (DLP) data [21]. The DLP, in this case, means transmitting single linear polarization and receiving two orthogonal linear polarizations coherently, containing HH/VH or VV/HV scattering matrix coefficients.

First, we describe the depolarized received wave state by the outer product of the Jones vector with ensemble averaging called 2×2 wave coherency matrix

$$
\left[J_2 \right] = \left\langle E_r \cdot E_r^{*T} \right\rangle \qquad \text{(Equation 2.78)}
$$

where $\left[J_2 \right]$ is the complex Hermitian positive semidefinite matrix. Let us consider the H-polarization transmission, where the received field can be expressed by

$$
E_{rH} = \begin{bmatrix} E_{rHH} \\ E_{rVH} \end{bmatrix} = \begin{bmatrix} S_{HH} & S_{HV} \\ S_{VH} & S_{VV} \end{bmatrix} \begin{bmatrix} 1 \\ 0 \end{bmatrix} = \begin{bmatrix} S_{HH} \\ S_{VH} \end{bmatrix} \qquad \text{(Equation 2.79)}
$$

Hence, we can form $\left\lfloor J_2 \right\rfloor$ as

$$
\left[J_2 \right] = \begin{bmatrix} \left\langle S_{HH} S_{HH}^* \right\rangle & \left\langle S_{HH} S_{VH}^* \right\rangle \\ \left\langle S_{VH} S_{HH}^* \right\rangle & \left\langle S_{VH} S_{VH}^* \right\rangle \end{bmatrix} \qquad \text{(Equation 2.80)}
$$

which is related to four stokes parameters,

$$[J_2] = \begin{bmatrix} \langle g_0 \rangle + \langle g_1 \rangle & \langle g_2 \rangle - \langle g_3 \rangle \\ \langle g_2 \rangle + \langle g_3 \rangle & \langle g_0 \rangle - \langle g_1 \rangle \end{bmatrix} \qquad \text{(Equation 2.81)}$$

It is important to note that the coherent dual-polarization (DP) system is required to retain the relative phase between the two received polarizations; otherwise, the off-diagonal elements in (Equation 2.80) are not measurable [21]. For example, ENVISAT's ASAR "alternating polarization" mode is not coherent DP [22] while recently available European Space Agency (ESA) Sentinel-1 dual-polarization data satisfy the coherent DP.

We apply eigen decomposition to the $[J_2]$ of coherent DP system,

$$[J_2] = [U_2] \begin{bmatrix} \lambda_1 & 0 \\ 0 & \lambda_2 \end{bmatrix} [U_2]^{*T} \qquad \text{(Equation 2.82)}$$

where $[U_2]$ is unitary matrix defined by wave polarization ellipse parameters α_w and relative phase δ in (Equation 2.13),

$$[U_2] = \begin{bmatrix} \cos\alpha_w & -\sin\alpha_w e^{-j\delta} \\ \sin\alpha_w e^{j\delta} & \cos\alpha_w \end{bmatrix} \qquad \text{(Equation 2.83)}$$

The wave entropy H_w that measures the degree of depolarization is defined by

$$H_w = \sum_{i=1}^{2} P_i \left(-\log_2 P_i \right) \qquad \text{(Equation 2.84)}$$

where probabilities are

$$P_i = \frac{\lambda_i}{\lambda_1 + \lambda_2} (i = 1, 2) \qquad \text{(Equation 2.85)}$$

The H_w is closely related to the degree of polarization m defined as the anisotropy using two eigenvalues, shown as

$$m = \frac{\sqrt{g_1^2 + g_2^2 + g_3^2}}{g_0} = \frac{\lambda_1 - \lambda_2}{\lambda_1 + \lambda_2} \qquad \text{(Equation 2.86)}$$

A zero m corresponds to maximum wave entropy $H_w = 1$ and vice versa.

Calculation of $\overline{\alpha_w}$ is performed by the following procedure,

$$\overline{\alpha_w} = P_1 \alpha_w + P_2 \left(\frac{\pi}{2} - \alpha_w \right) \qquad \text{(Equation 2.87)}$$

Let us evaluate the α_w for basic canonical targets. For odd-bounce, dipole scattering (wire target), and even-bounce scattering as tabulated in Table 2.1, we obtain the α_w when the received field is completely polarized wave,

$$\text{odd}: \alpha_w = 0,$$

$$\text{dipole}: \alpha_w = \frac{1}{2}\cos^{-1}\left(\frac{4\cos^4\theta - \sin^2(2\theta)}{4\cos^4\theta + \sin^2(2\theta)}\right),$$

$$\text{even}: \alpha_w = \frac{1}{2}\cos^{-1}(\cos(4\theta)) \qquad \text{(Equation 2.88)}$$

Notice that the α_w $(0 \le \alpha_w \le \pi/2)$ is determined by the target orientation angle θ $(-\pi/2 < \theta \le \pi/2)$; thus, they are no longer roll-invariant parameters.

We can derive α_w by the first and second component of the Stokes vector in (H-V) basis

$$\cos(2\alpha_w) = \frac{\langle g_1 \rangle}{\langle g_0 \rangle} = \frac{\left\langle |S_{HH}|^2 \right\rangle - \left\langle |S_{VH}|^2 \right\rangle}{\left\langle |S_{HH}|^2 \right\rangle + \left\langle |S_{VH}|^2 \right\rangle} \qquad \text{(Equation 2.89)}$$

The obtained pairs of H_w and $\overline{\alpha_w}$ values are plotted on the $H_w / \overline{\alpha_w}$ 2D plane with some boundaries. Two bounding curves, I and II, are defined similarly with (Equation 2.75). In the DP case, 2×2 diagonal matrices as a function of depolarizing parameter m_d for two curves are

$$\text{Curve I} \begin{bmatrix} 1 & 0 \\ 0 & m_d \end{bmatrix} 0 \le m_d \le 1,$$

$$\text{Curve II} \begin{bmatrix} m_d & 0 \\ 0 & 1 \end{bmatrix} 0 \le m_d \le 1 \qquad \text{(Equation 2.90)}$$

Consequently, the two curves show asymmetry, that differs from the FP case, as shown in Figure 2.9.

Although we can derive parameters from coherent DP data, these do not differentiate the odd or even-bounce scattering mechanism as we see in (Equation 2.88). The short reason for this is that the odd-bounce classification requires the relative phase between S_{HH} and S_{VV}. Such a lack of information limits the DP decomposition, and an interpretation of derived parameters differs from those of familiar FP ones. For example, the DP yields α_w that is a ratio of wave components, while FP yields scattering alpha angle α that is a ratio of scattering coefficients [23].

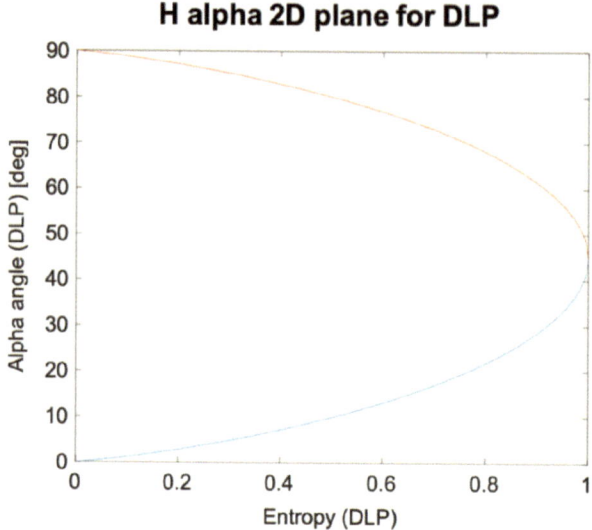

FIGURE 2.9 $H_w / \overline{\alpha}_w$ 2D plane for DLP data.

2.5   COMPACT POLARIMETRIC SAR

2.5.1   BASIC CONCEPT OF COMPACT POLARIMETRY

The data of FP mode, that is, to transmit and receive orthogonal two polarizations, based on a 3×3 polarimetric matrix gives us a complete polarimetric signature of the scene, which enhances the target parameter retrieval and polarimetric discrimination. However, inherent limitations such as reduction of swath width, an increase in system complexity, data volume, and power consumption should be compromised. On the other hand, DP measurement (system transmits on one polarization only) needs only half pulse-repetition frequency (double pulse repetition interval) relative to FP measurement; thus, swath width that is proportional to the receiving window can be more than double the width of the FP SAR, as demonstrated in Figure 2.10 (a) and (b). For this reason, DLP measurements are still the main operational mode in most spaceborne SAR missions. Nonetheless, the DLP mode does not afford complete information regarding the polarization state of the targets. For example, the DLP results do not give the possibility to separate surface scattering and double bounce scattering as we demonstrated before.

In recent years, to circumvent the aforementioned drawbacks of the DP mode, DP with circular polarization, or 45° tilted linear polarization in transmission has emerged with the so-called compact polarimetric (CP) SAR [24]–[26]. In CP, the transmission of both H and V polarizations done in FP is realized by circular or 45° tilted linear polarization transmission, which is equivalent to simultaneous transmission of H and V (with 90° of δ for circular polarization). The objective of CP is to realize better polarimetric decomposition and maximize the polarimetric

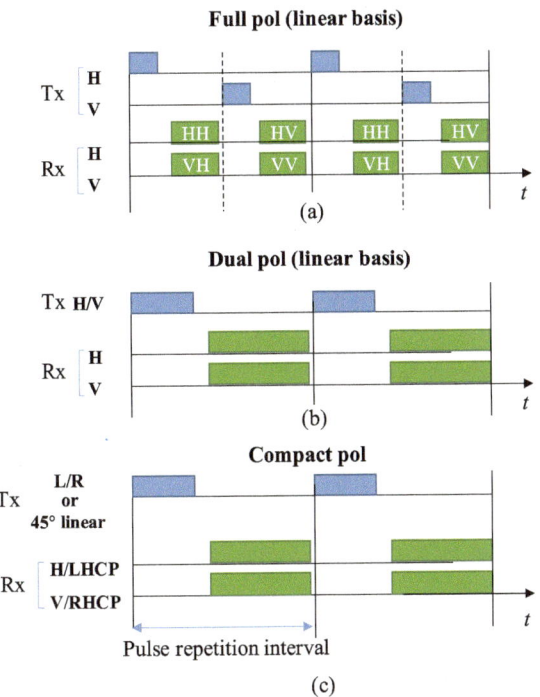

FIGURE 2.10 Tx and Rx timing of three polarimetric modes. (a) FP on (H-V) basis, (b) DP on (H-V) basis, (c) CP.

information of the target with single-transmitted polarization. Since it transmits single-polarization, wider swath width can then be retained better than in FP mode. Therefore, the CP is considered a balanced strategy to obtain reasonable swath width and polarimetric information. Three modes have been discussed, that is, the $\pi/4$ mode [25], the dual-circular polarimetric (DCP) mode [26], and the circular transmit while in linear receive (CTLR) mode [24]. These different polarimetric SAR modes are categorized in Figure 2.11.

The CP with circular transmits and linear receive was first introduced in the lunar-orbiting satellites as an imaging radar. Mini-SAR on India's Chandraayan-1 [27] and Mini-RF on NASA's LRO [28], both launched in 2008, adopted the CP architecture. Afterward, CP mode was implemented on RISAT-1 [29] and ALOS-2 [30] (as an experimental mode) which are Earth-observing SAR. Furthermore, most of the imaging mode in RADARSAT Constellation Mission (RCM) as a three-satellite constellation and launched in 2019, uses CP mode [31].

2.5.2 COMPACT POLARIMETRY DATA ANALYSIS

One requirement of the CP data is that the radar system should receive coherent orthogonal polarizations to retain the relative phase between two received channels;

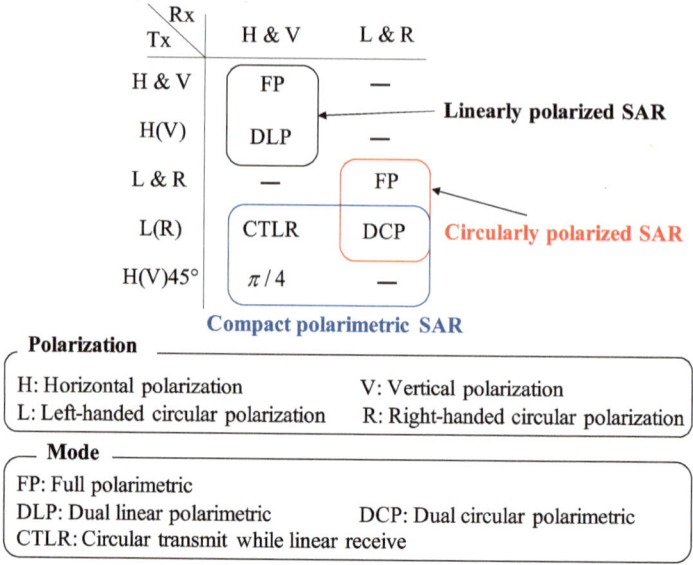

FIGURE 2.11 Categorization of various polarimetric SAR modes.

otherwise, the wave coherency matrix is not sufficiently derived. We see the same requirement in the DLP decomposition case, as demonstrated.

Let us demonstrate the scattering vector of three CP modes, that is, CTLR, DCP, and $\pi/4$. In the case of linear polarization receiving, the expression in (Equation 2.50) can be generalized on an arbitrary polarization basis (A-B) of transmission, defined as

$$k = \begin{bmatrix} E_{rH} \\ E_{rV} \end{bmatrix} = \frac{1}{\sqrt{1+\rho\rho^*}} \begin{bmatrix} S_{HH} & S_{HV} \\ S_{VH} & S_{VV} \end{bmatrix} \begin{bmatrix} 1 & -\rho^* \\ \rho^* & 1 \end{bmatrix} \begin{bmatrix} e^{j\xi} & 0 \\ 0 & e^{-j\xi} \end{bmatrix} E_t(AB)$$

(Equation 2.91)

Based on the definition in (Equation 2.91), we can now obtain the received field vector representation for CTLR ($\rho = j$ and $\alpha = 0$) and $\pi/4$ ($\rho = 1$ and $\alpha = 0$) modes,

$$k_{CTLR} = \frac{1}{\sqrt{2}} \begin{bmatrix} S_{HH} + jS_{HV} \\ S_{VH} + jS_{VV} \end{bmatrix}, \ (\text{LHCP transmit})$$

(Equation 2.92)

$$k_{\frac{\pi}{4}} = \frac{1}{\sqrt{2}} \begin{bmatrix} S_{HH} + S_{HV} \\ S_{VV} + S_{VH} \end{bmatrix} \cdot (+\pi/4 \ \text{transmit})$$

(Equation 2.93)

On the other hand, we can transform the basis for both transmitting and receiving in the case of DCP modes. As a result, the scattering vector of DCP modes for the LHCP transmit case can be defined as

$$\mathbf{k}_{DCP} = \begin{bmatrix} E_{rL} \\ E_{rR} \end{bmatrix} = \begin{bmatrix} S_{LL} \\ S_{RL} \end{bmatrix} = \frac{1}{2} \begin{bmatrix} S_{HH} - S_{VV} + 2jS_{HV} \\ j\left(S_{HH} + S_{VV}\right) \end{bmatrix} \qquad \text{(Equation 2.94)}$$

where we use the basis relationship in (Equation 2.55).

We can then form the 2×2 wave coherency matrix defined in (Equation 2.80) by scattering vectors of CP modes. For DCP mode with LHCP transmit, we get

$$\left[J_2 \right]_{DCP} = \begin{bmatrix} \langle S_{LL} S_{LL}^* \rangle & \langle S_{LL} S_{RL}^* \rangle \\ \langle S_{RL} S_{LL}^* \rangle & \langle S_{RL} S_{RL}^* \rangle \end{bmatrix} = \begin{bmatrix} \langle g_o \rangle + \langle g_3 \rangle & \langle g_2 \rangle + j \langle g_1 \rangle \\ \langle g_2 \rangle - \langle j g_1 \rangle & \langle g_o \rangle - \langle g_3 \rangle \end{bmatrix} \qquad \text{(Equation 2.95)}$$

By substituting (Equation 2.94) into (Equation 2.95), we get

$$\left[J_2 \right]_{DCP} = \frac{1}{4} \begin{bmatrix} \left| \left(S_{HH} - S_{VV}\right) \right|^2 & -j\left(S_{HH} - S_{VV}\right) \cdot \left(S_{HH} + S_{VV}\right)^* \\ j\left(S_{HH} + S_{VV}\right) \cdot \left(S_{HH} - S_{VV}\right)^* & \left| \left(S_{HH} + S_{VV}\right) \right|^2 \end{bmatrix} \cdot$$
$$\frac{1}{4} \begin{bmatrix} 4\left|S_{HV}\right|^2 & 0 \\ 0 & 0 \end{bmatrix} + \frac{1}{4} \begin{bmatrix} 4\mathrm{Im}\left(\left(S_{HH} - S_{VV}\right) \cdot S_{HV}^*\right) & 2\left(S_{HH} + S_{VV}\right)^* \cdot S_{HV} \\ 2\left(S_{HH} + S_{VV}\right) \cdot S_{HV}^* & 0 \end{bmatrix}$$

$$\text{(Equation 2.96)}$$

The 2×2 wave coherency matrix for DCP modes results as a sum of three terms [32]. The first and second terms are expressed by Pauli basis coefficients, that are, $\left(S_{HH} - S_{VV}\right)$, $\left(S_{HH} + S_{VV}\right)$ and S_{HV}. On the other hand, the third term consists of co-pol/cross-pol correlation.

The parameter retrieval in the context of CP SAR often starts with the Stokes vector or 2×2 wave coherency matrix [33]. We have demonstrated that the four Stokes parameters lead to the polarimetric geometrical parameters τ, ξ, and A as well as relative phase δ.

We have already shown that the decomposition by DLP mode is not capable of discriminating the three basic scattering mechanisms, but only the received wave state can be estimated. On the other hand, when circular polarization is chosen as the transmission, four Stokes parameters can discriminate the three basic scattering types, as we will demonstrate next. This can be inferred from the $\left[J_2 \right]_{DCP}$ in (Equation 2.96) in that the wave coherency matrix in CP mode consists of Pauli basis coefficients. It is noted that since Stokes parameters are determined regardless of received polarization basis, as shown in (Equation 2.96), all CP modes are comparable to each other.

It is important to note that the information content between the CP and DLP are different in practice. The HV polarization of DLP is difficult to accurately reconstruct using CP data, which was mentioned in [34]. Therefore, the data obtained in CP mode is nothing more than complementary information to the DLP mode. The suitability of information content is highly dependent on the applications. For example, HV measurement is key to forest biomass observation.

Several target decomposition theories for CP data have been proposed. There are two main groups: decomposition deduced by the Stokes vector [35] and the 2×2 wave coherency matrix [36]. The two decomposition approaches carry identical information. In the following subsections, we specifically focus on the decomposition approach based on the data format from DCP mode.

2.5.3 TARGET DECOMPOSITION THEOREM FOR COMPACT MODE

The CP decomposition concept addresses classifying the target in terms of three basic scattering mechanisms, that are even-bounce, odd-bounce, and volume scattering.

In this context, Raney introduced the target characterization approach by secondary/child Stokes parameters [24], [37]. The $m - \delta$ decomposition was first proposed to perform the decomposition by means of the degree of polarization (DoP) m and relative phase δ derived from Stokes parameters [37]. The 2D plane using m and δ ($m - \delta$ plane) is employed to map the target scattering feature. The m is closely related with entropy that measures the randomness of backscatter. The δ is the relative phase between received horizontal and vertical E-field components; hence it can represent the rotational direction of the ellipse. For a perfect LHCP transmission, $\delta = -90°$ and $\delta = 90°$ correspond to dominant odd-bounce (surface) and even-bounce (double-bounce) scattering. The derived parameters yield three components of scattering power as [38]

$$\begin{bmatrix} P_{odd} \\ P_{vol} \\ P_{even} \end{bmatrix}_{m-\delta} = \begin{bmatrix} \frac{1}{2} g_0 m (1 - \sin \delta) \\ g_0 (1 - m) \\ \frac{1}{2} g_0 m (1 + \sin \delta) \end{bmatrix} \qquad \text{(Equation 2.97)}$$

where $P_{odd} + P_{vol} + P_{even} = g_0$. However, Raney et al. [37] indicated the ambiguity of δ for decomposition purposes when the transmitted circular polarization is not perfectly circular and suggested employing an ellipticity angle χ instead of δ for target decomposition, called $m - \chi$ decomposition [37]. The $m - \chi$ decomposition is performed by

$$\begin{bmatrix} P_{odd} \\ P_{vol} \\ P_{even} \end{bmatrix}_{m-\delta} = \begin{bmatrix} \frac{1}{2} g_0 m (1 - \sin \chi) \\ g_0 (1 - m) \\ \frac{1}{2} g_0 m (1 + \sin \chi) \end{bmatrix} \qquad \text{(Equation 2.98)}$$

Decomposition using three parameters (m,χ,τ) could offer further insight into the reflecting surfaces with known ellipticity of the transmitted field [37].

On the other hand, Cloude et al. proposed the compact polarimetric model-based decomposition by relating the Stokes vector observables obtained from CP to classical full-pol decomposition methods [35]. More specifically, the Stokes vector is linked to scattering coefficients of the medium and not to the properties of the propagation wave. Following this procedure, the scattering mechanism α can then be mapped into Stokes vector elements.

We have already shown the relationship between the wave coherency matrix and scattering matrix coefficients in (Equation 2.96). By using the relationship between the 2×2 wave coherency matrix and the Stokes vector in (Equation 2.95) as well as the definition of the 3×3 coherency matrix $\left[T_3(HV) \right]$ in (Equation 2.57), we obtain

$$
\mathbf{g} = \begin{bmatrix} \langle g_0 \rangle \\ \langle g_1 \rangle \\ \langle g_2 \rangle \\ \langle g_3 \rangle \end{bmatrix} = \begin{bmatrix} \frac{1}{2}\left(t_{11} + t_{22} + t_{33}\right) \pm \mathrm{Im}\left(t_{23}\right) \\ \mathrm{Re}\left(t_{12}\right) \pm \mathrm{Im}\left(t_{13}\right) \\ \mathrm{Re}\left(t_{13}\right) \pm \mathrm{Im}\left(t_{12}\right) \\ \mathrm{Im}\left(t_{23}\right) \pm \frac{1}{2}\left(t_{22} + t_{33} - t_{11}\right) \end{bmatrix}.
\qquad \text{(Equation 2.99)}
$$

This relationship can map the Stokes vector into the coherency matrix elements.

Here the compact decomposition theory considers that the partially polarized wave consists of a noise signal $(g_1 = g_2 = g_3 = 0)$ and a completely polarized wave of the Stokes vector. The model-based decomposition was proposed in this framework where two models: single (rank-1 symmetric) scattering mechanism (ground) and randomly oriented dipole scattering (volume) are involved, namely random volume over ground (RVOG). Of these, the volume component represents the noise signal. Each of them has the form of 3×3 coherency matrices as follows

$$
\left[T_3 \right]_{ground} = m_s \begin{bmatrix} \cos^2 \alpha_s & \cos \alpha_s \sin \alpha_s \, e^{j\phi} & 0 \\ \cos \alpha_s \sin \alpha_s \, e^{-j\phi} & \sin^2 \alpha_s & 0 \\ 0 & 0 & 0 \end{bmatrix},
$$

$$
\left[T_3 \right]_{volume} = m_v \begin{bmatrix} 2 & 0 & 0 \\ 0 & 1 & 0 \\ 0 & 0 & 0 \end{bmatrix}
\qquad \text{(Equation 2.100)}
$$

where α_s and ϕ account for rotation invariant scattering alpha angle for dominant scattering and phase of t_{12}, the m_s and m_v represent the power of ground and volume scattering, respectively. The corresponding Stokes vector is derived via the relationship in (Equation 2.99)

$$g_{ground} = \frac{m_s}{2} \begin{bmatrix} 1 \\ \sin 2\alpha_s \cos \phi \\ \pm \sin 2\alpha_s \sin \phi \\ \pm \cos 2\alpha_s \end{bmatrix},$$

$$g_{volume} = 2m_v \begin{bmatrix} 1 \\ 0 \\ 0 \\ 0 \end{bmatrix} \qquad \text{(Equation 2.101)}$$

The received wave of the RVOG scattering model involves the sum of the ground component and the random volume component; hence the Stokes vector has the form

$$g_{RVOG} = 2m_v \begin{bmatrix} 1 \\ 0 \\ 0 \\ 0 \end{bmatrix} + \frac{m_s}{2} \begin{bmatrix} 1 \\ \sin 2\alpha_s \cos \phi \\ \pm \sin 2\alpha_s \sin \phi \\ \pm \cos 2\alpha_s \end{bmatrix} \qquad \text{(Equation 2.102)}$$

It is noted that this decomposition is invertible where we have four observables with four unknowns under the assumption of the volume scattering model, that is, the randomly oriented dipole scattering. Consequently, we obtain four parameters

$$\begin{cases} m_s = 2g_0 m \\[2mm] m_v = \dfrac{1}{2} g_0 (1-m) \\[2mm] \alpha_s = \dfrac{1}{2} \tan^{-1} \left(\dfrac{\sqrt{g_1^2 + g_2^2}}{\pm g_3} \right) \\[2mm] \phi = \arg\left(g_1 + jg_2\right) \end{cases} \qquad \text{(Equation 2.103)}$$

Those parameters lead to the pseudo-three-component decomposition defined by

$$
\begin{bmatrix} P_{odd} \\ P_{vol} \\ P_{even} \end{bmatrix}_{Cloude} = \begin{bmatrix} \frac{1}{2} g_0 m \left(1 - \cos 2\alpha_s\right) \\ g_0 \left(1 - m\right) \\ \frac{1}{2} g_0 m \left(1 + \cos 2\alpha_s\right) \end{bmatrix}
\qquad \text{(Equation 2.104)}
$$

2.5.4 Eigenvector-Based Decomposition for Compact Modes

Eigenvector-based decomposition of the 2×2 wave coherency matrix for DCP data $\left[J_2 \right]_{DCP}$ defined in (Equation 2.95) has the form

$$
\left[J_2 \right]_{DCP} = \left\langle k_{DCP} \cdot k_{DCP}^{*T} \right\rangle = \begin{bmatrix} u_1 & u_2 \end{bmatrix} \begin{bmatrix} \lambda_1 & 0 \\ 0 & \lambda_2 \end{bmatrix} \begin{bmatrix} u_1 & u_2 \end{bmatrix}^{*T}
\qquad \text{(Equation 2.105)}
$$

where u_i are the orthogonal eigenvectors of the unitary matrix,

$$
u_i = e^{j\phi_i} \begin{bmatrix} \cos \alpha_{DCPi} e^{j\phi_i} & \sin \alpha_{DCPi} \end{bmatrix}^T
\qquad \text{(Equation 2.106)}
$$

The H_{DCP} and mean $\alpha_{DCP} (\overline{\alpha_{DCP}})$ are then given by

$$
H_{DCP} = \sum_{i=1}^{2} P_i \left(-\log_2 P_i \right),
\qquad \text{(Equation 2.107)}
$$

$$
\overline{\alpha_{DCP}} = \sum_{i=1}^{2} P_i \alpha_{DCPi}
\qquad \text{(Equation 2.108)}
$$

where scattering probabilities are

$$
P_i = \frac{\lambda_i}{\lambda_1 + \lambda_2} (i = 1,2)
\qquad \text{(Equation 2.109)}
$$

Let us consider the scattering from a single canonical target to evaluate α_{DCP}. Under such a condition, the unit eigenvector u_1 can be expressed by a normalized single target vector k_{DCP}, where we have the relationship of $\left[J_2 \right] = \lambda_1 u_1 \cdot u_1^{*T} = k_{DCP} \cdot k_{DCP}^{*T}$, defined as

$$
\begin{aligned}
u_1 &= e^{j\phi_1} \begin{bmatrix} \cos \alpha_{DCP1} e^{j\phi_1} \\ \sin \alpha_{DCP1} \end{bmatrix} = \frac{1}{\left| k_{DCP} \right|} k_{DCP} \\
&= \frac{1}{2 \left| k_{DCP} \right|} \begin{bmatrix} S_{HH} - S_{VV} + 2 j S_{HV} \\ j \left(S_{HH} + S_{VV} \right) \end{bmatrix}
\end{aligned}
\qquad \text{(Equation 2.110)}
$$

From the relationship in (Equation 2.110), we can relate the α_{DCP} to the elements of $\boldsymbol{k}_{DCP}$ for a single canonical target

$$\cos^2 \alpha_{DCP} = \frac{\left|S_{HH} + S_{VV}\right|^2 + 4\left|S_{HV}\right|^2}{\left|S_{HH} + S_{VV}\right|^2 + \left|S_{HH} - S_{VV}\right|^2 + 4\left|S_{HV}\right|^2} \qquad \text{(Equation 2.111)}$$

Notice that when comparing with alpha angle α for FP defined in (Equation 2.73), we can obtain the following relationship between α_{DCP} and α

$$\alpha_{DCP} = 90° - \alpha \qquad \text{(Equation 2.112)}$$

Thus, three α_{DCP} of the basic scatterers are straightforwardly derived as

$$\text{odd}: \alpha_{DCP} = 90°, \text{dipole}: \alpha_{DCP} = 45°, \text{even}: \alpha_{DCP} = 0° \qquad \text{(Equation 2.113)}$$

Therefore, α_{DCP} is a roll-invariant parameter too, same as α of FP. Note again that the above relationship is valid only for the single canonical target. For distributed targets, the exact correspondence between FP and DCP may fail [39].

The obtained pairs of H_{DCP} and α_{DCP} values are then plotted on the $H_{DCP} / \overline{\alpha}_{DCP}$ 2D plane with some boundaries to clarify feasible regions. Note that the feasible region of the H_{DCP} / α_{DCP} plane for the CP mode is the same as that of DLP, as shown in Figure 2.9. To distinguish the physical scattering mechanism on the

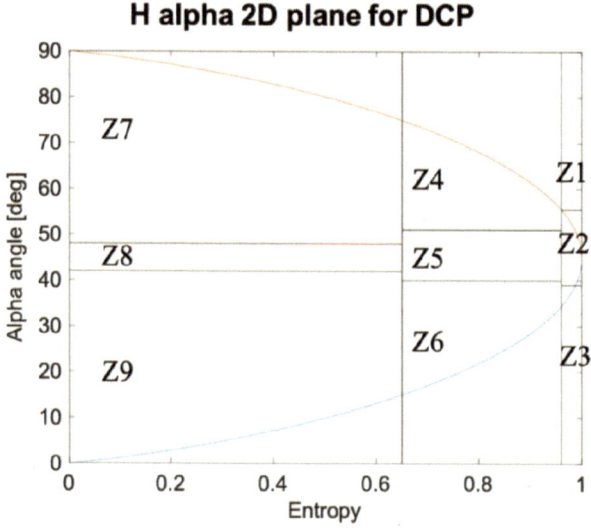

H alpha 2D plane for DCP

FIGURE 2.12 $H/\overline{\alpha}$ 2D plane for DCP data. The descriptions of each zone are given in Figure 2.8.

2D plane, Zhang et al. proposed the classification space for DCP mode through the investigation of distribution centers and densities of different scattering mechanisms [40]. We display Zhang's classification space in Figure 2.12 but we apply the $\alpha_{DCP} = 90° - \alpha_{DCP}$ condition to compare it to the FP classification space where the descriptions of each zone are the same, as shown in Figure 2.8.

2.6 SUMMARY

In this chapter, we have demonstrated the basic theory of, and calculations for, SAR polarimetry. We have shown the scattering and polarimetric decomposition theory for circularly polarized SAR data. The polarimetric theory has then been extended to CP SAR modes, especially for the DCP mode, and we highlighted the advantage of using circular polarization in the DP system. The knowledge demonstrated in this chapter is fundamental for subsequent circularly polarized SAR data analysis in the following chapters.

REFERENCES

[1] G. Sinclair, "The transmission and reception of elliptically polarized waves," Proc. IRE, vol. 38, no. 2, pp. 148–151, 1950.

[2] S. R. Cloude and E. Pottier, "A review of target decomposition theorems in radar polarimetry," IEEE Trans. Geosci. Remote Sens., vol. 34, no. 2, pp. 498–518, 1996.

[3] Y. Yamaguchi, Polarimetric SAR Imaging Theory and Applications. Boca Raton, FL: CRC Press, 2020.

[4] L. C. Shen and J. A. Kong, Applied Electromagnetism. Boston, MA: PWS Publishing Co. 1985.

[5] C. Balanis, Antenna Theory: Analysis and Design, Fourth Edition. Hoboken, NJ: John Wiley & Sons, 2005.

[6] J. S. Lee and E. Pottier, Polarimetric Radar Imaging: From Basics to Applications. Boca Raton, FL: CRC Press, 2009.

[7] S. R. Cloude, Polarisation: Applications in Remote Sensing. Oxford, UK: Oxford University Press, 2009.

[8] W. M. Boerner, Direct and Inverse Methods in Radar Polarimetry, Proceedings of the NATO-ARW, September 18–24, 1988, 1987–91, NATO ASI Series C: Math & Phys. Sciences, vol. C-350, Parts 1 & 2, Kluwer Academic Publications, 1992.

[9] W. L. Stutzman, Polarization in Electromagnetic Systems, vol. 53, no. 9. Norwood, MA: Artech House, 2018.

[10] S. S. Gao, Q. Luo, and F. Zhu, Circularly Polarized Antennas. West Sussex, UK: John Wiley & Sons, Ltd, 2013.

[11] A. Freeman and S. S. Saatchi, "On the detection of Faraday rotation in linearly polarized L-band SAR backscatter signatures," IEEE Trans. Geosci. Remote Sens., vol. 42, no. 8, pp. 1607–1616, 2004.

[12] G. Singh, S. Mohanty, Y. Yamazaki, and Y. Yamaguchi, "Physical scattering interpretation of POLSAR coherency matrix by using compound ccattering phenomenon," IEEE Trans. Geosci. Remote Sens., vol. 58, no. 4, 2020.

[13] J. W. Goodman, "Some fundamental properties of speckle," J. Opt. Soc. Am., vol. 66, no. 11, 1976.

[14] K. Tragl, E. Lueneburg, A. Schroth, and V. Ziegler, "Polarimetric covariance matrix concept for random radar targets," in Seventh Int. Conf. Antennas Propagation, ICAP 91, 1991.

[15] J. R. Huynen, *Phenomenological Theory of Radar Targets*, The Netherlands: Technical University of Delft, 1970.

[16] A. Freeman and S. L. Durden, "A three-component scattering model for polarimetric SAR data," IEEE Trans. Geosci. Remote Sens., vol. 36, no. 3, pp. 963–973, May 1998.

[17] Y. Yamaguchi, T. Moriyama, M. Ishido, and H. Yamada, "Four-component scattering model for polarimetric SAR image decomposition," IEEE Trans. Geosci. Remote Sens., vol. 43, no. 8, pp. 1699–1706, 2005.

[18] S. R. Cloude, "Uniqueness of target decomposition theorems in radar polarimetry," Direct Inverse Methods Radar Polarim., vol. 350, pp. 267–296, 1992.

[19] S. R. Cloude and E. Pottier, "Concept of polarization entropy in optical scattering," Opt. Eng., vol. 34, no. 6, pp. 1599–1611, 1995.

[20] S. R. Cloude and E. Pottier, "An entropy-based classification scheme for land applications of polarimetric SAR," IEEE Trans. Geosci. Remote Sens., vol. 35, no. 1, 1997.

[21] S. Cloude, "The dual polarization entropy/alpha decomposition: A PALSAR case study," in Proc. 3rd PolInSAR Work., 2007.

[22] R. K. Raney, "Dual-polarized SAR and Stokes parameters," IEEE Geosci. Remote Sens. Lett., vol. 3, no. 3, pp. 317–319, 2006.

[23] L. Mascolo, S. R. Cloude, and J. M. Lopez-Sanchez, "Model-based decomposition of dual-pol SAR data: Application to Sentinel-1," IEEE Trans. Geosci. Remote Sens., vol. 60, pp. 1–19, 2022.

[24] R. K. Raney, "Hybrid-polarity SAR architecture," IEEE Trans. Geosci. Remote Sens., vol. 45, no. 11, pp. 3397–3404, 2007.

[25] J.-C. Souyris, P. Imbo, R. Fjortoft, S. Mingot, and J.-S. Lee, "Compact polarimetry based on symmetry properties of geophysical media: the π/4 mode," IEEE Trans. Geosci. Remote Sens., vol. 43, no. 3, pp. 634–646, 2005.

[26] S. N. and P. M., "Compact polarimetric analysis of X-band SAR data," in Proceeding Eur. Conf. Synth. Aperture Radar, 2006.

[27] P. Spudis *et al.*, "Mini-SAR: an imaging radar experiment for the Chandrayaan-1 mission to the Moon," Curr. Sci., vol. 96, no. 4, pp. 533–539, Jan. 2009.

[28] R. K. Raney *et al.*, "The lunar mini-RF radars: Hybrid polarimetric architecture and initial results," Proc. IEEE, vol. 99, no. 5, pp. 808–823, 2011.

[29] T. Misra *et al.*, "Synthetic aperture radar payload on-board RISAT-1: configuration, technology and performance," Curr. Sci., vol. 104, no. 4, pp. 446–461, Jan. 2013.

[30] Y. Yokota *et al.*, "Evaluation of compact polarimetry and along track interferometry as experimental mode of PALSAR-2," in 2015 IEEE Int. Geosci. Remote Sens. Symp., 2015, pp. 4125–4128.

[31] G. Kroupnik, D. De Lisle, S. Côté, M. Lapointe, C. Casgrain, and R. Fortier, "RADARSAT constellation mission overview and status," in 2021 IEEE Radar Conf., 2021, pp. 1–5.

[32] M. E. Nord, T. L. Ainsworth, J.-S. Lee, and N. J. S. Stacy, "Comparison of compact polarimetric synthetic aperture radar modes," IEEE Trans. Geosci. Remote Sens., vol. 47, no. 1, pp. 174–188, 2009.

[33] R. K. Raney, "Hybrid dual-polarization synthetic aperture radar," Remote Sens., vol. 11, no. 13. 2019.

[34] R. Touzi, "Hybrid versus matched antenna for dual- and fully polarimetric SAR," in PolInSAR 2013, Aug. 2013, p. 5.

[35] S. R. Cloude, D. G. Goodenough, and H. Chen, "Compact decomposition theory," IEEE Geosci. Remote Sens. Lett., vol. 9, no. 1, pp. 28–32, 2012.

[36] R.Guo, Y.-B. Liu, Y.-H. Wu, S.-X. Zhang, M.-D. Xing, and H. He, "Applying H/α decomposition to compact polarimetric SAR," IET Radar, Sonar Navig., vol. 6, no. 2, pp. 61–70(9), Feb. 2012.

[37] R. K. Raney, J. T. S. Cahill, G. W. Patterson, and D. B. J. Bussey, "The m-chi decomposition of hybrid dual-polarimetric radar data with application to lunar craters," J. Geophys. Res. Planets, vol. 117, no. E12, Dec. 2012.

[38] F. J. Charbonneau et al., "Compact polarimetry overview and applications assessment," Can. J. Remote Sens., vol. 36, no. sup2, pp. S298–S315, Jan. 2010.

[39] S. Ghods, S. V. Shojaedini, and Y. Maghsoudi, "A modified H/α classification method for DCP compact polarimetric mode by reconstructing quad H and α parameters from dual ones," IEEE J. Sel. Top. Appl. Earth Obs. Remote Sens., vol. 9, no. 6, pp. 2233–2241, 2016.

[40] H. Zhang, L. Xie, C. Wang, F. Wu, and B. Zhang, "Investigation of the capability of $H - \alpha$ decomposition of compact polarimetric SAR," IEEE Geosci. Remote Sens. Lett., vol. 11, no. 4, pp. 868–872, 2014.

[41] T. Fukusako, "Basic of circularly polarized antenna the 56th antenna and propagation design and analysis method workshop", Proc. Inst. Electron. Inform. Comm. Eng., pp. 24–30, Nov. 2016.

3 System Design

Josaphat Tetuko Sri Sumantyo

3.1 INTRODUCTION

The parameter design of hardware and image processing of synthetic aperture radar (SAR) is very important to estimate the performance of the targeted system. First, we need to consider the platform on which the SAR system will be installed. There are several platforms for the SAR system, that is, unmanned aerial vehicles or drones, aircraft, high-altitude platform systems or stratosphere platforms, microsatellites, and so forth. Each platform has specific characteristics, for example, available payload (weight), flying altitude, space for installation, power supply (frequency, voltage, power, connector type), flight duration, and memory, among others. In this book, we will focus on the design of airborne circularly polarized SAR for CN235MPA aircraft [1]–[2].

3.2 MISSION

This book discusses the design of the SAR system for CN235MPA aircraft as shown in Figures 3.1 and 3.2. Figure 3.1 shows the size of the CN235MPA aircraft with (a) side view, (b) bird view, (c) top view, and (c) front view. The aircraft is 23.40m in length, has a wingspan of 25.81m, and is 8.18m in height. The specification of this aircraft is shown in Table 3.1 [1]–[2].

This aircraft has several pods for surveillance missions, such as maritime, land observation, and moving target monitoring, among others. We employed the nose radome to install the circularly polarized antennas as shown in Figure 3.2. This figure shows the radome, antenna, and radar system in CN235MPA aircraft, where (a) CN235MPA aircraft, (b) nose radome, (c) circularly polarized SAR system, (d) front view of circularly polarized antenna's panel, and (e) back view of the antenna's panel. The service ceiling of this aircraft is 7,620m, where we design the system to operate at an altitude of 1,000m to 4,000m. The maximum cruise speed v_p is 450km/h or 125m/s, and we designed the system by considering the cruise speed v_p of 300km/h or 83.33m/s at an altitude of 1,000m.

DOI: 10.1201/9781003282693-3

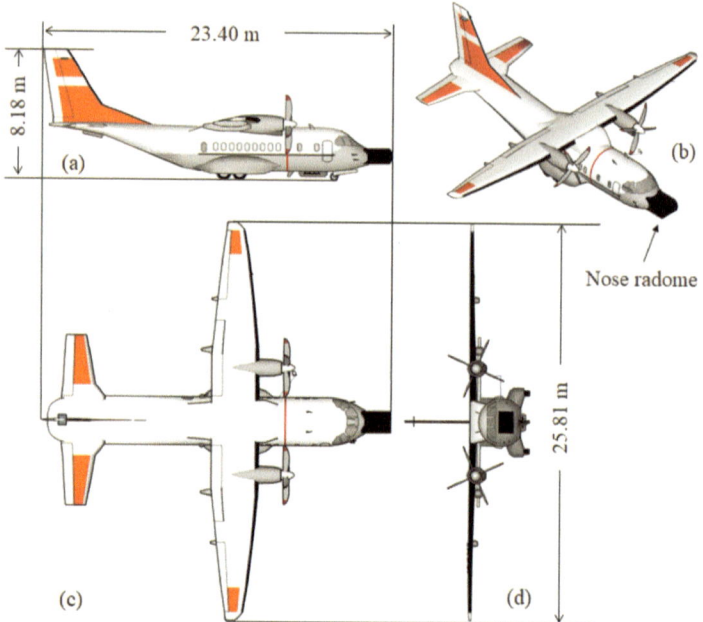

FIGURE 3.1 The size of the CN235MPA Aircraft. (a) side view, (b) bird view, (c) top view, and (d) front view.

FIGURE 3.2 Radome, antenna, and a radar system in CN235MPA aircraft. (a) CN235MPA aircraft, (b) nose radome, (c) circularly polarized SAR system, (d) front view of circularly polarized antenna's panel, and (e) back view of the antenna's panel.

TABLE 3.1

Specifications of CN235MPA Aircraft [1]–[2]

Parameter	Value	Units	Remarks
	General Characteristics		
Crew	2	Person	Pilot and co-pilot
Payload capacity	51	Passengers	
	6,000	kg	13,100 lb
Length	23.40	m	70 ft 2.5 in
Wingspan	25.81	m	84 ft 8 in
Height	8.18	m	26 ft 10 in
Wing area	59.10	m²	636.1 sq ft
Aspect ratio	11.27:1		
Airfoil	NACA 653-218		
Empty weight	9,800	kg	21,605 lb
Max takeoff weight	16,100	kg	35,420 lb
Powerplant	2 × General Electric CT7-9C3 turboprops		1,305 kW (1,750 hp) each (take-off)
	Performance		
Cruise speed	450 (at 4,575 m)	km/h	286 mph, 248 kn (at 15,000 ft)
Stall speed	156	km/h	97 mph, 84 kn (flaps down)
Range	4,355	km	2,706 mi, 2,350 nmi
Service ceiling	7,620	m	25,000 ft
Rate of climb	7.8	m/s	1,780 ft/min

3.3 PARAMETER DESIGN

3.3.1 OBJECTIVE AND SPECIFICATION OF SAR

This chapter discusses how to design an airborne circularly polarized SAR with a payload of about 100kg of a CN235MPA aircraft. The optimum operation altitude is 1,000m to 4,000m. The principal concept of airborne circularly polarized SAR is shown in Figure 3.3. We will derive the basic specification of airborne circularly polarized SAR that works on a frequency of 5.3GHz, or C-band, with targeted ground resolution of about 1m, look angle of 20° to 70°, antenna size of 1.2m x 0.4m (4 panels of 0.6m x 0.2m) for left-handed circular polarization (LHCP) and right-handed circular polarization (RHCP) or azimuth beamwidth of 5° and range beamwidth of 14°, antenna radiation efficiencyof >80%, pulse repetition frequency (PRF) of 100 to 10,000Hz. The circularly polarized SAR has a receiver antenna composed of LHCP and RHCP antennas. This image is used to retrieve the physical information on Earth's surface, that is, soil moisture, biomass, cryosphere, agriculture, ocean dynamics, land deformation, disaster monitoring, and digital elevation model (DEM), among others.

In this section, we discuss the derivation of the specifications of an airborne circularly polarized SAR system for CN235MPA aircraft. The circular polarized SAR has

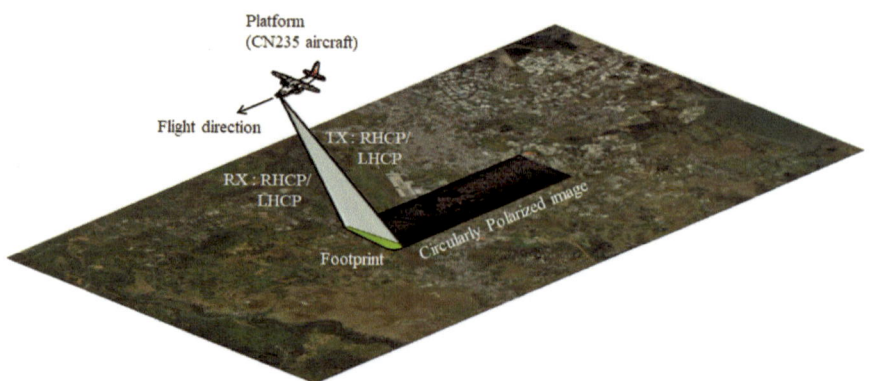

FIGURE 3.3   The concept of airborne circularly polarized SAR.

specific characteristics of polarization. This chapter starts with discussing the polarization and characteristics of circularly polarized SAR.

In general, the basic polarization is elliptical polarization, which includes linear and circular polarization. All these types of microwave polarization are classified based on the value of their axial ratio (AR). From [3], the AR parameter can be defined as

$$\varepsilon = \cot^{-1}(-R) \quad [\text{degree}] \qquad \text{(Equation 3.1)}$$

where ε is the ellipticity angle (-45°≤ ε≤45°) and R is the value of the axial ratio (AR). Here, R is equal to 1 and infinite for perfect circular polarization and linear polarization (LP), respectively. The electromagnetic wave is categorized as elliptical polarization when R is between 1 and infinite. In terms of polarization sense, the sign of R is positive for right-handed (RH) polarization and negative for left-handed (LH) polarization. The R parameter is also commonly stated in the decibel (dB) unit as $20 \log_{10}|R|$.

In the circularly polarized SAR antenna realization, the circularly polarized microwave is generated by feeding the circularly polarized SAR antenna system so that the radiated electromagnetic wave will have a 90° phase difference (δ) between the vertical component (E_y) and the horizontal component (E_x). In the case of generating LHCP, the value of δ equals -90° and for RHCP, the value of δ equals 90°. In the circularly polarized SAR system, to be called the circularly polarized antenna, it is targeted that the antenna should have AR characteristics up to 3dB in its radiation pattern, as explained in the next section. In our experiment, both RHCP and LHCP are employed in the transmitter (TX) and receiver (RX) of the circularly polarized SAR antenna system.

3.3.2   The Geometry of the Platform and SAR

In a conventional linearly polarized (LP) SAR system, the 3dB half-power beamwidth of the antenna is used to determine its basic parameters. The 3dB half-power beamwidth is defined as angular separation inside the main lobe of the antenna gain pattern in elevation or azimuth direction, by which the magnitude of the radiation pattern decreases -3dB from the peak.

The 3dB half-power (HP) beamwidth in elevation direction ($\theta_{el\text{-}HP}$) and azimuth direction ($\theta_{az\text{-}HP}$) can be estimated as [4]–[5].

$$\theta_{el-HP} = 0.886\frac{\lambda}{W} \quad [\text{degree}] \qquad \text{(Equation 3.2a)}$$

$$\theta_{az-HP} = 0.886\frac{\lambda}{L} \quad [\text{degree}] \qquad \text{(Equation 3.2b)}$$

where W is the physical antenna width (elevation); L is the physical antenna length (azimuth), and λ is the radar wavelength in center frequency. The design parameter in this book is listed in Table 3.2. The value of $\theta_{el\text{-}HP}$ determines the ground swath width that can be covered by the LP-SAR system [4]–[5]. On the other hand, the $\theta_{az\text{-}HP}$ is related to the synthetic aperture length ($L_{sa\text{-}LP}$) and the estimated azimuth resolution $\delta_{az\text{-}LP}$ of the LP-SAR system which can be expressed as [6]

TABLE 3.2
Design Parameters of Airborne Circularly Polarized SAR

Parameter	Value
Altitude	1,000 m to 4,000 m
Center Frequency, f_c (Wavelength, λ)	5,300 MHz (57 cm)
Bandwidth, B	400 MHz
Baseband Range	DC to 200 MHz
Peak Power	400 W (56 dBm)
Pulse Width, τ_p	5 µs
Duty Circle, D_c	0.5%
Polarization	TX & RX: RHCP+LHCP
Off Nadir, θ_o	20° up to 70°
Resolution	Up to 1 m
Antenna Size	Length l 0.6 m x Width w 0.2 m (4 panels)
Antenna Efficiency, η	80%
Antenna Gain, G	26 dBic
Axial Ratio, AR	≤3 dB
System Losses, Ls	3 dB
Noise Figure, F	3 dB
Noise Temperature, T	300 K
Thermal Noise Equivalent ($\sigma_{NE}°$)	-50 dB
Pulse Repetition Frequency (PRF$_{op}$) (Operational)	1,000 Hz
Signal-to-Noise Ratio (SNR)	20 dB
Speed of Platform, v_p	300 km/h (83.33 m/s)

$$L_{sa-LP} = R_o \theta_{az-HP} \quad [\text{m}] \qquad \text{(Equation 3.3)}$$

$$\delta_{az-LP} = \frac{R_o \lambda}{2L_{sa-LP}} \approx \frac{L}{2} \quad [\text{m}] \qquad \text{(Equation 3.4)}$$

where R_o is the nearest range distance to the target in azimuth plane view; as described in Figure 3.4. The gain of antenna boresight G is shown as

$$G = \eta 4\pi A/\lambda^2 \quad [\text{dB}] \qquad \text{(Equation 3.5)}$$

where η is antenna efficiency, and A is antenna area. From Equations (3.2), (3.3), (3.4), and (3.5), it can be concluded that in the airborne SAR system, most of its parameters are related to the physical size of the SAR sensor antenna (L and W).

Compared to the LP-SAR system, besides the 3dB half-power beamwidth, the antenna beamwidth of a circularly polarized SAR system is determined by the AR characteristic of the antenna. Here, within the targeted beamwidth (i.e., within the 3dB half-power beamwidth), the antenna should have AR $\leq$ 3dB. Although, theoretically, the perfect circularly polarized wave is achieved when AR is 0dB, from the experiment [7]–[9] it is shown that it is very difficult to be realized practically. The 3dB AR beamwidth for an airborne Circularly polarized SAR antenna can be expressed as

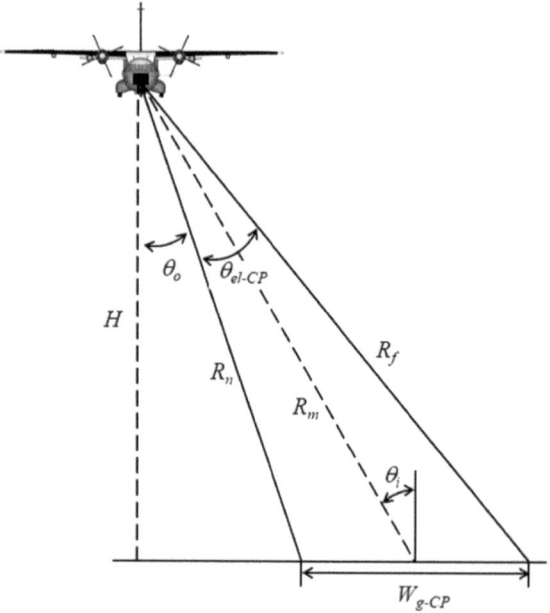

FIGURE 3.4   The geometry of the slant range of the airborne circularly polarized SAR.

$$\theta_{el-CP} \le \theta_{el-HP} \quad \text{[degree]} \qquad \text{(Equation 3.6a)}$$

$$\theta_{az-CP} \le \theta_{az-HP} \quad \text{[degree]} \qquad \text{(Equation 3.6b)}$$

where θ_{el-CP} is 3dB AR beamwidth in the elevation direction, and θ_{az-CP} is 3dB AR beamwidth in the azimuth direction in circular polarization. Here, both θ_{el-HP} and θ_{az-HP} are explained in Equation (3.2).

The range value of θ_{el-CP} and θ_{az-CP} that can be used in the circularly polarized SAR system that AR ≤3dB, only a small fraction of θ_{el-HP} that can be used in the circularly polarized SAR system. Referring to Figure 3.4, we see a slant range distance at far range (R_f), middle range (R_m), and near range (R_n). The airborne circularly polarized SAR slant range geometry can be expressed in terms of θ_{el-CP} as

$$R_n = \frac{H}{\cos\left(\theta_o - \left(\theta_{el-CP}/2\right)\right)} \quad \text{[m]} \qquad \text{(Equation 3.7)}$$

$$R_m = \frac{H}{\cos\theta_o} \quad \text{[m]} \qquad \text{(Equation 3.8)}$$

$$R_f = \frac{H}{\cos\left(\theta_o + \left(\theta_{el-CP}/2\right)\right)} \quad \text{[m]} \qquad \text{(Equation 3.9)}$$

Here, H is the aircraft platform altitude; θ_o is the off-nadir angle. The incident angle could be derived as

$$\theta_i = sin^{-1}\left(sin\theta_o \frac{R_e + H}{R_e}\right) \quad \text{[degree]} \qquad \text{(Equation 3.10)}$$

where R_e is the Earth radius (≈6371 km). In the case of the airborne system, the value of H is relatively very small as compared to R_e ($H \ll R_e$) and hence $\theta_i \approx \theta_o$.

Figure 3.5 shows the relationship between off-nadir angle and slant range. This figure shows that the near range and far range increase when the off-nadir angle increases. The targeted area could be observed by adjusting the altitude and the off-nadir angle. If the altitude of the platform is adjusted between 1,000m to 4,000m and the off-nadir angle between 20° to 70°, the near range varies between 1,026m to 8,756m, and the far range varies between 1,124m to 18,032m.

The circularly polarized SAR ground swath width (W_{g-CP}) can be expressed as

$$W_{g-CP} = \sqrt{R_f^2 - H^2} - \sqrt{R_n^2 - H^2} \quad \text{[m]} \qquad \text{(Equation 3.11)}$$

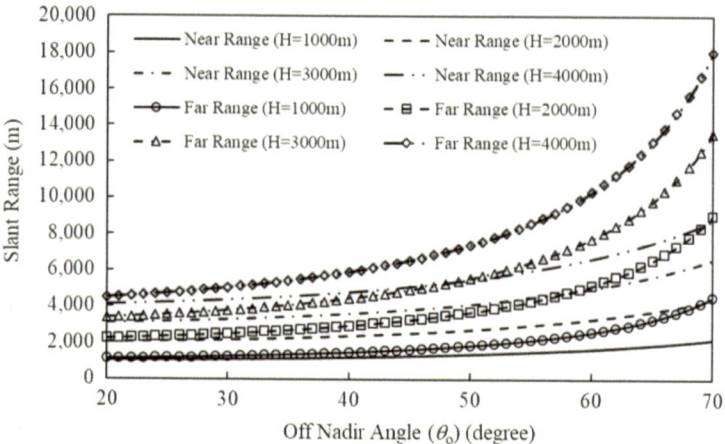

FIGURE 3.5 Slant range.

By substituting Equations (3.7) and (3.9) into Equation (3.11), we obtain

$$\sqrt{\left(\frac{1}{\cos\left(\theta_o + \left(\theta_{el-CP}/2\right)\right)}\right)^2 - 1} - \sqrt{\left(\frac{1}{\cos\left(\theta_o - \left(\dfrac{\theta_{el-CP}}{2}\right)\right)}\right)^2 - 1} = \frac{W_{g-CP}}{H}$$

(Equation 3.12)

Figure 3.6 shows the relationship between off-nadir angle and ground swath width. This figure shows the ground swath width increases drastically in off-nadir angles larger than 60°, where we consider the altitude between 1,000m to 4,000m. This slant range will determine the distance and area to be observed by the sensor, and the power consumed by the SAR system, especially peak power and average power will be discussed further.

3.3.3 Resolution

The SAR image is composed of pixels, where a pixel is generally thought of as the smallest single component of a digital image. In this sub-section, we will discuss how to determine the resolution of a pixel at the range and the azimuth directions. Figure 3.7 shows the azimuth plane view of the airborne circularly polarized SAR. As shown in this figure, the synthetic aperture length of the circularly polarized SAR (L_{sa-CP}) in the azimuth direction is determined by the θ_{az-CP} beamwidth. Then the azimuth resolution of the airborne circularly polarized SAR system, δ_{az-CP}, can be obtained as

$$L_{sa-CP} = R_o \theta_{az-CP} \quad [m]$$

(Equation 3.13)

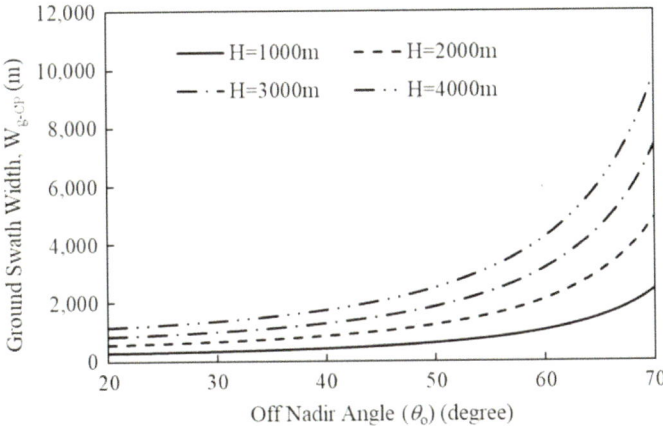

FIGURE 3.6 Ground swath width.

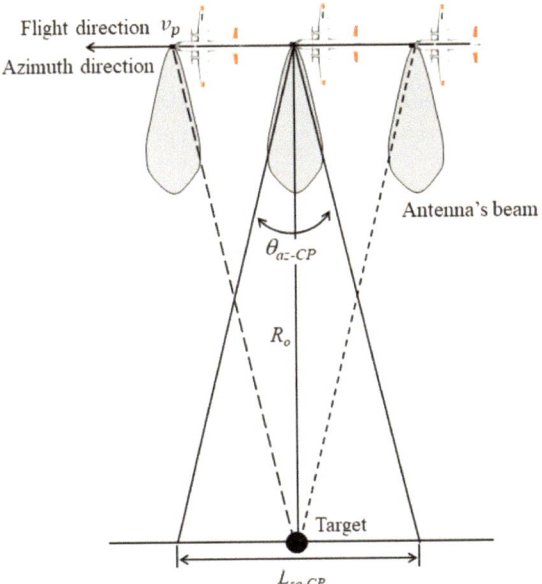

FIGURE 3.7 Azimuth plane view of the airborne circularly polarized SAR.

$$\delta_{az-CP} = \frac{R_o \lambda}{2L_{SA-CP}} = \frac{\lambda}{2\theta_{az-CP}} \quad [m] \qquad \text{(Equation 3.14)}$$

Figure 3.8 shows the relationship between off-nadir angle and synthetic aperture length, where it shows the variation of synthetic aperture length at the near range and the far range with the off-nadir angle between 20° and 70°. The synthetic aperture

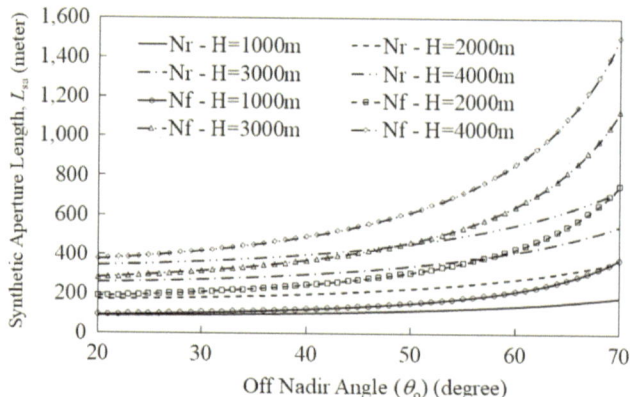

FIGURE 3.8   Synthetic aperture length.

length in the near range varies between 86m and 732m, and the far range varies between 94m and 1,507m. When calculating the resolution of real aperture using Equation (3.4) by considering the slant range length and synthetic aperture length, we obtain the estimated azimuth resolution δ_{az-LP} is about 0.4m, further if it is estimated by the half of antenna length (0.6m) it will obtain 0.3m. Then the derivation of azimuth resolution by using Equation (3.14) will obtain 0.34m at the near range and the far range.

The resolution in the slant range δ_{sr} and the ground range direction, δ_{gr}, is given by [5] [10], where the ground range resolution depends on the incident angle.

$$\delta_{sr} = \frac{c}{2B} \ [m]$$ (Equation 3.15)

$$\delta_{gr} = \frac{c}{2B \sin\theta_i} \ [m]$$ (Equation 3.16)

Where c is the velocity of light (3×10^8 m/s); B is the pulse bandwidth. Here we employed B as 400MHz in the airborne circularly polarized SAR system, therefore we obtain the best resolution when the slant range is 0.375m. The resolution of the ground range direction is around 1.1m at the near range and 0.4m at the far range. This ground resolution does not have a relation to the altitude and depends on the off-nadir angle. The resolution at the far range is better compared to the resolution at the near range.

3.3.4   SIGNAL-TO-NOISE RATIO

Signal-to-noise ratio (SNR or S/N) is a measure used in designing SAR systems that compares the level of the desired signal to the level of background noise. SNR is defined as the ratio of signal power to the noise power, often expressed in decibels (dB). A ratio higher than 1:1 (greater than 0dB) indicates more signal than noise. The overall SNR can be expressed as

$$SNR = \frac{P_t (G\eta)^2 \lambda^3 (PRF) \sigma^\circ c \tau_p}{(4\pi)^3 R_m^3 KTBF 4vL_s \sin\theta_i}$$
(Equation 3.17)

where P_t is transmitted peak power (400W); G is circularly polarized boresight antenna gain (25.7dB for antenna size of 0.6m x 0.2m); η is antenna efficiency (80% for microstrip array antenna); PRF is operating pulse repetition frequencies (1,000Hz); σ° is backscattering coefficient (targeted value is -13dB); τ_p is transmitted pulse length (5μs); K is Boltzmann coefficient (1.38x10^{-23} JK^{-1}); T is receiver temperature (in Kelvin, 300K); F is the receiver noise figure (3dB); v_p is the speed of platform or aircraft (300km/h or 83.33m/s); L_s is the circularly polarized SAR system loss (3dB). The incident angle θ_i is defined by Equation (3.10). Finally, the targeted SNR obtains 20dB.

3.3.5   SAMPLING TIME INTERVAL

The total length of flight (L) for the targeted image size (I_{CP}) is illustrated in Figure 3.9. Here, the required data take duration (t_D) can be estimated as

$$t_D = \frac{L}{v} \ [s]$$
(Equation 3.18)

$$L = L_o + L_{sa-CP} \ [m]$$
(Equation 3.19)

where L_o is the length of the image size that is targeted to be captured; and L_{sa-CP} is described in Equation (3.13). By referring to Figure 3.4, start sampling time t_s and stop sampling time t_f can be shown as

$$t_s = 2R_n / c \ [s]$$
(Equation 3.20)

$$t_f = 2R_f / c + \tau_p \ [s]$$
(Equation 3.21)

$$T_i = t_f - t_s \ [s]$$
(Equation 3.22)

where R_n, R_f, T_i, and τ_p are minimum return (near range), maximum return (far range), sampling time interval, and pulse length (setting value is 5μs), respectively. Figure 3.10 shows the sampling time interval to observe the target area with an altitude between 1,00m and 4,000m. This figure shows the sampling time interval with altitudes 1,000m to 4,000m is between 5.7μs to 66.8μs in the range of off-nadir angle from 20° to 70°.

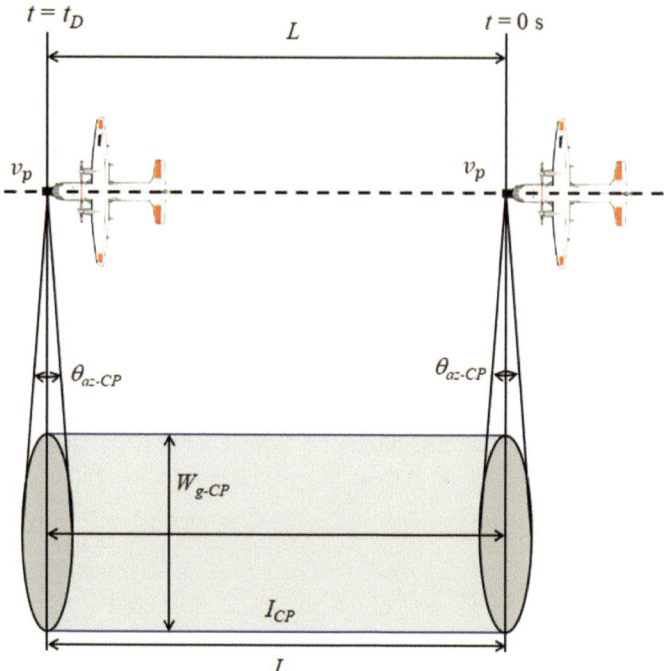

FIGURE 3.9   Illustration of airborne circularly polarized SAR data retrieving.

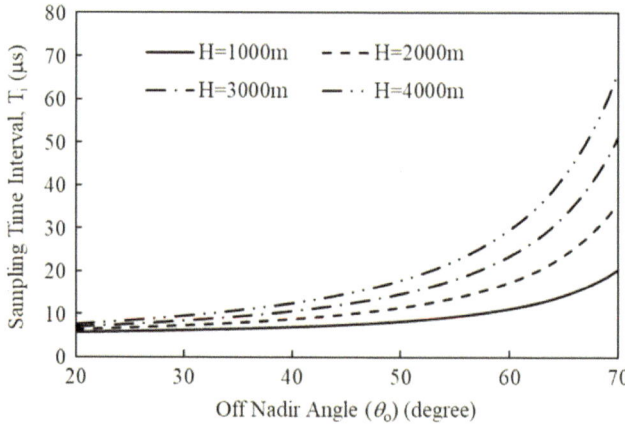

FIGURE 3.10   Sampling time interval.

3.3.6   PRF AND DUTY CYCLE

The pulse repetition frequency (PRF) is the parameter to determine the number of chirp pulses that must be transmitted and received by the SAR system. It is the number of pulses of a repeating signal in a specific time unit, normally measured in pulses per second. PRF is defined by minimum and maximum values as

$$PRF_{min} = B_p = 2v_p\theta_{az-CP} / \lambda \quad [\text{Hz}] \qquad \text{(Equation 3.23)}$$

$$PRF_{max} = c / 2\left(R_f - R_n\right) \quad [\text{Hz}] \qquad \text{(Equation 3.24)}$$

where the PRF is the number of pulses of a repeating signal in a specific time unit, normally measured in pulses per second. The minimum PRF depends on doppler bandwidth or the speed of the platform v_p, the 3dB AR beamwidth in the azimuth direction, and wavelength. Then the maximum PRF depends on slant range length at the near range and the far range. The minimum PRF obtains 246Hz by substituting the parameters listed in Table 3.2, and Figure 3.11 shows the maximum PRF. We employed 1,000 PRF that satisfied the minimum and the maximum PRF. The duty cycle is obtained at 0.5% by using the equation

$$D_c = PRF_{op} \times \tau_p \times 100 \quad [\%] \qquad \text{(Equation 3.25)}$$

A duty cycle or power cycle is the fraction of one period in which a signal or system is active. The duty cycle is commonly expressed as a percentage (%) or a ratio. A period is the time it takes for a signal to complete an on-and-off cycle.

3.3.7 BANDWIDTH

Minimum pulse bandwidth is obtained by using ground swath width W_{g-CP} of Equation (3.11) and the incident angle θ_i of Equation (3.10) that shown as

$$B = c / 2W_{g-CP}\sin\theta_i \quad [\text{Hz}] \qquad \text{(Equation 3.26)}$$

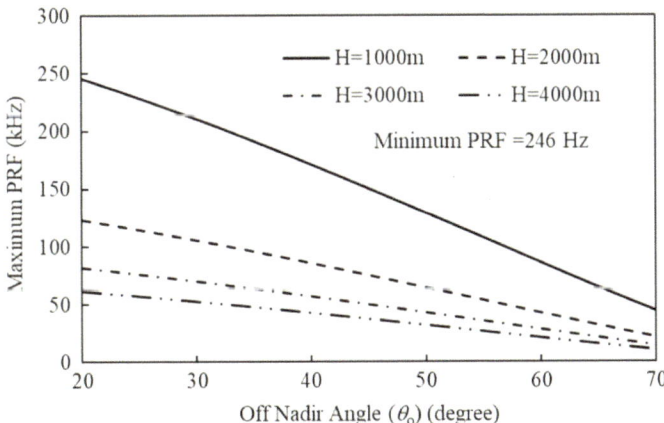

FIGURE 3.11   The pulse repetition frequency (PRF).

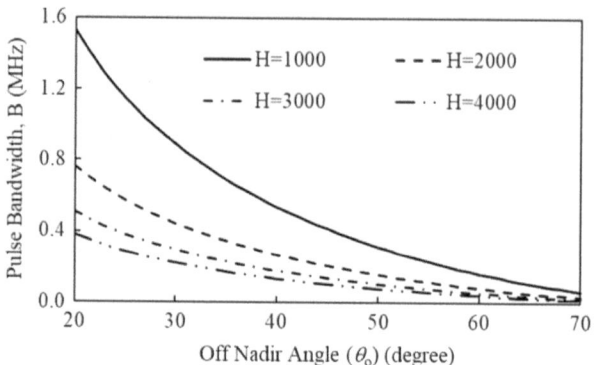

FIGURE 3.12 Pulse bandwidth.

This selected bandwidth will determine the best slant range and ground range resolutions, time bandwidth product (TBP), compressed pulse length, noise power, sampling frequency, peak output power, and backscattering coefficient. Figure 3.12 shows the (minimum) pulse bandwidth by considering the design parameters of Table 3.2. This figure shows the required (minimum) pulse bandwidth has a range of 0.016MHz to 1.6MHz. We could consider employing higher bandwidth to improve the performance of the SAR system, especially ground range resolution. In this book, we employed 400MHz of pulse bandwidth to obtain the ground range of about 1m.

3.3.8 TRANSMIT POWER

In the RF transmitter design, the most important parameter is the peak transmit (output) power P_t, because this parameter is used to choose the filter, modulator, pre-amplifier, solid-state power amplifier (SSPA), and so forth. The peak transmit power is derived by radar equation and can be obtained by

$$P_t = \frac{4(4\pi)^3 R_m^3 FKTBL_s v_p \sin\theta_i}{\sigma_{NE}^o PRF_{op} \tau_p G^2 \lambda^3 c} \quad [\text{W}] \qquad \text{(Equation 3.27)}$$

where R_m is slant range at the center (middle) of the swath width. We could also consider it by using the slant range at the far range, R_f to get a guaranteed peak to transmit power, but empirically R_m is enough. L_s is system loss including the loss by increasing the temperature inside the system, attenuation, among others. v_p is the speed of the platform. θ_i is the incident angle. σ_{NE}^o is thermal noise equivalent. PRF_{op} is operational pulse repetition frequency. τ_p is pulse length (5μs). G is the gain of the antenna. λ is the wavelength of the center frequency, while c is the speed of light. F, T, K, and B are noise figure, noise temperature, Boltzmann constant, and pulse bandwidth, respectively, where the multiplication of these four parameters is

$$N_o = FTKB \qquad \text{(Equation 3.28)}$$

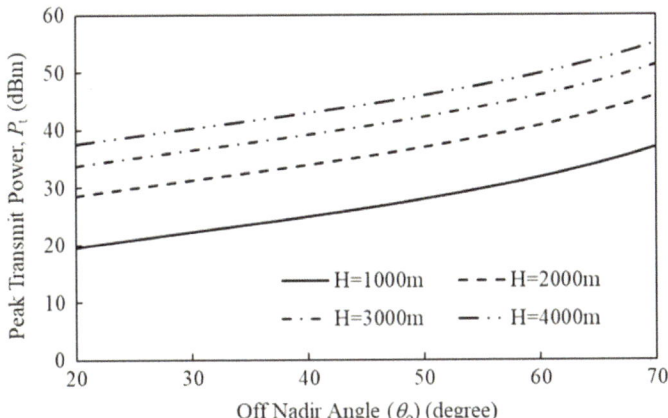

FIGURE 3.13   Peak transmit power.

Noise power indicates the total noise power in a given bandwidth at the input or output of a device in the SAR system when the signal is not present; the integral of noise spectral density over the bandwidth.

Equations (3.27) and (3.28) show the peak transmit power is proportional to noise power. The peak transmit power is the maximum power that the output power supply can sustain for a short time and is sometimes called the peak surge power. Peak power differs from continuous power which refers to the amount of power that the supply can provide continuously. The peak power is always higher than the continuous power and is only required for a limited amount of time. Figure 3.13 shows the peak transmit power. It shows the maximum power needed at an off-nadir angle of 70° at an operating altitude of 4,000m is about 323W (55.1dBm). By consideration of other loss and attenuation (cable, temperature, mismatched, etc.), we fabricated the airborne SAR with a peak transmit power is 400W.

The average output power is the time average of the instantaneous power. The average transmit power P_{avg} is shown as

$$P_{avg} = P_t \times PRF_{op} \times \tau_p \quad [\text{W}] \qquad \text{(Equation 3.29)}$$

This means the average transmit power is proportional to the peak transmit power, pulse repetition frequency, and pulse length. Figure (3.14) shows the average power for SAR with an off-nadir angle of 20° to 70° with the maximum value being about 2W (32.1dBm).

3.3.9   TIME-BANDWIDTH PRODUCT

In the parameter design, the time bandwidth product (TBP or TBWP) is also an important parameter. The TBP is the product of pulse width τ_p and the receiver's minimum bandwidth B as

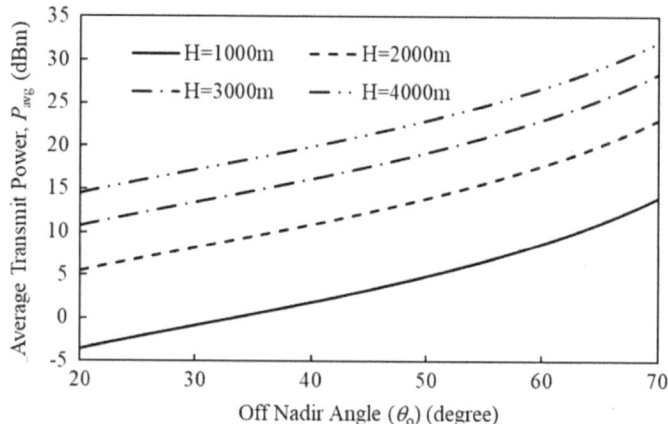

FIGURE 3.14   Average transmit power.

$$TBP = \tau_p B \qquad \text{(Equation 3.30)}$$

This is an important parameter for designing the radar to measure the possible pulse compression rate and the expectant time-side-lobes. The radar receiver should have a bandwidth as small as possible to avoid receiving additional noise and disturbances. But the bandwidth should be wide enough to let the desired echo signals pass through. The shorter the pulses, the wider the required bandwidth. The wider the bandwidths, the better range resolution can be achieved. The length of compressed pulse width t_c can be calculated by

$$t_c = \frac{1}{B} \; [s] \qquad \text{(Equation 3.31)}$$

The pulse width τ_p is improved to be the compressed pulse width t_c, hence we can determine the signal power improvement as

$$r_{sp} = \frac{\tau_p}{t_c} \qquad \text{(Equation 3.32)}$$

3.3.10   DOPPLER BANDWIDTH

The Doppler history of a target in a SAR image, while traversing the beam, is used to focus the data in the azimuth direction. Here, the Doppler frequency or Doppler centroid is a measure of the effective antenna squint angle. It is used to adjust the bandpass characteristics of the azimuth compression filter to the location of the signal spectrum. An inaccurate Doppler centroid will affect the resolution and SNR, and also generate the aliased azimuth frequency components to fall within the passband of the compression filter, and thus reduce the signal-to-ambiguity ratio [14]–[16]. The

Doppler effect, named after Austrian physicist Christian Doppler who proposed it in 1842, is the difference between the observed frequency and the emitted frequency of a wave for an observer (aircraft or other platforms) moving relative to the source of the electromagnetic waves. Here, the lower boundary of the PRF is determined by the Doppler bandwidth of the SAR as Equation (3.23) derives by the speed of the aircraft v_p and beamwidth θ_{az} at azimuth direction. If we consider the parameters in Table 3.2, we obtain the Doppler bandwidth is 246Hz.

$$B_D = \frac{2v_p \theta_{az}}{\lambda} \quad [\text{Hz}] \qquad \text{(Equation 3.33)}$$

3.3.11 SYNTHETIC APERTURE

By referring to Figures 3.4, 3.7, and 3.9, the integration time to observe an instantaneous footprint at the maximum range and the speed direction at azimuth direction is

$$T_{syn} = \frac{\gamma_b \lambda R_f}{2v_p d_y} \quad [\text{s}] \qquad \text{(Equation 3.34)}$$

where γ_b is the azimuth broaden coefficient (%). R_f, v_p, and d_y are the far-range distance, the speed of the platform or aircraft, and the azimuth resolution, respectively. The azimuth resolution d_y is the same as the best azimuth resolution:

$$R_{az-best} = d_y = l/2 \quad [\text{m}] \qquad \text{(Equation 3.35)}$$

where l is the length of the antenna in the azimuth direction. Then the integration time is employed to calculate the required synthetic aperture length as

$$L_{syn} = v_p T_{syn} \quad [\text{m}] \qquad \text{(Equation 3.36)}$$

Maximum synthetic aperture length can be derived by

$$L_{max} = R_{max} \theta_{az} \quad [\text{m}] \qquad \text{(Equation 3.37)}$$

where R_{max} is the maximum slant range that we could employ the far range distance R_f and the required synthetic aperture length L_{syn} as Equations (3.9) and (3.36), that could be derived by

$$R_{max} = \sqrt{R_f^2 + \left(L_{syn}/2\right)^2} \quad [\text{m}] \qquad \text{(Equation 3.38)}$$

Then the required circularly polarized azimuth 3dB beamwidth θ_{az} is

$$\theta_{az} = \frac{\gamma_b \lambda}{2R_{az-best}} \quad [\text{degree}] \qquad \text{(Equation 3.39)}$$

Further, the quality or density of the image in the azimuth direction could be derived from the azimuth compression ratio η_a as

$$\eta_a = \frac{L_{syn}}{d_y} \ [\%] \qquad\qquad \text{(Equation 3.40)}$$

and synthetic antenna beamwidth β_s as

$$\beta_s = \frac{\lambda}{2L_{syn}} \ [\text{degree}] \qquad\qquad \text{(Equation 3.41)}$$

By considering the parameters in Table 3.2, we acquired the synthetic aperture beamwidth at an off-nadir angle of 20° to 70° to vary from 0.02° to 0.004°. It means the synthetic aperture beamwidth at the far range is better than at the near range.

3.3.12 MEMORY ESTIMATION

In SAR image digital signal processing, downsampling is the important process of resampling in a multi-rate digital signal processing system, where the downsampling is synonymous with compression to describe an entire process of bandwidth reduction (filtering) and sample-rate reduction. When the process is performed on a sequence of samples of a signal or a continuous function, it produces an approximation of the sequence that would have been obtained by sampling the signal at a lower rate. In the design of the SAR system, we define the maximum down-sampling rate as

$$K_{pmax} = \frac{PRF_{op}}{1.2B_D} \qquad\qquad \text{(Equation 3.42)}$$

where PRF_{op} is the operational PRF and B_D is the Doppler bandwidth using Equation (3.33). Then the down-sampling rate in integers is determined as

$$K_p = rounddown\left(K_{pmax}\right) \qquad\qquad \text{(Equation 3.43)}$$

Further, the sampling number before prefilter N_{syn} could is defined as

$$N_{syn} = L_{syn} PRF / v_p \qquad\qquad \text{(Equation 3.44)}$$

and after prefiltering m_s is

$$m_s = N_{syn} / K_p \qquad\qquad \text{(Equation 3.45)}$$

These N_{syn} and m_s are variables to estimate the number of pixels of the SAR image. By referring to Table 3.2, we obtained the N_{syn} ranging from 1,128 to 4,522 samples, and m_s ranging from 376 to 1,507 samples for off-nadir angles of 20° to 70°, respectively.

The size of each pixel in the azimuth direction could be derived by the azimuth sample spacing in the aperture domain (D_u) as

$$D_u = v_p / PRF_{op} \quad [\text{m}] \qquad \text{(Equation 3.46)}$$

then the azimuth sample spacing after Presummer Filter (d_u) will be

$$d_u = K_p D_u \quad [\text{m}] \qquad \text{(Equation 3.47)}$$

By considering parameters in Table 3.2, D_u and d_u are obtained at 0.083m and 0.25m, respectively.

Azimuth range half size Y_o is defined by

$$Y_o = I_s / 2W_{g-CP} \quad [\text{m}] \qquad \text{(Equation 3.48)}$$

where I_s is the size of the area of observed area; for example, we set 10,000,000m² if we assume to observe the area with this size, we will obtain Y_o from 17,478 to 2,042 at the off-nadir angle of 20° to 70°. W_g is the ground swath width as determined by Equation (3.11).

Slant range half size R_o is derived by

$$R_o = W_{g-CP} / 2 \quad [\text{m}] \qquad \text{(Equation 3.49)}$$

where W_{g-CP} is designed swath width from Equation (3.11). Then the azimuthal range size R_{az} is

$$R_{az} = 2Y_o \quad [\text{m}] \qquad \text{(Equation 3.50)}$$

By referring to Figure 3.9, the total flight distance of $2L$ is

$$2L = 2Y_o + L_{syn} \quad [\text{m}] \qquad \text{(Equation 3.51)}$$

Radar cross section σ (distributed target) is defined as

$$\sigma = \sigma^o r_g dy \quad [\text{dB}] \qquad \text{(Equation 3.52)}$$

where σ^o, r_g, and dy are the backscattering coefficient, the ground range resolution, and the azimuth resolution, respectively. The number of azimuth samples (total) S_{az} (distributed target) will be obtained as

$$S_{az} = 2L / du \qquad \text{(Equation 3.53)}$$

The recording data duration t_{DT} is obtained below by referring to Figure 3.9.

$$t_{DT} = 2L / v_p \quad [\text{s}]$$

(Equation 3.54)

3.3.12.1 Real Signal

Firstly, the minimum analog-to-digital converter (ADC) sampling frequency (f_s) for the real signal method is shown as

$$f_r = 2B \times 1.2 \quad [\text{Hz}]$$

(Equation 3.55)

Further, the sampling frequency f_r is employed to calculate the sample spacing in the time domain (dt) as

$$dt = 1 / f_r \quad [\text{s}]$$

(Equation 3.56)

The number of range samples per pulse will be

$$dt = 2T_i / 2dt \quad [\text{s}]$$

(Equation 3.57)

where T_i is the sampling time interval as explained in Equation (3.22). Data volume V_d of each channel will be

$$V_d = N_b n S_{az} \quad [\text{bits/channel}]$$

(Equation 3.58)

where N_b, n, and S_{az} are the number of bits (8 bits), the number of range samples per pulse, and the number of azimuth samples (total) as calculated by Equation (3.53). Finally, the data rate d_r could be obtained below to estimate the speed of data communication of each channel.

$$d_r = V_d / t_{DT} \quad [\text{bps/channel}]$$

(Equation 3.59)

By considering the parameter in Table 3.2, we obtain the d_r from 14.5Mbps to 52.4Mbps/channel at the off-nadir angle of 20° to 70°.

3.3.12.2 Complex Signal

In the same manner as the sub-section before for the real signal, the minimum analog-to-digital converter (ADC) sampling frequency (f_s) for the complex signal method is derived as

$$f_c = B \times 1.2 \quad [\text{Hz}]$$

(Equation 3.60)

Then the sample spacing in the time domain (dt) to be

$$dt = 1 / f_c \quad [\text{s}]$$

(Equation 3.61)

and the number of range samples per pulse as

$$dt = 2T_i / 2dt \quad [\text{s}] \qquad \text{(Equation 3.62)}$$

where T_i is the sampling time interval as explained in Equation (3.22). Data volume V_d is obtained as

$$V_d = N_b n S_{az} \quad [\text{bits/channel}] \qquad \text{(Equation 3.63)}$$

where N_b, n, and S_{az} are the number of bits (8 bits), the number of range samples per pulse, and the number of azimuth samples (total) as determined by Equation (3.53). Finally, the data rate d_r is acquired as

$$d_r = V_d / t_{DT} \quad [\text{bps/channel}] \qquad \text{(Equation 3.64)}$$

By considering the parameter in Table 3.2, we obtain the d_r from 7.2Mbps to 26.2Mbps/channel at the off-nadir angle of 20° to 70°.

3.4   SUMMARY

This chapter discussed the parameter design for airborne circularly polarized synthetic aperture radar (SAR) and the implementation of CN235MPA onboard Circularly

---

TABLE 3.3
Specifications of C-band Circularly Polarized SAR System

Parameter	Value	Units	Remarks
Mode of operation	Stripmap		
Carrier/Center frequency	5.3	GHz	C-band
Modulation format	Pulsed-Linear Frequency Modulation (LFM)		
Bandwidth	100 to 400	MHz	Adjustable
Slant range resolution	0.375 to 1.5	meters	
Pulse width	0.1 to 10	µs	
Pulse repetition frequency	100 to 10,000	Hz	Adjustable
Peak transmit power	400	Watts	Adjustable
Receiver gain	50	dB	
Receiver noise figure (NF)	3	dB	
Antenna polarization	Circular polarization (LL, RR, LR, RL)		Axial ratio ≤ 3 dB
Antenna isotropic gain	22	dBic	
Antenna beamwidth	Range: 13	Degrees	
	Azimuth: 6	Degrees	
Incident angle	20 to 70	Degrees	
Platform height	1,000 to 4,000	Meters	
Observation time	65.536	seconds	

polarized SAR that works in the center frequency of 5.3GHz (C-band) and 400MHz of bandwidth for operation altitude 1,000m. Based on the parameter design in this chapter, we summarize the results in Table 3.3 as the specifications of the C-band airborne Circularly Polarized SAR system [11]–[13]. The radiofrequency (RF) hardware and flight test will be discussed in the next chapters.

REFERENCES

[1] https://web.archive.org/web/20140326120344/http://www.airbusmilitary.com/Aircraft/CN235/CN235About.aspx (Accessed on 30 April 2022)

[2] www.indonesian-aerospace.com/aircraft/detail/22_cn235-220+mpa (Accessed on 30 April 2022)

[3] W.L. Stutzman., *Polarization in Electro-magnetic System*, Boston, MA, Artech House, 1993.

[4] A. Freeman, W.T.K. Johnson, B. Huneycutt, R. Jordan, S. Hensley, P. Siqueira, and J. Curlander, "The 'Myth' of the Minimum SAR Antenna Area Constraint," IEEE Trans. Geosci. Remote Sens., Vol.38, pp.320–324, 2000.

[5] J.T.S. Sumantyo, "Circularly Polarized Synthetic Aperture Radar Onboard Unmanned Aerial Vehicle, Series: Intelligent Systems, Control, and Automation: Science and Engineering," in *Autonomous Control Systems and Vehicles: Intelligent Unmanned System*, N. Kenzo, M. Kartidjo, K.J. Yoon, and A. Budiyono (eds), UAV Books, Vol.65, 2013, ISBN 978-4-431-54275-9.

[6] I.G. Cumming and F.H. Wong, *Digital Processing of Synthetic Aperture Radar Da-ta*, Boston, MA: Artech House, 2005.

[7] C.E. Santosa, J.T.S. Sumantyo, K. Urata, M.Y. Chua, K. Ito, and S. Gao, "Development of a Low Profile Wide-Bandwidth Circularly Polarized Microstrip Antenna for C-Band Airborne CP-SAR Sensor," Progress in Electromagnetics Research C, Vol.81, pp.77–88, January 2018, DOI:10.2528/PIERC17110901.

[8] C.E. Santosa, J.T.S. Sumantyo, M.Y. Chua, K.U.Nagamine, K. Ito, and S. Gao, "Subarray Design for C-Band Circularly-Polarized Synthetic Aperture Radar Antenna onboard Airborne," Progress in Electromagnetics Research, Vol.163, pp.107–117, September 2018, DOI:10.2528/PIER18060602.

[9] C.E. Santosa, J.T.S. Sumantyo, S. Gao, and K. Ito, "Broadband Circularly Polarized Microstrip Array Antenna with Curved-Truncation and Circle-Slotted Parasitic," IEEE Transactions on Antennas and Propagation (TAP), Vol.69, No.9, pp.5524–5533, September 2021 DOI: 10.1109/TAP.2021.3060122.

[10] Chan, Y.K., Koo, V.C., Lim, T.S., "Conceptual Design of a High Resolution, Low Cost X-band Airborne Synthetic Aperture Radar System, " Progress in Electromagnetic Research Vol. 3, pp.943–947, 2007.

[11] J.T.S. Sumantyo, M.Y. Chua, C.E Santosa, G.F Panggabean, T. Watanabe, B. Setiadi, F.D. Sri Sumantyo, K. Tsushima, K. Sasmita, A. Mardivanto, E. Supartono, E.T Rahardjo, G. Wibisono, M.A. Marfai, R.H. Jatmiko, Sudaryatno, T. H. Purwanto, B.S. Widartono, M. Kamal, D. Perissin, S. Gao, and Koichi Ito, "Airborne Circularly Polarized Synthetic Aperture Radar," IEEE Selected Topics in Applied Earth Observations and Remote Sensing (JSTARS), Vol.14, pp.1676–1692, January 2021, DOI:10.1109/JSTARS.2020.3045032.

[12] M.Y. Chua, J.T.S. Sumantyo, C.E Santosa, G.F Panggabean, F.D. Sri Sumantyo, T.Watanabe, Y.Q. Ji, P.P Sitompul, M. Nasucha, F.Kurniawan, B. Purbantoro, A. Awaludin, K. Sasmita, E.T Rahardjo, G. Wibisono, R.H. Jatmiko, Sudaryatno, T. H.

Purwanto, B.S. Widartono, and M. Kamal, "The Maiden Flight of Hinotori-C: The First C Band Full Polarimetric Circularly Polarized Synthetic Aperture Radar in the World," IEEE Aerospace and Electronic Systems Magazine, Vol.34, No.2, pp.24–35, February 2019, DOI:10.1109/MAES.2019.180120.

[13] J.T.S. Sumantyo, H.Wakabayashi, A. Iwasaki, F. Takahashi, H. Ohmae, H.Watanabe, R.Tateishi, F. Nishio, M. Baharuddin, and P.R.Akbar, "Development of Circularly Polarized Synthetic Aperture Radar Onboard Microsatellite," in Proceedings of the Progress in Electromagnetics Research Symposium (PIERS), 382–385, 2009.

[14] R. Bamler, "Doppler Frequency Estimation and the Cramer-Rao Bound," IEEE Transactions on Geoscience and Remote Sensing, Vol.29, No.3, pp.385–390, May 1991.

[15] F.K. Li and W.T.K Johnson, "Ambiguities in Spaceborne Synthetic Aperture Radar Systems," IEEE Transactions on Aerospace and Electronic Systems, Vol.AES-19, pp.389–397, 1983.

[16] F.K. Li, D.N. Held, J. Curlander, and C. Wu, "Doppler Parameter Estimation for Synthetic Aperture Radar," IEEE Transactions on Geosciences and Remote Sensing, Vol.GE-23, pp.47–56, 1985.

4 RF System

Ming Yam Chua and Josaphat Tetuko Sri Sumantyo

4.1 INTRODUCTION

Hinotori-C is a pulsed Linear Frequency Modulated (Pulsed-LFM) imaging radar. The sensor is designed to operate in stripmap SAR mode with the major technical specifications tabulated in Table 4.1 [1].

The Hinotori-C is an SAR sensor operating in the microwave C-band region, with the carrier frequency fixed at 5.3GHz. The bandwidth of the baseband chirp of Hinotori-C SAR can be configurable in real-time, in the range of 10MHz to 400MHz. While the sensor is operating at its maximum bandwidth of 400MHz, achieving a best slant range resolution of 37.5cm.

The Hinotori-C SAR sensor can work at a maximum altitude of 4,000m. The operating incident angle of the sensor can be in the range of 25° to 60°. The sensor's transmitter can generate 280W peak power of RF chirps with 10% of the maximum operating duty cycle. For a low-altitude flight mission, the output power level of the transmitter unit can be regulated to a lower level using the built-in front-end digitally tunable attenuator that can operate in the range of 0dB to 31dB with the step of 1dB. the pulse width and the pulse repetition frequency (PRF) of the Hinotori-C SAR sensor are re-configurable on the fly with an operating range of 0.1μs to 100μs and 100Hz to 10kHz, respectively to maximize the signal processing gain from the pulse compression processing.

In terms of the data acquisition capabilities of the sensor, the onboard digitizer can simultaneously record a large continuous stream of echoes for as many as 65,536 records of azimuth bins (corresponding to 65.536 seconds of acquisition time at the PRF rate of 1kHz) in a single acquisition. The streaming capability of the onboard digitizer supports 'unlimited' echo acquisition (practically, the available amount of onboard storage in the system limits the total recordable number of echoes). Meanwhile, the receiver chain of the sensor has around 50dB of total gain and about 4.5dB of noise figure (NF). Figure 4.1 shows the Hinotori-C SAR electronics assemblies installed in the cabin of CN-235 maritime patrol aircraft (MPA) during its maiden flight mission in March 2018 in Makassar, Indonesia [2]–[5].

4.2 SYSTEM BLOCK DIAGRAM

The SAR electronics of the Hinotori-C were integrated primarily using commercially off-the-shelf available components. The development time of the entire SAR electronics system was as short as six months, with the majority of time spent on the system integration, system testing, and the validation of the system's performance.

DOI: 10.1201/9781003282693-4

TABLE 4.1

Technical Specifications of Hinotori-C SAR

Parameter	Value
Operating mode	Stripmap
Carrier frequency	5.3 GHz (C-band)
Modulation format	Pulsed-LFM
Slant range resolution	1.5 – 0.375 m
Bandwidth	10 – 400 MHz
Pulse width	0.1 – 100 μs
Pulse repetition frequency	100 Hz – 10 kHz
Peak transmit power	280 W with front-end 31 dB adjustable attenuation (1 dB step size)
Receiver gain	50 dB
Receiver noise figure	4.5 dB
Receiver dynamic range	-55 dBm – -85 dBm
Antenna polarization	Full Circular (LL, RR, LR, RL)
Antenna beamwidth	13° (Range) 6° (Azimuth)
Incident angle	25° – 60°
Platform height	500 – 4000 m
Observation time	65.536 s

FIGURE 4.1   Hinotori-C SAR electronics assemblies.

The hardware is comprised of several subsystems, as depicted in Figure 4.2. In general, the subsystems are namely,

   i.   The Radio Frequency Unit (RFU)
  ii.   Antenna (discussed in Chapter 6)
 iii.   The Baseband and Control Unit (BCU) (Detail in Chapter 5)
 iv.   The Inertial Navigation Unit (INU) (Detail in Chapter 5)
  v.   The Power Unit (PU) (Detail in Chapter 5)

Table 4.2 summarizes the primary functions of each subcomponent in each subsystem.

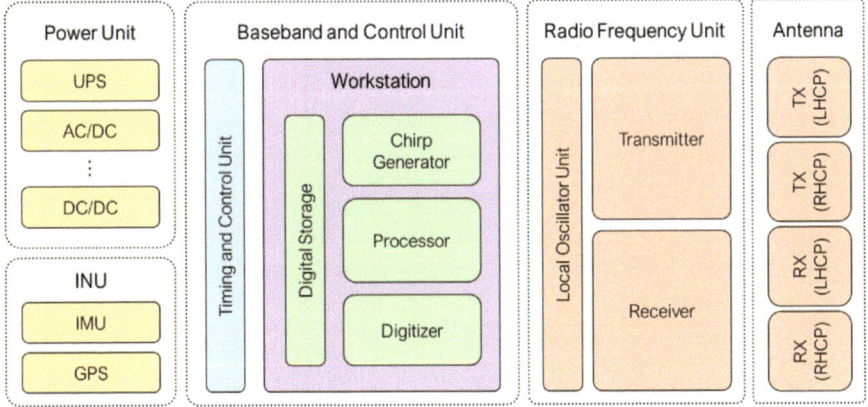

FIGURE 4.2   System block diagram of Hinotori-C SAR electronics.

TABLE 4.2
Primary Functions of the Subsystems in the Hinotori-C SAR Electronics

Subsystem	Subcomponent	Primary Function
Radio Frequency Unit (RFU)	Transmitter	The transmitter modulates the baseband radar waveform into a higher frequency and higher power electrical signal to ease the illumination by the transmitting antenna. The transmitter for radar is usually similar to the transmitter for a radio communication system.
	Receiver	The receiver collects, amplifies, and converts the high-frequency electrical signal collected by the receiving antenna into a suitable form for recording.
	Local Oscillator	The local oscillator generates the synchronized radio clocks to the rest of the subcomponents in the radar system, such as the transmitter, receiver, chirp generator, digitizer, and timing unit.

(continued)

TABLE 4.2 (Continued)
Primary Functions of the Subsystems in the Hinotori-C SAR Electronics

Subsystem	Subcomponent	Primary Function
Antenna	TX & RX Antenna	The antenna radiates electromagnetic waves to the target area for illumination. For reception, the antenna intercepts some of the power of the electromagnetic waves and produces an electrical signal.
Baseband and Control Unit (BCU)	Chirp generator	The chirp generator produces the modulated baseband waveform used in radar. The most commonly used waveform in SAR is the linear frequency modulated (LFM) waveform, more widely known as "chirp".
	Digitizer	The digitizer records the SAR echoes in digital form and stores the recorded waveform in digital storage.
	Processor	The processor controls all of the digitally controllable subcomponents such as the transmitter, receiver, chirp generator, digitizer, and the timing and control unit.
	Digital Storage	The digital storage temporarily stores the digitized stream of echoes during the mission. The engineer/scientist will then retrieve the data for SAR further post-mission processing.
	Timing and Control Unit (TCU)	The TCU generates high-precision timing signals to control the relevant subcomponents.
Inertial Navigation Unit (INU)	Inertial Measurement Unit (IMU)	The IMU measures and records the SAR payload's specific force, angular rate, and sometimes the body's orientation, using a mixture of sensors, such as accelerometers, gyroscopes, and sometimes magnetometers. The recorded navigations would be helpful in the post-processing of SAR raw data, particularly for the processing that involves motion compensation.
	Global Positioning System (GPS)	The GPS records the vehicle's position and speed during the flight mission. The time resolution of GPS data is usually much lower than IMU data.
Power Unit (PU)	Uninterruptible Power Supply System (UPS)	The UPS regulates and supplies stable electrical energy to all subcomponents. It also serves as a backup power supply to prevent short-duration power shortages.
	AC/DC & DC/DC converter	The converters (AC/DC & DC/DC) convert the electrical energy from one form the another. The subcomponents in different subsystems require different supply ratings.

4.3 RADIO FREQUENCY UNIT

The primary functions of the radio frequency unit (RFU) in the Hinotori-C sensor are,

   i)   To up-convert the baseband chirp waveform into the carrier band using an
        upconverter (one or more upconversion mixers)
  ii)   To amplify the carrier band chirp signal into a higher power level for a more
        extended range illumination using high power amplifier (HPA)
 iii)   To amplify the collected weak echoes into a higher power level for downcon-
        version using low noise amplifier (LNA)
  iv)   To downconvert the echoes in carrier band to baseband using a downconverter
        (one or more down-conversion mixers)
   v)   To generate synchronized clocks for the rest of the units using one or more
        local oscillators (LO)

There are three subsystems in the RFU; namely, (i) the transmitter, (ii) the receiver, and (iii) the local oscillator unit. These subsystems carry out all the functions mentioned above.

The RFU is the most critical hardware to design in SAR electronics. A poorly designed radio frequency front-end will produce degraded signals, significantly affecting the final SAR image's quality. In SAR, or radio, the two commonly adopted transmitter/receiver architectures are (i) superheterodyne, and (ii) homodyne (sometimes also called direct conversion or zero-IF) [6].

Figure 4.3(a) shows the architecture of a transmitter that uses homodyne architecture. A similar frequency conversion architecture applies to a homodyne receiver using components that perform frequency down-conversion. The homodyne architecture is easy to implement since it only has one frequency translation. However, the local oscillator (LO) leakage suppression highly relies on the mixer used, which typically ranges 30–50dB. The LO drive's power level in these mixers is usually in the range of +10dB to +20dBm. For a baseband waveform with a power level of -10dBm, the transmitter will generate an RF signal with the power ratio of the signal to its carrier signal of -10dBc to -40dBc (decibels relative to the carrier), which is insufficient for an SAR. Not only that, an SAR usually uses a frequency-modulated waveform in which the center frequency of the modulated waveform is centered at its carrier frequency. The LO frequency used in the transmitter is usually the same as the carrier frequency in a single-stage frequency translation architecture such as homodyne architecture, making the LO removal harder and less efficient [7]–[8].

Considering a slightly more complicated transmitter architecture with more than one frequency translation stage, such as a superheterodyne architecture, as shown in Figure 4.3(b), the leaked LO signals from the frequency translation devices can be removed easily by filtering. In a superheterodyne transmitter, the inevitable LO component at the output of the mixer can be suppressed by carefully choosing the frequencies of its LO signals. The rule of thumb is to avoid using the same or very close frequency for these LO with the frequency of the radar's carrier, making the filters and the LO suppression capability of the mixers in the transmitter can effectively suppress the leaked LO signals from the mixers. For better LO leakage suppression,

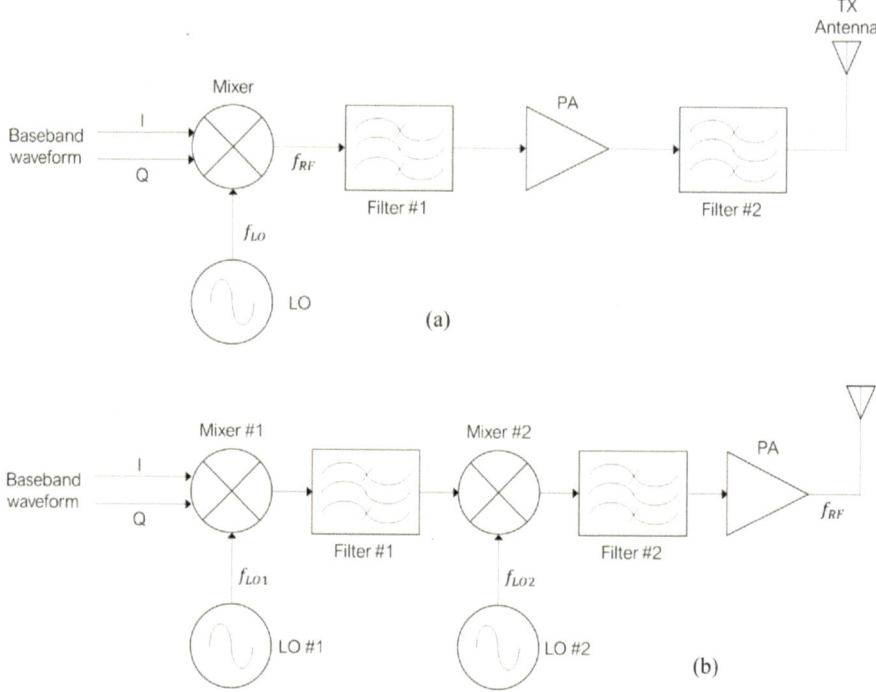

FIGURE 4.3 Transmitter architecture; (a) homodyne transmitter, (b) superheterodyne transmitter.

Hinotori-C SAR adopted the superheterodyne architecture in its transmitter and receiver.

4.3.1 SAR TRANSMITTER

The SAR transmitter generates a high-power chirp located at its intended carrier band. The transmitting antenna then converts the high-frequency radio signals generated by the transmitter into electromagnetic waves and illuminates the area under observation. Figure 4.4 shows the pictorial view of the transmitter and the solid-state power amplifier (SSPA) module, while Figure 4.5 depicts the hardware architecture of the transmitter. The transmitter consists of,

   i)   An upconverter circuit
   ii) A gating and amplification circuit
   iii) A DC biasing circuit
   iv) An embedded controller

4.3.1.1 The Upconverter Circuit

The upconverter circuit (Figure 4.6) produces the high-frequency chirp signals by modulating and mixing the baseband chirp with the synchronized LO signals using

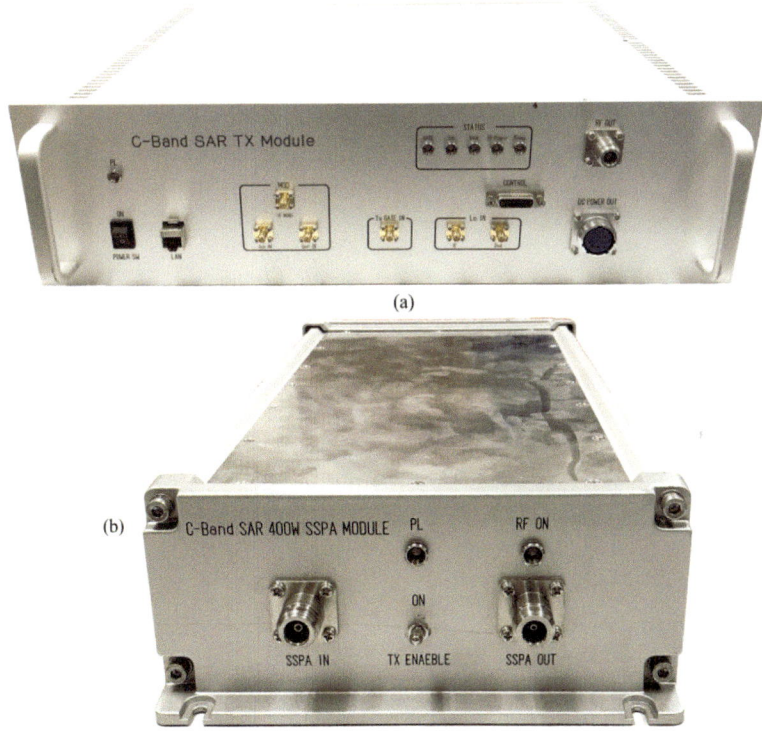

FIGURE 4.4   (a) Hinotori-C SAR transmitter; (b) Hinotori-C SAR solid-state power amplifier module.

RF mixers. In the general description, an RF mixer is an electrical circuit that produces signals with new frequencies from two signals applied to it. Technically, a mixer performs frequency translation using the nonlinear characteristic of the devices, changing the signal's frequency of its input into another frequency, which could be higher for the case of transmission (up-conversion in the transmitter) or lower for the ease of signal processing or recording (down-conversion in the receiver). This process is essential in a transmitter to scale down the size of the RF electronic components such as antenna, mixer, amplifier, filter, circulator, and attenuator for a more practical implementation since the physical dimension of these devices is proportional to the operating wavelength, λ, of the device.

The upconverter in Hinotori-C SAR's transmitter performs the frequency translation twice to produce the chirp signal at its final center frequency of 5,300MHz. The two mixers in the upconverter perform the frequency translation by mixing their input with their respective LO signals, one with 2,850MHz and another with 2,450MHz, to produce an output frequency of 5,300MHz. The upconversion mixer used in the transmitter can be either (i) quadrature image rejection mixer for Mixer #1 or (ii) single-sideband or image rejection mixer for Mixer #2. These mixers can suppress the image frequency by 30–40dB.

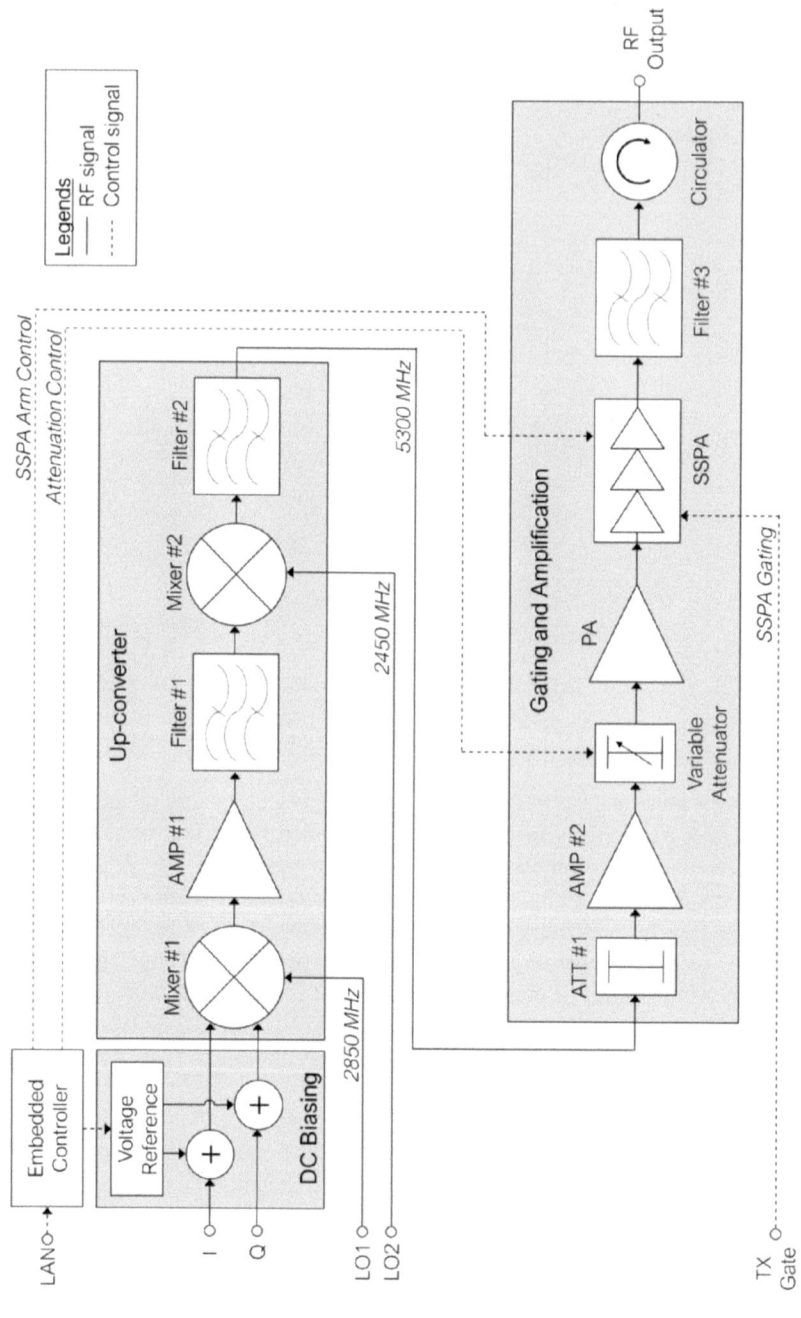

FIGURE 4.5   The transmitter architecture of Hinotori-C SAR.

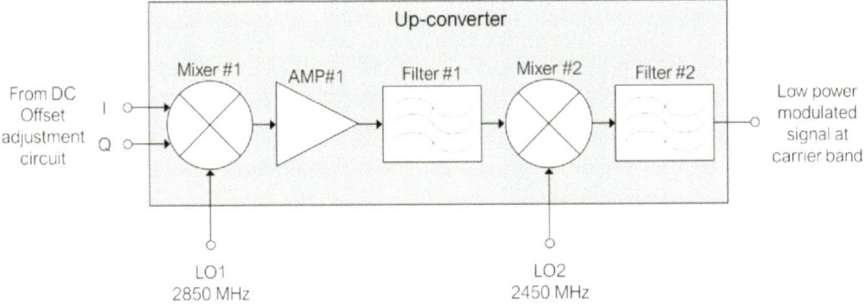

FIGURE 4.6 The upconverter circuit in the transmitter of Hinotori-C SAR.

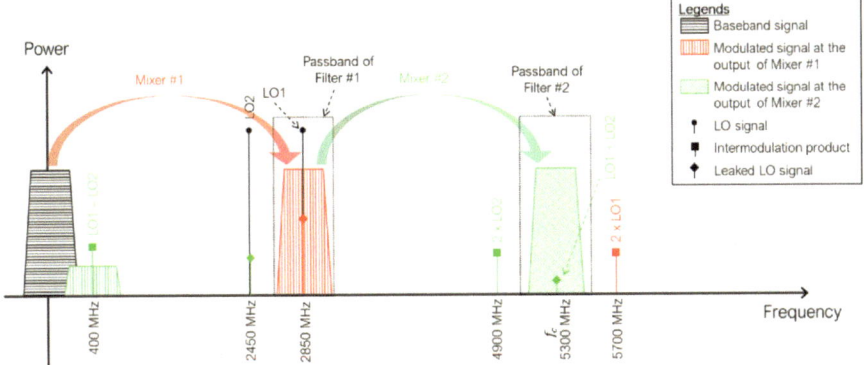

FIGURE 4.7 The mixing products from the mixers in Hinotori-C SAR transmitter.

The mixers in the upconverter produce several unwanted signals at their output port by their nature during the mixing process. The unwanted signals are the intermodulation products, the higher-order harmonics from the applied fundamental signals at its input, and the leaked LO signals, as illustrated in Figure 4.7. In the diagram, the red-labeled signals are produced by Mixer #1, while the green-labeled signals are produced by Mixer #2.

The carefully chosen bandpass filters (Filter #1 and #2 in Figure 4.6) placed at the output of these mixers suppress the power level of these unwanted signals to a significantly lower level. The additional filtering process ensures these unwanted signals will not degrade the signal quality of the intended modulated signal. Filter #1 in the upconverter suppresses the unwanted products produced by Mixer #1 so that the next mixer will not further mix these unwanted signals with its input signals. Similarly, Filter #2 in the upconverter removes the unwanted products produced by Mixer #2, retaining only the modulated signals centered at the 5,300 MHz carrier frequency. The frequencies chosen for the two LO shall not be too close to each other. Otherwise, it will not be easy to remove the intermodulation products if they are within the operating bandwidth of the transmitter.

Mixer #1 and #2 are passive mixers with about 10dB of conversion loss. The power level of the quadrature input signals to the first mixer is about 0dBm to -10dBm. The power level of the chirp produced by the upconverter is around -20dBm to -30dBm, which is low and insufficient for the gating and amplification circuit to produce the required 280W peak power signal. Thus, the low power amplifier (AMP #1 in Figure 4.6) after Mixer #2 provides the first signal amplification stage by amplifying the weak modulated signals from the upconverter into a more suitable power level for the coming stages of amplification. Placing the amplifier in between Mixer #1 and #2 gives an additional advantage of creating a higher power level signal at the input of Mixer #2 to ensure a better mixing process in the mixer.

4.3.1.2 Gating and Amplification Circuit

The modulated signals produced by the upconverter are usually insufficient for use in the SAR sensors to detect targets even as near a few hundred meters away. The signals must be further amplified to the expected power level before the antenna can transmit them. The amplifiers in the gating and amplification circuit do the required job. Figure 4.8 shows the diagram of the gating and amplification circuit in the Hinotori-C SAR's transmitter. The components that amplify the signal are the small-signal amplifier (AMP #2), the Power Amplifier (PA), and the solid-state power amplifier (SSPA).

To do the amplification better, the selection of the power amplifiers in the transmitter has to meet the following requirements:

i)   Sufficient **Gain** to amplify the input signal to the required power level (Refer to Figure 4.9 for an illustration of the amplifier's Gain)
ii) Sufficient **Power Handling Capability** (1dB compression point, P_{1dB}) to handle the high-power signal produced by the amplifier without damaging itself (Refer to Figure 4.9 for an illustration of the amplifier's P_{1dB})

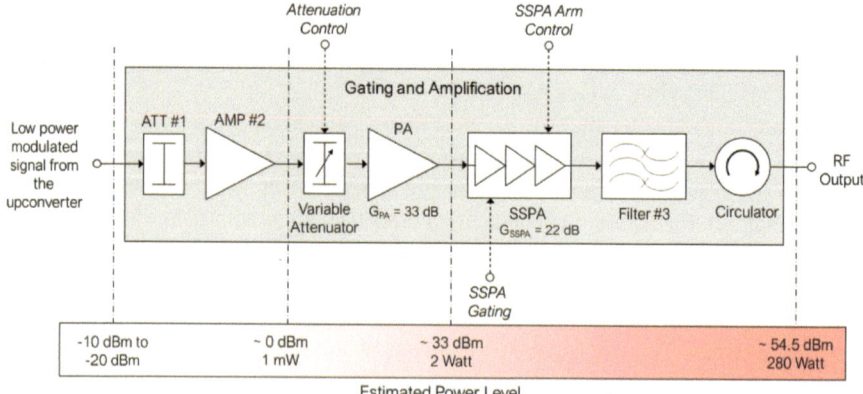

FIGURE 4.8   The gating and amplification circuit in the transmitter of Hinotori-C SAR.

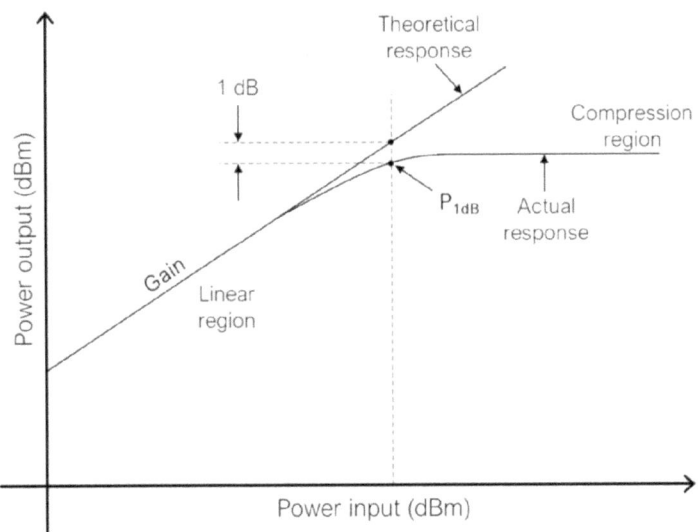

FIGURE 4.9   Gain (G) and 1dB compression point (P1dB) of an amplifier.

iii) **Stable** to avoid additional oscillations caused by the changes in the conditions of the amplifier, such as changes in temperature, frequency, load, and power level

iv) **Efficient** enough to turn the supplied energy into the output signal without generating much heat

The first amplifier (AMP #2) in the gating and amplification circuit boosts the signals to a power level G_{PA} dB below the P1dB of the power amplifier (PA), where G_{PA} is the gain of the PA amplifier in decibel. Since AMP #2 has a fixed gain, to avoid saturating the PA amplifier while maximizing the output power of the PA amplifier, the attenuation level of the fixed attenuator at the input of the PA (ATT #1) has to be tuned manually for every system. The output signal from the PA amplifier is then further amplified by the SSPA to the desired power level (280Watt). The gain and P1dB of the SSPA has to be chosen carefully during the design stage, taking into account the gain and the P1dB of the PA amplifier. Since SAR uses pulsed chirp, to optimize the transmitter's power consumption and to prevent components heating issues in the SSPA, the SSPA chosen shall be the pulsed type that supports high-speed drain pulsing. Some examples of practical implementation of these amplifiers for C-band SAR are:

i)   Mini Circuits' ZVE-3W-83+ 2000 MHz to 8000 MHz High Power Amplifier or Pasternack's PE15A4055 2 GHz to 12 GHz Medium Power Amplifier for PA amplifier

ii)   Wolfspeed's GaN on SiC solutions such as CGHV59350F 350W 5200MHz to 5900MHz GaN HEMT amplifier

FIGURE 4.10   The RF cavity bandpass filter used in Hinotori-C SAR.

The variable attenuator in the gating and amplifier circuit allows the user to adjust the output power level of the system by feeding a lower power level into the PA amplifier. The attenuation level of the attenuator is controllable using the developed graphical user interface (GUI) software through the embedded controller. The front-end circulator functions to alleviate the impedance mismatch between the transmitter's output and the transmitting antenna's input port to minimize the signals' reflection when the transmitting antenna connects to the system. Meanwhile, Filter #3 provides the final filtering stage in the transmitter. The filter used in this stage is usually the type of RF cavity bandpass filter with a very high roll-off rate. Figure 4.10 shows the image of the RF cavity bandpass filter used in the Hinotori-C SAR transmitter.

4.3.1.3   DC Biasing Circuit

The SAR transmitter takes two baseband signals at its input. These input signals are pairs of SAR baseband waveforms with one named in-phase signal, or the I, and the other named quadrature signal, or the Q. The two signals are the same amplitude sinusoid with the same frequency at the same time instance, but 90° out of phase (orthogonal) with the I leading the Q. Mixer #1 in the upconverter circuit takes these signals to perform quadrature modulation and at the same time upconvert it to its intermediate radio frequency at 2850MHz. The mixer used here is usually the quadrature upconversion single sideband/image rejection mixer with its general architecture illustrated in Figure 4.11.

An ideal quadrature mixer produces only its desired product. The phase and amplitude balance in the mixer vectorially cancels the LO signals and the image frequency. Nevertheless, in practice, there are imperfections in the phase and amplitude balance in the mixer circuit, as illustrated in Figure 4.11. The imperfections in the baluns of the mixer, or in other words, the poor port isolation, will cause increased LO feedthrough, RF/IF feedthrough, and spurious products. Meanwhile, the imperfections in the phase balance of the LO or the amplitude or phase balance of the I/Q channels will cause imperfect rejection of the unwanted products such as image frequency or sidebands in the mixer.

There are many ways to improve the rejection level of the LO and the sideband in a practical mixer. A common technique for reducing the LO leakage is applying a quadrature bias to the mixer I and Q inputs. In this technique, an offset voltage is added to each IF signal (I & Q signals for quadrature mixer) before mixing it with the LO signals [9]. The offset voltage applied negates specific imbalances in the mixer

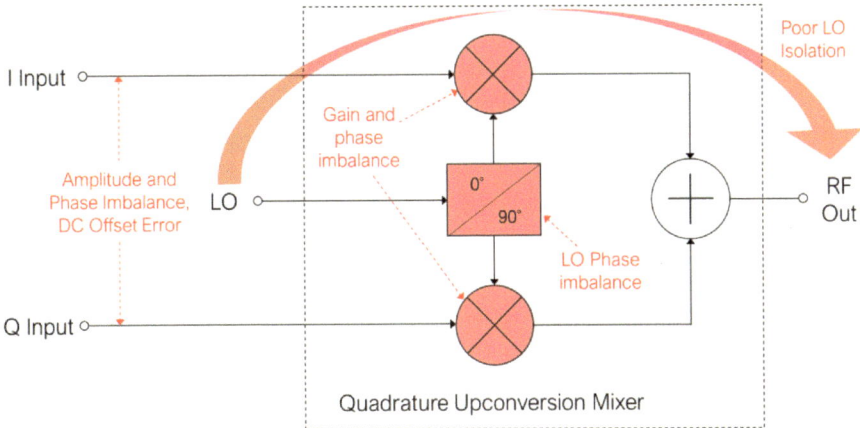

FIGURE 4.11 Block diagram of a quadrature upconversion image rejection mixer and the source of error in the mixer (highlighted in red).

and thus biases the mixer to operate at its LO minimum region. Figure 4.12 shows the plot of LO leakage of an image rejection mixer with zero phase imbalance over a range of I and Q offset voltages. The blue contour in the plot indicates the LO minima region of the mixer.

The offset voltages required to tune the mixer to its LO minima differ since each mixer has different DC and frequency characteristics. The position of the LO minima is sensitive to several factors. The operating frequency, amplitude, and temperature changes can degrade (increase) the power level of the LO leakage. In other words, these offset voltages need to be re-determined sporadically to recalibrate the mixer to maintain the lowest possible LO leakage. Hinotori-C applied two automated methods of locating the transmitter's LO minima – gradient search and linear search methods.

Several types of electronic devices can do the required biasing. One popular device is a diplexer, such as an RF bias tee. A bias tee is a three-port network commonly used to alter the DC bias point of electronic components. Abstractly, as illustrated in Figure 4.13, the bias tee can be viewed as an ideal capacitor that allows AC through but blocks the DC bias and an ideal inductor that blocks AC but allows DC. The offset voltage could be added to the I or Q signal by applying the offset voltage onto the DC port and the I or Q signal onto the RF port of the bias tee.

Alternatively, an operational amplifier in the summing amplifier configuration can do the biasing. Figure 4.14 shows an example of a non-inverting summing operational amplifier circuit. The amplifier adds the I or Q signal with the offset voltage and, at the same time, can amplify or attenuate the combined signal. The resistors R_f and R_i in the amplifier control the amount of amplification and attenuation of the amplifier. Since the typical bandwidth of the chirp in SAR is a few hundred megahertz, the chosen operational amplifier should have a good Gain Bandwidth Product (GBP) and slew rate.

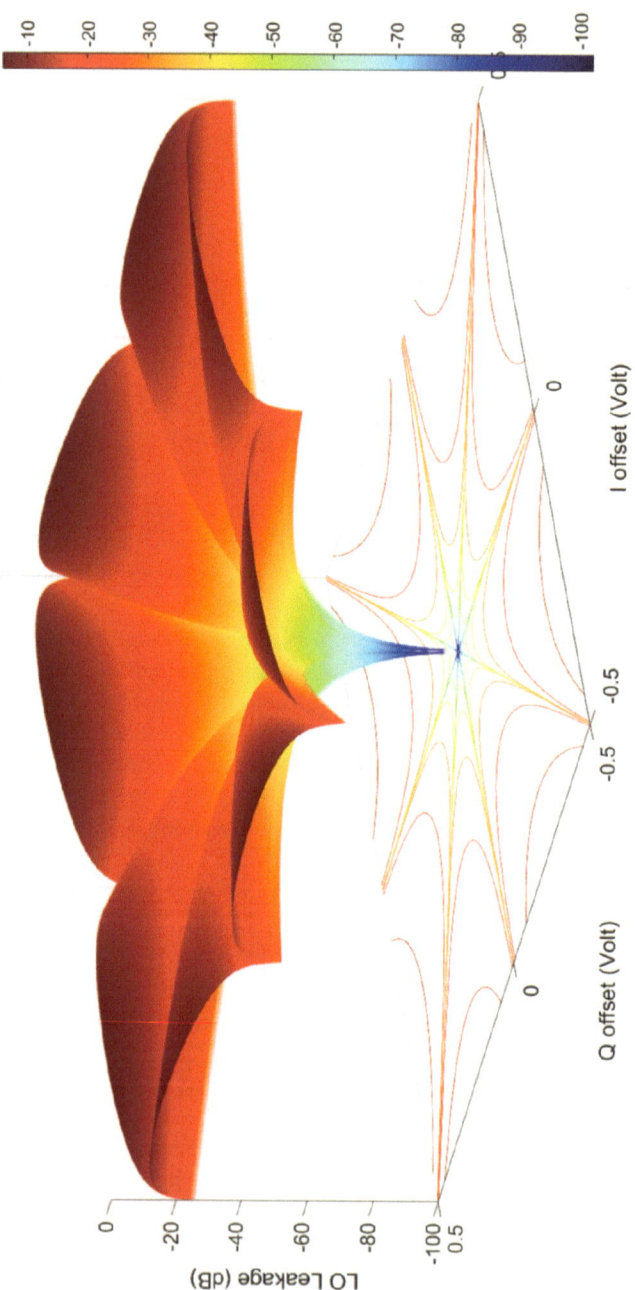

FIGURE 4.12   Quadrature mixer LO leakage vs. I and Q bias.

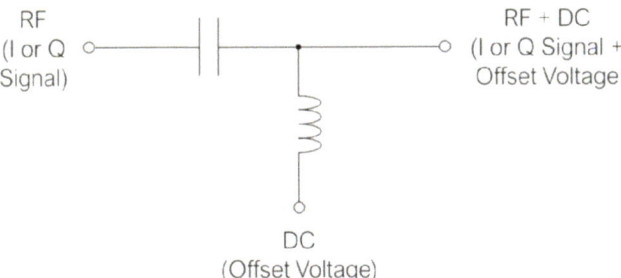

FIGURE 4.13   Equivalent circuit of a bias tee.

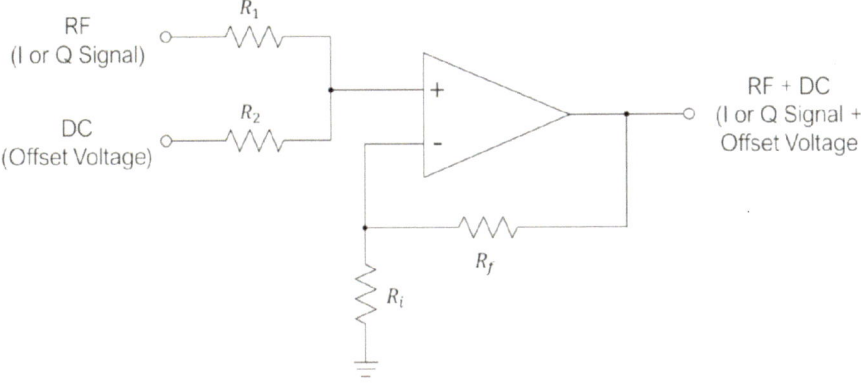

FIGURE 4.14   A summing operational amplifier.

4.3.1.4   Embedded Controller

The embedded controller in the SAR transmitter controls several devices in the transmitter. The controller can connect to the host software through an Ethernet interface, receives the command from the host PC, and,

- Interface to the in-system Digital-to-Analog Converter (DAC) to generate the offset voltages required for the DC offset adjustment circuitry
- Issue commands through Serial Peripheral Interface (SPI) to configure the attenuation level of the variable attenuator
- Controls the power supplies of the Solid-State Power Amplifier (SSPA) module to arm/disarm the SSPA module
- Interface to the onboard Analog-to-Digital Converter (ADC) to digitize the current and temperature signals and send them back to the host PC for current and temperature monitoring

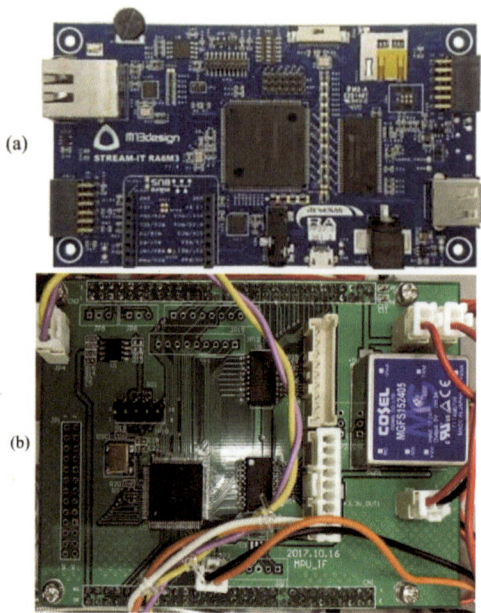

FIGURE 4.15   (a) The Renesas RA6M3 evaluation board, (b) the custom design intel MAX CPLD board.

Commonly, the excellent choice of embedded controllers is a microcontroller or programmable logic devices such as Field Programmable Gate Array (FPGA) and Complex Programmable Logic Device (CPLD). Hinotori-C SAR transmitter uses Renesas RA6M3 microcontroller for interface and control, and Intel (previously known as Altera) MAX V CPLD for custom logic implementation. The receiver unit uses a similar embedded controller but out carry control tasks in the receiver unit, see Figure 4.15.

4.3.2   SAR Receiver

The SAR receiver receives the weak echoes through the receiving antenna, amplifies the echoes to a higher power level, and converts the echoes to baseband for digitization. Figure 4.16 shows the pictorial view of the Hinotori-C SAR receiver, while Figure 4.17 depicts the hardware architecture of the receiver. The receiver consists of:

  i)   A gating and amplification circuit
 ii)   A downconverter circuit
iii)   A DC biasing circuit
 iv)   An embedded controller

4.3.2.1   Gating and Amplification Circuit

The signals from the transmitter propagate to the targets through the air in the form of electromagnetic waves. The targets reflect a portion of the signal and propagate

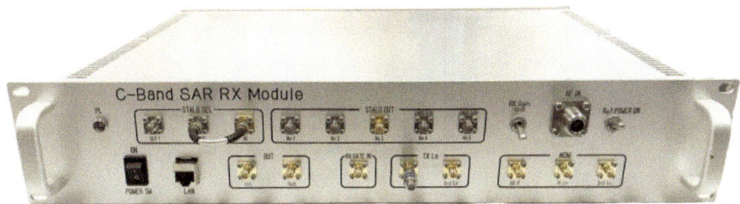

FIGURE 4.16   Hinotori-C SAR receiver.

back to the sensor as echoes. During wave propagation, electromagnetic waves suffer a tremendous amount of magnitude degradation from several sources of impairment, for example, two-ways path loss and reflectivity loss. Thus, the power level of the received echoes is usually very weak, typically smaller than -100dBm, which is insufficient to be detected.

The gating and amplification circuit in the Hinotori-C SAR's receiver amplifies these echoes to a more sophisticated level for downconversion. There are two groups of electronic components in the Gating and Amplification circuit, illustrated in Figure 4.18, precisely the gating circuit and the amplification circuit. The echoes first passage through the gating circuit for several signal regulations and then pass through the amplification circuits for signal amplification. Similar to the circulator in the transmitter, the front-end circulator here alleviates the impedance mismatch between the receiving antenna and the receiver to minimize the signals' reflection in the system. Filter #1, which is a type of Cavity Bandpass Filter, acts as the frequency selector that tunes the receiver to the center frequency and bandwidth of the SAR system.

Meanwhile, the two cascaded low noise amplifiers (LNAs) in the amplification circuit (LNA #1 and LNA #2) produce more than 70dB of power gain in the receiver. Contrary to the amplifiers in the transmitter, which require high power handling capability (1dB compression point, P_{1dB}), the amplifier used in the receiver shall possess a low noise figure (NF), for example, an LNA. In case the power level of the cascaded LNA is too high, the variable attenuator in the amplification circuit allows the user to manually adjust the power level before feeding the signal into the downconverter. The attenuation level of the variable attenuator is controllable using the developed graphical user interface (GUI) software through the embedded controller. Meanwhile, the switch, which is controllable by the embedded controller in the receiver, "mutes" the receiver to prevent over-amplification of the antenna coupling signal when the transmitter is transmitting. Filter #2 removes the unwanted frequency products produced by the LNAs.

A noise figure (*NF*) is an indicator that specifies the noise performance of a system. Essentially, the noise figure defines the amount of noise a specific component adds to the system with the noise power from a simple load with an amount of

$$P_n = kTB$$ (Equation 4.1)

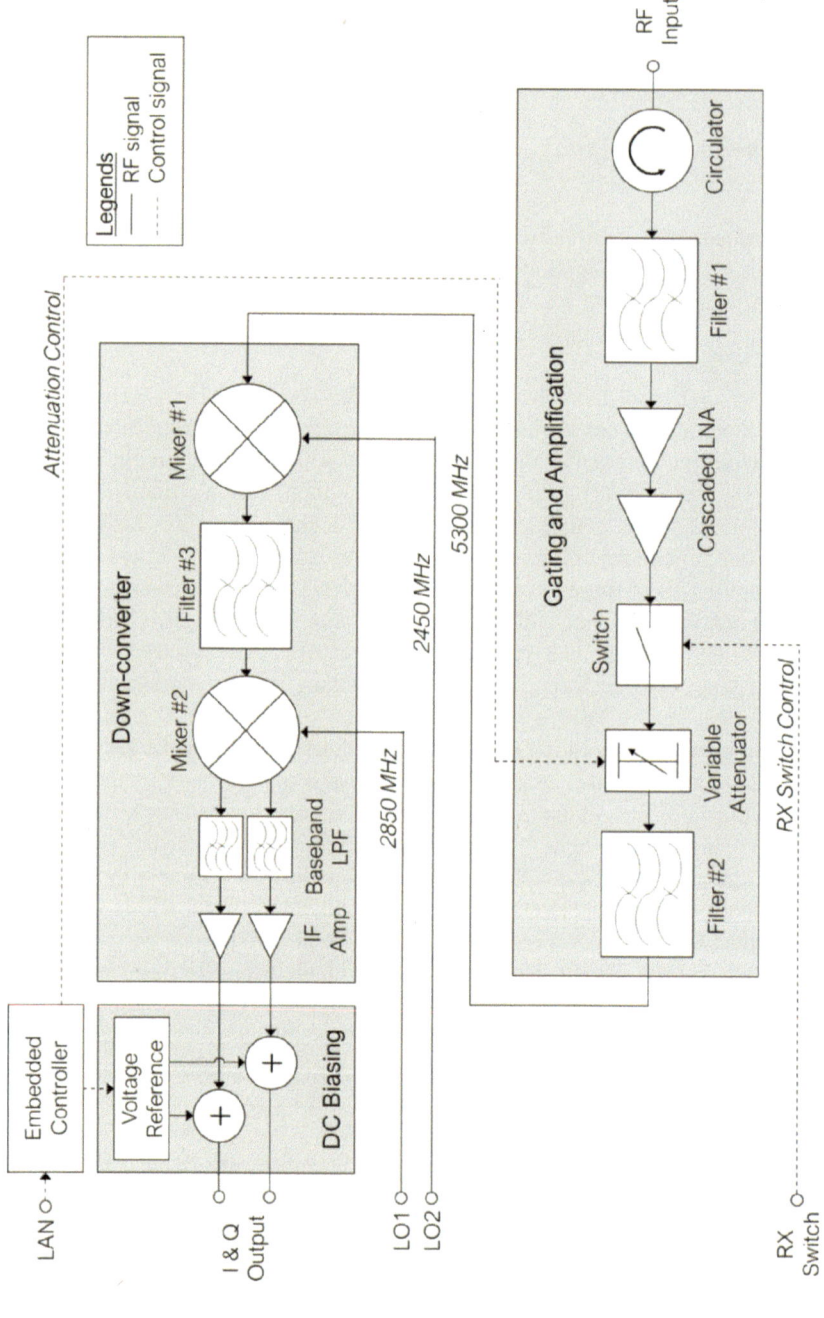

FIGURE 4.17 The receiver architecture of Hinotori-C SAR.

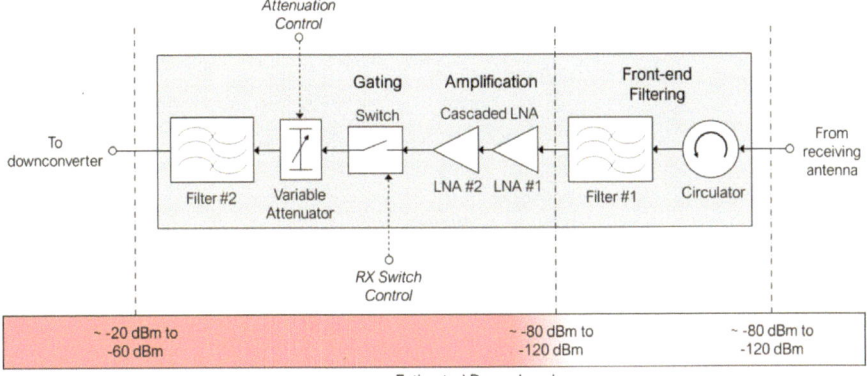

FIGURE 4.18   The gating and amplification circuit in the receiver of Hinotori-C SAR.

where

k= Boltzman's constant
T= absolute temperature of the load
B= measurement bandwidth

Every electronics component has a different amount of noise figure. The noise figure of the particular component is equal to

$$NF = 10\log_{10}\left(\frac{SNR_i}{SNR_o}\right) = SNR_{i,dB} - SNR_{o,dB} \qquad \text{(Equation 4.2)}$$

where

SNR_i = input signal-to-noise ratio (SNR)
SNR_o = output signal-to-noise ratio (SNR)
$SNR_{i,dB}$ = input signal-to-noise ratio (SNR) in decibel
$SNR_{o,dB}$ = output signal-to-noise ratio (SNR) in decibel

A noise figure is a good figure of merit to indicate the performance of a receiver. It specifies the amount of the system's SNR degradation caused by the receiver. An ideal receiver has a 0dB noise figure. Nevertheless, the inherent noise from the semiconductor devices in a receiver, for example, thermal noise, shot noise, and flicker noise results in a larger total noise figure in the receiver. The noise figure of a receiver with multiple cascaded components is not the summation of the noise figure of its every component, but it is equal to

$$NF = 10\log_{10}\left(F_1 + \sum_{i=2}^{n}\frac{F_i - 1}{\left(\prod_{j=1}^{i-1}G_j\right)}\right) \qquad \text{(Equation 4.3)}$$

where

NF= the noise figure of the receiver
F_i = the noise factor of the ith the component in the receiving chain
G_j = the gain of the jth the component in the receiving chain
n = number of components in the receiving chain
$i = i^{th}$components in the receiving chain

The equation above implies that it is crucial to place a component with a low noise figure and high gain as the first component in the receiving chain (after the receiving antenna) to minimize the receiver's noise figure. Table 4.3 lists the noise figure and the gain for every component in the receiver. By applying the data from Table 4.3 to Equation (4.3) for estimation, the noise figure of the Hinotori-C SAR receiver is about 4.5dB. Meanwhile, the gain of the receiver is about 50dB.

4.3.2.2 Downconverter Circuit

The echoes amplified by the amplification circuit reside at the carrier band. Although some ADC can sample these echoes, the signal's frequency is too high, and it is not feasible to convert these analog signals to digital data at this frequency. The amount of raw data generated would be tremendously large, causing subsequent problems.

i.   Large storage space is required to store these data.
ii. The data recorder requires an ultra-high-speed interface (> 20 GT/s) to transfer the digitized data to the data storage device.
iii. A large amount of raw data makes SAR data processing difficult.

The information carried by the echoes lies within the operating frequency band (chirp bandwidth centered at carrier frequency) of the signal, which is relatively smaller (nominally less than 10%) than the carrier frequency. A conventional

TABLE 4.3
Receiver Noise Figure

Component	Noise Figure (dB)	Gain (dB)
Circulator	0.4	-0.4
Filter #1	0.6	-0.6
LNA #1	3.5	35
LNA #1	3.5	35
Switch	1.5	-1.5
Variable attenuator	3.5	-3.5
Filter #2	0.2	-0.2
Mixer #1	8	-7.5
Filter #3	0.2	-0.2
Mixer #2	6	-5.5
Total	**4.5**	**50.6**

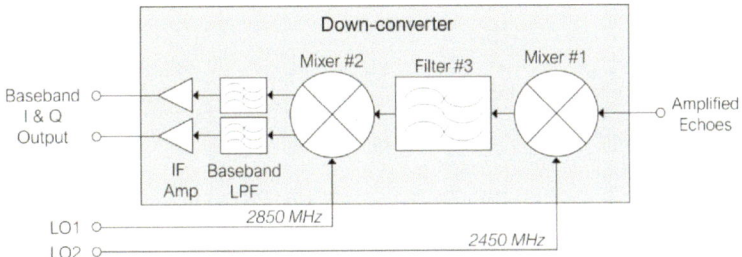

FIGURE 4.19 The downconverter circuit in the receiver of Hinotori-C SAR.

approach that could resolve the issues described above is to translate the signal's center frequency to a much lower frequency band called the baseband, which in most cases is 0Hz. The downconverter in the Hinotori-C SAR receiver does the desired frequency translation.

Figure 4.19 shows the circuit of the downconverter. Similar to the upconverter, the downconverter performs the frequency translation twice to shift the frequency of the amplified echoes from 5,300MHz to the baseband. Mixer #1 first mixes the amplified echoes with a 2,450MHz LO signal (LO2) to translate the frequency of the amplified echoes to the IF-band, which has a center frequency of 2,850MHz. Next, Mixer #2 moves the shifted amplified echoes from the IF-band to the baseband by mixing the signal with a 2,850MHz LO signal (LO1). The type of downconversion mixer used here can be (i) single-sideband or image rejection mixer for Mixer #1 and (ii) quadrature mixer for Mixer #2.

Same as the mixers in the upconverter, these mixers also produce several unwanted signals by their nature during the mixing process. These unwanted signals are the leaked LO signals, the intermodulation products, and the higher-order harmonics from the signals at the mixers' input, as illustrated in Figure 4.20. In that figure, the green-labeled signals are produced by Mixer #1, while the red signals are produced by Mixer #2. The power level of these unwanted signals is suppressed to a significantly low level by Filter #3 and the baseband low pass filter (LPF). The roll-off rate of these baseband low pass filters should be high enough to suppress one of the intermodulation products from Mixer #2 at 400MHz. The baseband filters are lumped-element filters built using capacitors and inductors since the cut-off frequency of these filters is a couple of hundred MHz. The pair of IF amplifiers amplify the voltage level of these baseband signals before the digitizer records them. Some IF amplifiers built using operational amplifiers have additional signal filtering capability.

4.3.2.3 DC Biasing Circuit

In a downconversion mixer, the LO signal leaked into the RF path will self-mix with its LO signal and produce a DC offset. If the LO-RF isolation of the mixer is insufficient, the produced DC offset could be severe enough to saturate the ADC in the digitizer. The receiver has a similar DC biasing circuit as the transmitter located after the downconverter. However, rather than using it to minimize the LO leakage in the

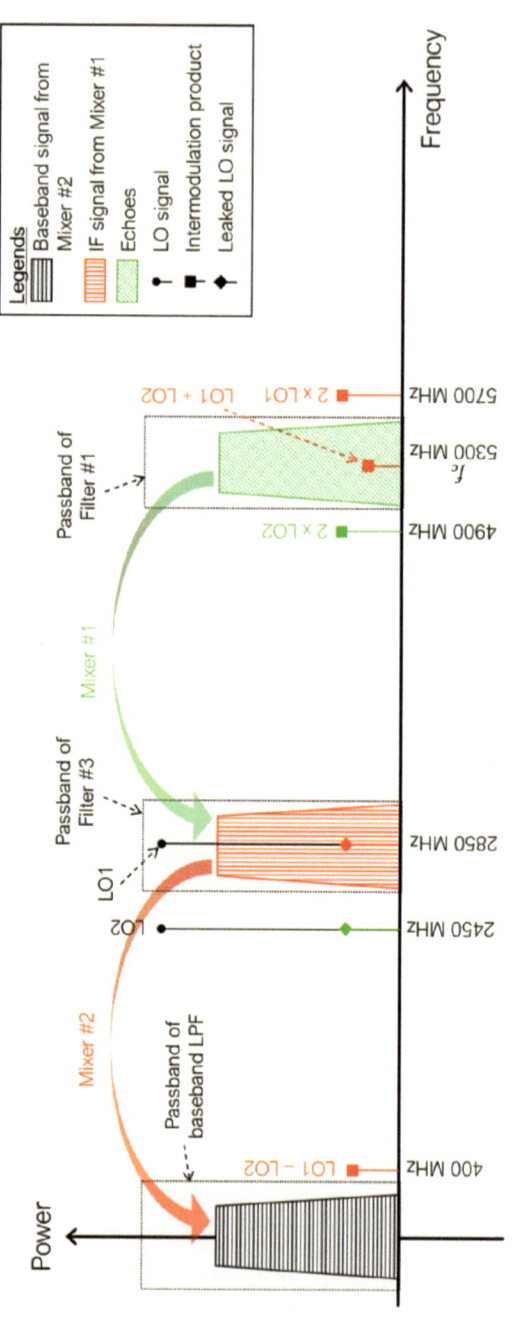

FIGURE 4.20   The mixing products from the mixers in Hinotori-C SAR receiver.

mixer as in the transmitter, this biasing circuit nulls the DC component generated by the mixer in the downconverter before the digitizer records the signals. The in-system DAC in the receiver controlled by the receiver's embedded controller generates these offset voltages. One easy way to figure out the voltage level needed is by first terminating the receiver's input and then tuning the DC level of the receiver's output as close to 0V.

4.3.3 LOCAL OSCILLATOR UNIT

Coherency in a pulsed radar system describes the phase relationships between the transmitted and the received pulses. Oscillations and electromagnetic waves are said to be coherent if their phase relationships are constant. All derivative signals in a fully coherent radar system, such as clocks, pulses, and gate signals, are derived from a master oscillator synchronous with its oscillation, thus having a fixed phase relationship to the master oscillator. Synthetic aperture radar is a fully coherent radar. It employs coherency in the system to improve the resolution of its images by processing successive radar echoes recorded from multiple antenna positions using a special signal processing algorithm to synthesize a very large antenna aperture.

The RF hardware in the transmitter and receiver, and the baseband electronics in the timing and control unit (TCU), digitizer, and chirp generator in the Hinotori-C SAR system require clock sources with different oscillation frequencies, waveform types, and signaling levels. The clock sources needed are in two groups, which are, (i) Very Stable Local Oscillator (STALO) signals with the frequency of 2,850MHz and 2,450MHz, and (ii) several coherent 10MHz reference clocks. The independent local oscillator unit (LOU) in the radio frequency unit (RFU) generates these clock signals.

LOU has two modules, the local oscillator (LO) module synthesizing high-frequency LO signals and the Reference Clock (REFCLK) module, which generates multiple 10MHz synchronized reference clock signals. There are two STALOs in the LO module, with each generating LO #1 (2,450MHz) and LO #2 (2,850MHz) signals, respectively. These STALO oscillations are synchronized to the reference clock module's reference signals to retain the clock coherency of the transmitter and the receiver with the master clock. Figure 4.21(a) depicts the block diagram of a LO module. An example of LO module implementation is shown in Figure 4.21(c) using a dual-frequency synthesizer module produced by Valon Technology.

On the other hand, the REFCLK module generates all reference clock signals distributed to the LO module, chirp generator, digitizer, and timing and control unit (TCU). The onboard oven-controlled crystal oscillator (OCXO) generates a master 10MHz sinusoid reference clock. The master clock is split into several analog outputs and each is fed into a voltage follower for impedance matching. The module also produces several digital reference signals with 3.3V transistor-transistor logic (TTL) level for devices that require a digital reference clock by converting some of the analog reference signals to digital form using TTL gate buffer integrated circuits. Figure 4.21(b) depicts the block diagram of the REFCLK module. Figure 4.21(d) shows the REFCLK module with five analog and four digital outputs.

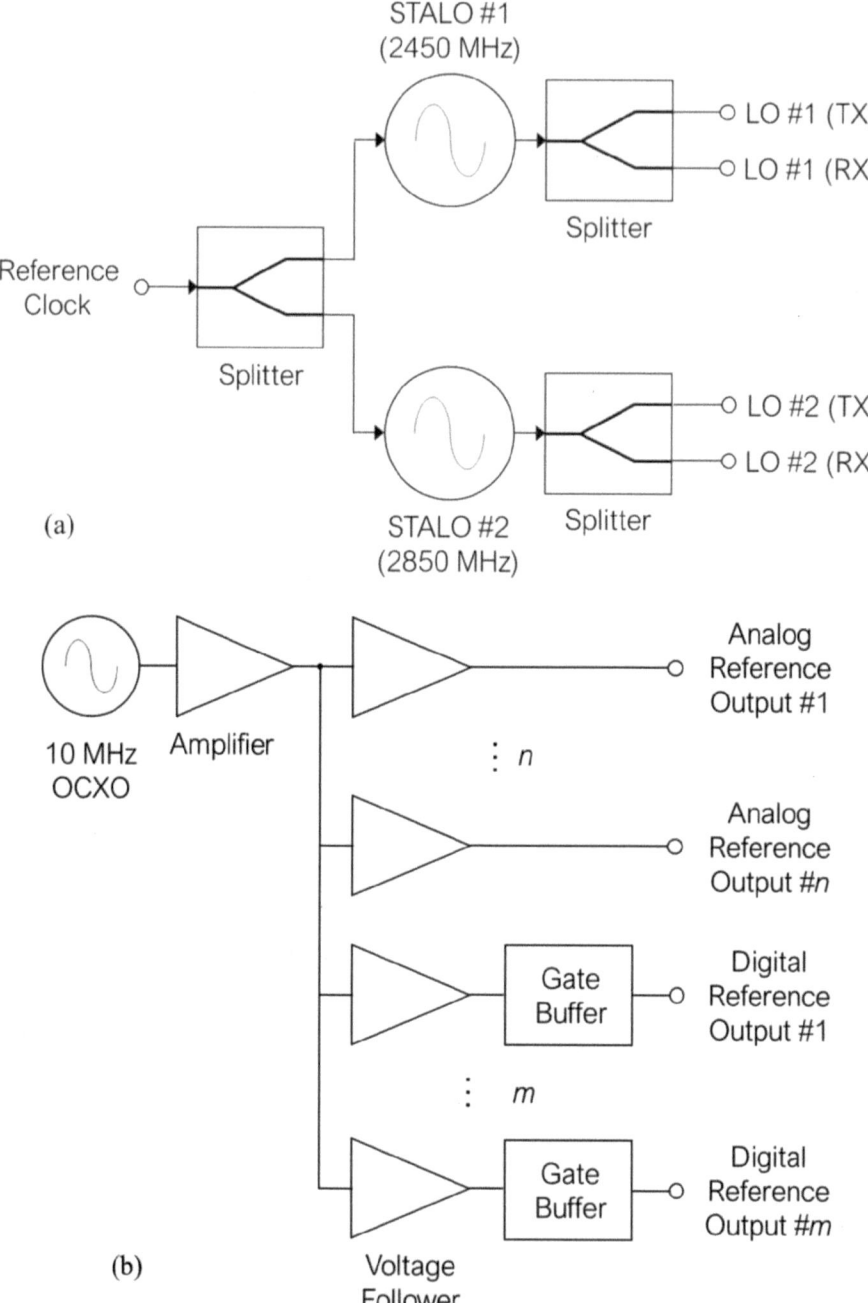

(a)

(b)

FIGURE 4.21 The local oscillator unit (LOU). System block diagram of (a) LO module; (b) REFCLK module, snapshot; (c) LO module; (d) REFCLK module.

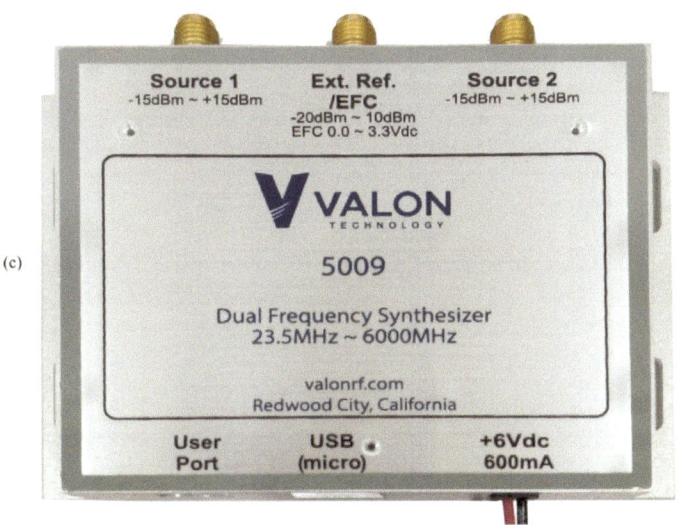

FIGURE 4.21   (Continued)

REFERENCES

[1] JT Sri Sumantyo, MY Chua, CE Santosa, GF Panggabean, T Watanabe, B Setiadi, FD Sri Sumantyo, K Tsushima, K Sasmita, A Mardiyanto, E Supartono, ET Rahardjo, G Wibisono, MA Marfai, RH Jatmiko, Sudaryatno, TH Purwanto, BS Widartono, M Kamal, D Perissin, S Gao, and Ki Ito, "Airborne Circularly Polarized Synthetic Aperture Radar," IEEE Selected Topics in Applied Earth Observations and Remote Sensing (JSTARS), Vol.14, pp.1676–1692, January 2021, DOI:10.1109/JSTARS.2020.3045032.

[2] JT Sri Sumantyo, MY Chua, CE Santosa, GF Panggabean, T Watanabe, B Setiadi, K Tsushima, FD Sri Sumantyo, K Sasmita, A Mardiyanto, E Supartono, ET Rahardjo, G Wibisono, RH Jatmiko, TH Purwanto, BS Widartono, M Kamal, RH Triharjanto, S Gao, K Ito, "Hinotori-C2 mission: CN235MPA Aircraft Onboard Circularly Polarized Synthetic Aperture Radar (CP-SAR)", 2019 IEEE International Geoscience and Remote Sensing Symposium (IGARSS2019), pp. 8538–8541, July 2019.

[3] JT Sri Sumantyo, MY Chua, CE Santosa, A Takahashi, and K Ito, "Aircraft and High Altitude Platform System Onboard Circularly Polarized Synthetic Aperture Radar (CP-SAR)", 2021 IEEE International Geoscience and Remote Sensing Symposium (IGARSS2021), pp. 8515–8518, July 2021.

[4] JT Sri Sumantvo, MY Chua, CE Santosa, GF Pariguabean, K Taushima, T Watanabe, K Sasmita, A Mardiyanto, FD Sri Sumantyo, ET Rahardjo, G Wibisono, E Supartono, S Gao, SP Parulian, M Nasucha, F Kurniawan, A Awaludin, B Purbantoro, YQ Ji, and N Imura, "Hinotori-C: A Full Polarimetric C Band Airborne Circularly Polarized Synthetic Aperture Radar for Disaster Monitoring", 2018 Progress in Electromagnetics Research Symposium (PIERS-Toyama), pp. 1466–1473, August 2018.

[5] MY Chua, JT Sri Sumantyo, CE Santosa, GF Panggabean, FD Sri Sumantyo, T Watanabe, YQ Ji, Peberlin, P Sitompul, M Nasucha, F Kurniawan, A Babag B Purbantoro, A Awaludin,, K Sasmita, ET Rahardjo, W Gunawan, RH Jatmiko, S Sudaryatno, TH Purwanto, BS Widartono, and M Kamal, "The Maiden Flight of Hinotori-C: The First C Band Full Polarimetric Circularly Polarized Synthetic Aperture Radar in the World", IEEE Aerospace and Electronic Systems Magazine, Vol. 34, Issue 2, pp. 24–35, June 2019.

[6] SA Maas, "Microwave Mixers," Artech House, Boston, 1986.

[7] VC Koo, YK Chan, G Vetharatnam, MY Chua, CH Lim, CS Lim, CC Thum, TS Lim, Z Ahmad, KA Mahmood, MH Shahid, CY Ang, WQ Tan, PN Tan, KS Yee, WG Cheaw, HS Boey, AL Choo, BC Sew, "A New Unmanned Aerial Vehicle Synthetic Aperture Radar for Environmental Monitoring", Progress in Electromagnetics Research, Vol. 122, pp. 245–268, 2012.

[8] KC Yee, CK Voon, YA Chin, SY Kuo, MY Chua, "Design and Development of a C-band RF Transceiver for UAVSAR", Progress in Electromagnetics Research C, Vol. 24, pp. 1–12, 2011.

[9] JG Baldwin and DF Dubbert, "Quadrature Mixer LO Leakage Suppression Through Quadrature DC Bias", Sandia Report (SAND2002–1316), May 2002.

5 Baseband and Control Unit

Ming Yam Chua and Josaphat Tetuko
Sri Sumantyo

5.1 INTRODUCTION

The hardware for an airborne synthetic aperture radar (SAR) system is comprised of several subsystems, as depicted in Figure 5.1. In general, the subsystems are:

- The Radio Frequency Unit (RFU) or SAR Electronics (Detail in Chapter 4)
- Antenna (Detail in Chapter 6)
- The Baseband and Control Unit (BCU)
- The Inertial Navigation System (INS)
- The Power Unit (PU)

The baseband and control unit (BCU) contains several critical subsystems to support several processes at the baseband level [1]–[5]. These subsystems are:

- Chirp generator
- Digitizer
- Timing and Control Unit (TCU)
- Processor and Digital Storage

5.2 CHIRP GENERATOR

The chirp generator produces the modulated baseband waveform used in radar. The most commonly used waveform in SAR is the linear frequency modulated (LFM) waveform, more widely known as "chirp".

5.2.1 THE CHIRP SIGNAL

Most SAR systems use the frequency modulated (FM) pulse for its pulse compression ratio (PCR) to improve the radar's range resolution [6]–[7]. There are two types of commonly employed frequency modulation schemes, namely the linear FM (LFM or widely knowns as "chirp") pulse, and the non-linear-FM (NLFM) pulse. LFM is the preferred choice due to the ease in its practical implementation on waveform generation and signal processing afterward. On the other hand, well-designed NLFM

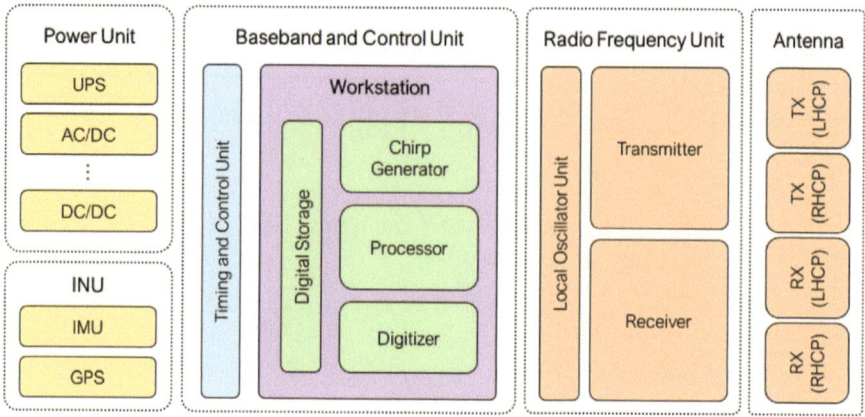

FIGURE 5.1   System block diagram of Hinotori-C SAR electronics.

waveforms exhibit a lower peak side-lobes ratio (PSLR) than LFM waveforms without compromising their impulse response width (IRW).

The chirp at the carrier band is centered at its carrier frequency, f_c, and sweeps $f_c - \frac{B}{2}$ Hz to $f_c + \frac{B}{2}$ Hz within T seconds. For a SAR transmitter adopting heterodyne architecture, the instantaneous frequency, $f(\tau)$, of the baseband chirp pulse, has to sweep starting $-\frac{B}{2}$ Hz to $\frac{B}{2}$ Hz, which can be formulated as

$$f(\tau) = \frac{B}{T}\tau - \frac{B}{2} \qquad \text{(Equation 5.1)}$$

where

τ = time variables in seconds
B = total bandwidth of LFM signal
T = pulse duration

With $\alpha = \frac{B}{T}$, which is the chirp rate in hertz per second, while the instantaneous phase of the baseband chirp pulse, $\phi_b(t)$ can be derived from Equation (5.1), formulated as

$$
\begin{aligned}
\phi_b(t) &= 2\pi \int_0^t f(\tau)\, d\tau \\
&= 2\pi \int_0^t \left(\alpha\tau - \frac{B}{2} \right) d\tau \\
&= 2\pi \left[\frac{1}{2}\alpha\tau^2 - \frac{B}{2}\tau \right]_0^t \\
&= \pi \left(\alpha t^2 - Bt \right)
\end{aligned}
\qquad \text{(Equation 5.2)}
$$

Hence, the baseband chirp signal, $x_b(t)$, is

$$x_b(t) = A \cdot \Pi\left(\frac{t}{T}\right) \cdot e^{j\phi_b(t)}$$

$$= A \cdot \Pi\left(\frac{t}{T}\right) \cdot e^{j\pi(\alpha t^2 - Bt)}$$

(Equation 5.3)

where

A = peak-to-peak amplitude of the chirp signal

$$\Pi\left(\frac{t}{T}\right) = \begin{cases} 1 & for\ 0 \leq t \leq T \\ 0 & elsewhere \end{cases}$$

It should be noted that $x_b(t)$ is a periodic function with a period of pulse repetition interval (PRI) seconds, which is the reciprocal of the pulse repetition frequency (PRF) in Hertz. Alternatively, the chirp signal can be written in the other form, which consists of cosine and sine terms. Using Euler's theorem, Equation (5.3) can be transformed to

$$x_b^e(t) = A \cdot \Pi\left(\frac{t}{T}\right)\left[\cos\left(\phi_b(t)\right) + j\sin\left(\phi_b(t)\right)\right]$$

(Equation 5.4)

The "real" part of Equation (5.4) is often called the in-phase chirp, while the "imaginary" part is called the quadrature chirp, constituting the baseband quadrature I & Q chirp waveform. A quadrature mixer produces a chirp signal at its carrier band by mixing the baseband I & Q chirp waveform with a LO signal at its designated carrier frequency. Mathematically, the chirp at the carrier band can be expressed as

$$x(t) = A \cdot \Pi\left(\frac{t}{T}\right) \cdot e^{j\pi(\alpha t^2 - Bt)} \cdot e^{j\pi f_c t}$$

(Equation 5.5)

where

f_c = the designated chirp center frequency

Figure 5.2(a) and Figure 5.2(b) plot the example of baseband LFM pulse waveform ($A = 1V$, $T = 10\mu s$, $A = 100\,MHz$) in time-domain and Figure 5.2(c) plots its spectrum. Meanwhile, Figure 5.2(d) and Figure 5.2(e) plot the chirp waveform's autocorrelated function (ACF), which is the same as the $\frac{\sin(X)}{X}$ function giving approximately -13.2dB peak-to-sidelobe ratio (PSLR). In other words, using this chirp signal for SAR will make the targets separated in time within the interval between the impulse and its sidelobe to appear as a single target with its attendant sidelobes. Nevertheless, the sidelobes can be suppressed effectively with a small cost of impulse

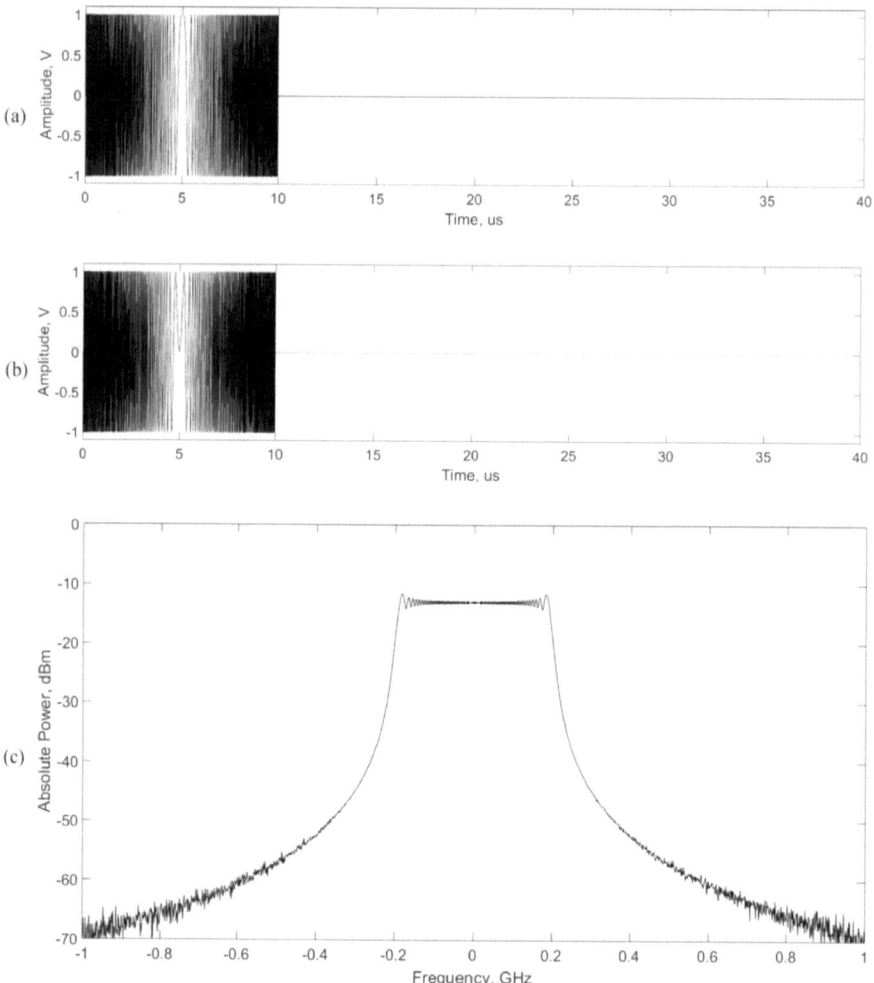

FIGURE 5.2 Chirp waveform; (a) real part, (b) imaginary part, (c) spectrum, (d) Auto-Correlation Function (ACF), (e) Ambiguity Function (AF).

response width (IRW) widening by applying amplitude-weighted functions such as Hamming, Hanning, Flat Top, Blackman, Gaussian, Dolph-Chebyshev window to its reference waveform during the matched filtering processing [8].

5.2.2 CHIRP WAVEFORM SYNTHESIS

There are various techniques to synthesize chirp signals. These techniques are generally in two categories; analog and digital approaches [9]–[13]. A voltage controlled oscillator (VCO) generates a chirp signal in an analog chirp generator. The slope of the linear ramp signal applied to the VCO controls the direction (up-chirp or

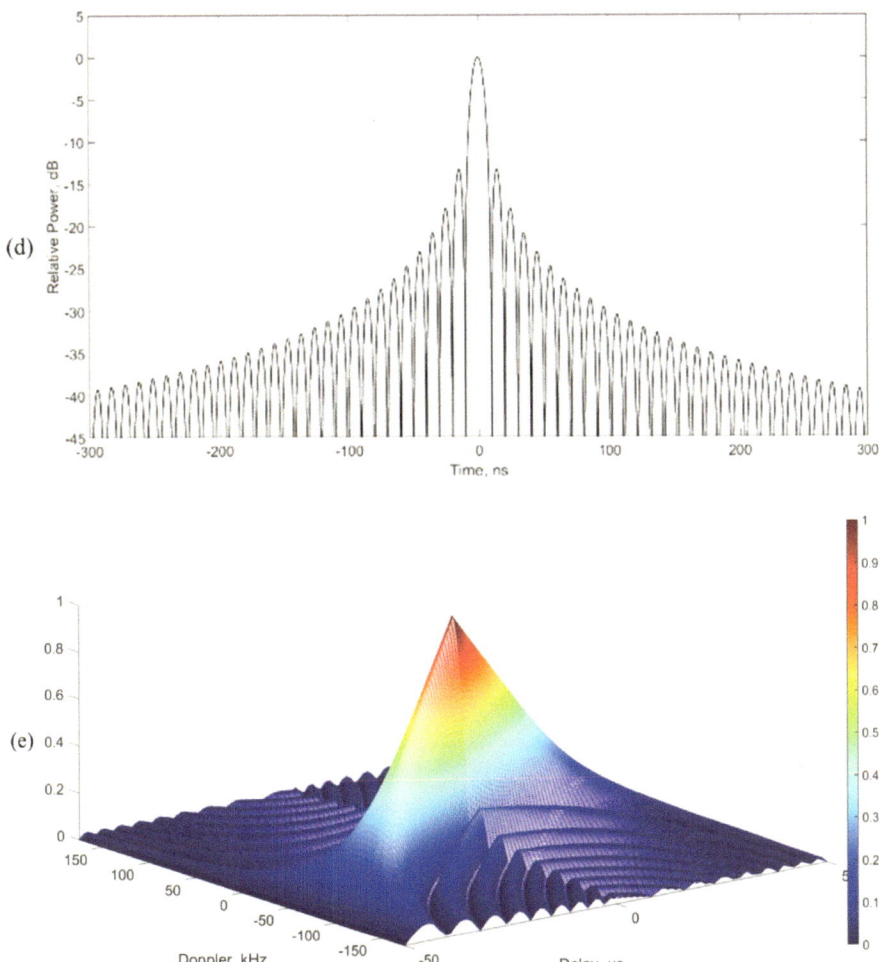

FIGURE 5.2   (Continued)

down-chirp) and the sweep rate of the chirp. However, due to several limitations of VCO, such as significant response or settling time and the non-linearity of VCO, the digital approach has become a more sophisticated technique to synthesize the chirp signals for SAR systems.

In the digital approach and digital hardware, such as gates, latches, and memory elements produce the chirp signal. The digital hardware generates the digital waveform data and is converted to analog signals using a digital-to-analog converter (DAC). Contrary to the analog chirp generator, the digital chirp generator can easily configure the waveform properties, such as modulation format, frequency, bandwidth, and waveform output duration by modifying the waveform data or the memory

TABLE 5.1

Analog vs. Digital Chirp Generator

Performance Indicator	Analog CG	Digital CG
Frequency response	Good	Excellent
Frequency linearity	Good but requires additional hardware	Excellent
Frequency/Phase stability	Good but requires additional hardware	Excellent
FM noise	Good	Good
Spurious/harmonics	Minimum filtering	Requires filtering
Hardware complexity	Low	High
Digital compatibility	Yes, but it requires additional hardware	Yes

contents. In terms of implementation, a digital chirp generator can be built using digital electronics or programmable logic devices, such as field programmable gate array (FPGA).

Table 5.1 summarizes the strengths and weaknesses of the digital chirp generator over the analog approach. Most of the modern SAR system's chirp generators are digital chirp generators and have the capability for reconfiguration and frequency/ phase stability. A high-speed DAC is the key hardware component in a digital chirp generator, as many of the DAC hardware specifications limit some signal parameters and the signal quality of the chirp produced. The following are some critical specifications for evaluating the performance of a high-speed DAC in a digital chirp generator.

5.2.2.1 Settling Time

The DAC settling time is an important criterion when choosing a DAC for high-speed waveform synthesis. It determines the time required to transition from an initial output voltage and to settle within a defined threshold of a desired final output voltage. The settling time of a DAC limits the maximum update/sampling rate of the DAC. Since the Nyquist sampling theorem states that the sampling rate of a sampling system must be at least twice its frequency to reproduce a pure sine wave accurately, the settling time usually determines the maximum frequency a DAC could produce. Figure 5.3 shows the fundamental definitions of a full-scale setting time for a DAC. The nominal settling time should be 5%–10% of the DAC's sampling time.

5.2.2.2 Glitch Impulse Area

An ideal DAC output should change from its old value to the new one monotonically. However, for a practical DAC, the output tends to undershoot or overshoot, or both. This phenomenon is known as a *glitch*. Glitch produces temporary spuriousness in the output. It arises from two causes: (i) the capacitive coupling effect during a digital to analog conversion, and (ii) the switching effect in the DAC. The *glitch impulse area* characterizes glitches with the unit of volt-seconds (usually in μV/sec

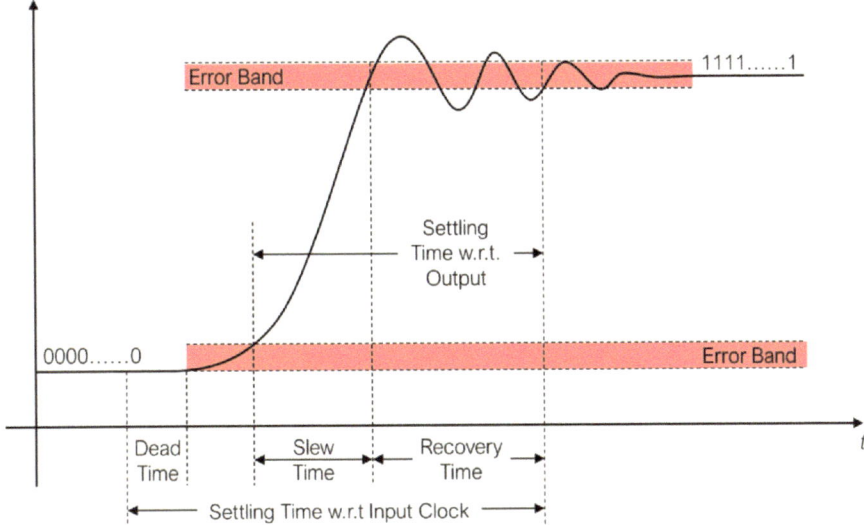

FIGURE 5.3   DAC settling time.

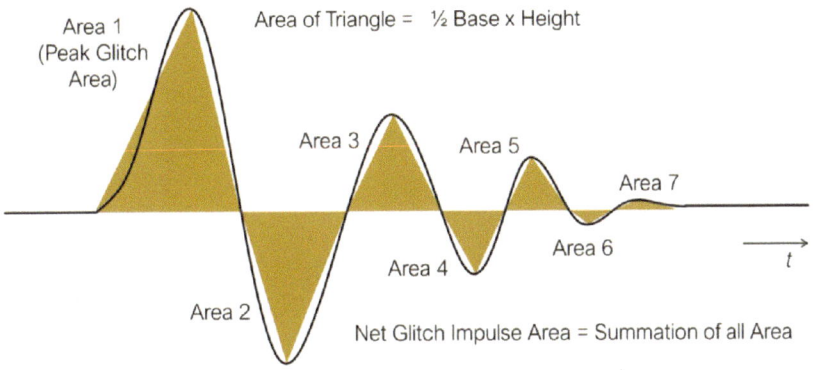

FIGURE 5.4   Glitch impulse area.

or pV/sec). It can be easily estimated from the mid-scale settling time waveform, as depicted in Figure 5.4.

5.2.2.3   Signal-to-Noise Ratio (SNR)

The signal-to-noise ratio (SNR) of a DAC is the ratio of the root mean square value for a fullscale sinusoid input to the root mean square value of the output noise in the specified bandwidth. The SNR of a DAC can be mathematically related to the DAC resolution as

$$SNR_{dB} = n \times 20\log_{10}(2) + 10\log_{10}\left(\frac{12}{8}\right)dB \qquad \text{(Equation 5.6)}$$

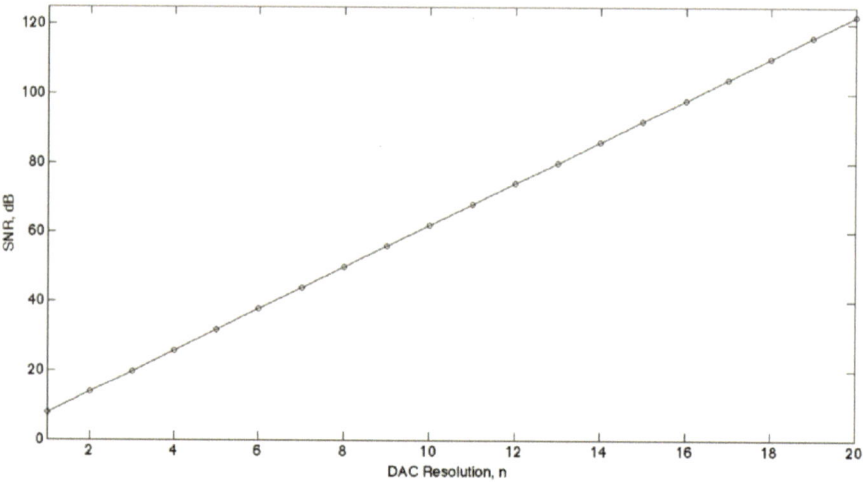

FIGURE 5.5   SNR vs. DAC resolution.

where

    n = the bit number of the DAC

Simplifying Equation (5.6) yields a more generalized equation to quantize the SNR of a DAC as

$$SNR_{dB} = 1.7609 + 6.0206 \times n \, dB \qquad \text{(Equation 5.7)}$$

Figure 5.5 depicts the SNR as a function of the resolution of a DAC.

Since an electronic system has a fixed amount of noise power, a higher resolution DAC can produce better quality signals. However, there are many factors that limit the usage of higher resolution DAC in a SAR system, such as

   i.   A higher resolution DAC needs a larger amount of data from the digital system that drives the DAC.

  ii.   There is a ceiling on the highest number of bits in commercial DACs.

 iii. The SNR requirement in the transmitter is not as critical as the receiver.

5.2.2.4   Spurious Free Dynamic Range (SFDR)

A spur is any observable frequency bin from the analog output of the DAC. The spurious free dynamic range (SFDR) is the available dynamic range of a DAC when spurious noise is not able to interfere with its fundamental signal. The SFDR can be measured by calculating the amplitude difference between its fundamental order to its spur up to its Nyquist bandwidth. Figure 5.6 shows several definitions of SFDR in the frequency domain.

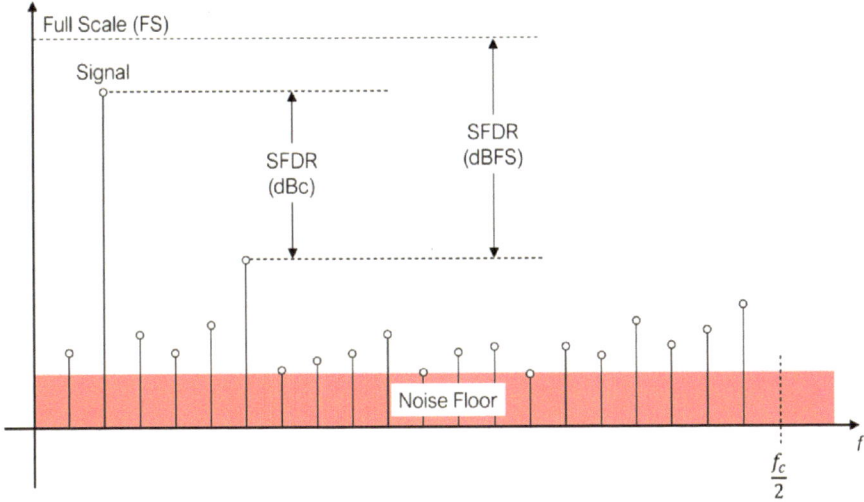

FIGURE 5.6 Spurious Free Dynamic Range (SFDR).

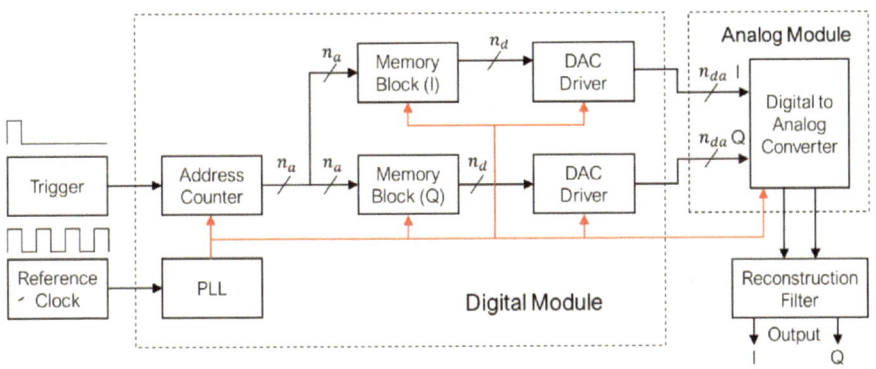

FIGURE 5.7 A general block diagram of an Arbitrary Waveform Generator (AWG).

The following section will discuss the two common types of digital chirp generator systems used in most SAR systems. The discussion will use the chirp generator built by JMRSL used in its circularly polarized SAR as an example of its implementation.

5.2.3 Chirp Signal Synthesis Using Look-Up Table (LUT) Technique

The look-up table (LUT) waveform synthesis technique uses more generalized hardware that can generate any waveform quickly and easily, as long as it is within its hardware capability. The technique is sometimes also called the memory-based waveform generation technique. Arbitrary waveform generator (AWG) is a popular waveform generator in the commercial field that generates waveforms using the same technique. Figure 5.7 shows the general functional block diagram of typical hardware that generates waveforms using the LUT technique.

There are two important hardware groups in the system, with one group responsible for generating digital waveform data and the other for producing the analog waveform. The practical realization requires several digital logic blocks, consisting of a binary counter as the vector address generator, two memory blocks (either RAM or ROM), a DAC driver, a DAC module, and a reconstruction filter. Since all logic blocks are sequential logics, the clock for each logic block is usually supplied internally in the synthesizer architecture. If the system requires a high-speed clock signal (in the order of multiple hundreds of MHz), a phase lock loop (PLL) multiplies the input reference clock to the desired frequencies.

In this technique, the amplitude level of a finite sample of the waveform to be produced is first pre-calculated and stored in a range of locations in the memory block. Once an external trigger signal is detected, the address counter starts to count from the first address of the waveform data in the memory block until its last. While the address counter counts, its output points to the memory element's location one by one to retrieve its memory content. The address counter's output width (n_a) is usually the same as the width of the memory block's address line (n_d). The data retrieved (memory block's output) is then converted to an analog level by the DAC. The number of data lines of the DAC (n_{da}) depends on the resolution of the chosen DAC. The lower significant bit has to be truncated if n_{da} is less than n_d.

During the conversion process, the produced analog waveform contains image frequencies (copies) resulting from the DAC's sampling process. A reconstruction filter (usually a low-pass filter that cut-off at the sampling frequency of the DAC) can effectively remove these image frequencies to reconstruct a smooth analog signal from the output of the DAC module. Assuming the LUT contains a complete cycle 1Hz sinusoid waveform, the output frequency (f_{out}) of the sinusoid tone is dependent on the frequency of the applied clock (f_c) and the size of the data in the memory block (n_{data}), which is

$$f_{out} = \frac{f_c}{n_{data}}$$ (Equation 5.8)

In short, chirp synthesis using the LUT technique plays back the waveform data pre-stored in the system. One of the chirp generators in Hinotori-C airborne circularly polarized SAR generates a chirp waveform utilizing the same technique using a commercial high-speed AWG from Signatec with the product name PXDAC4800 (Figure 5.8). The AWG can output data at a maximum real rate of 1.2 GSPS at 14-bit for two channels, 8-bit for four channels, or 600MSPS at 14-bit for four channels. A graphical user interface (GUI) developed inhouse functions to generate the chirp waveform data with the desired waveform parameters, downloads the data into the memory block of the AWG, and controls the playback of the waveform. Section 5.7 will discuss the developed GUI for the chirp generator. Figure 5.9 shows the capture chirp waveform (orange and blue plot) using a high-speed oscilloscope.

The LUT waveform synthesis technique offers an easy and reliable way to produce a waveform quickly. However, for it to generate different waveforms, it requires

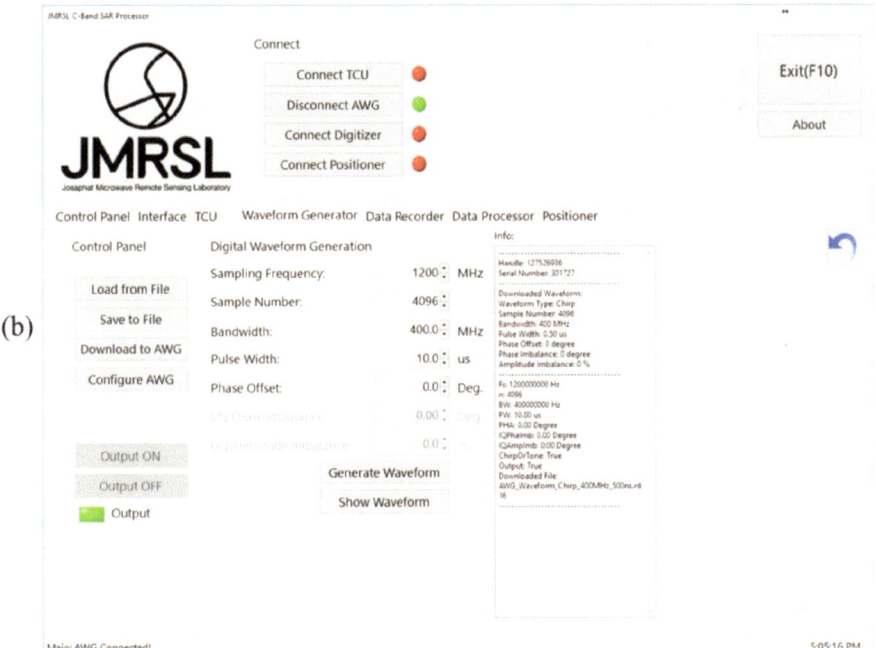

FIGURE 5.8   Hinotori-C Circularly Polarized-SAR chirp generator; (a) Signatec PXDAC4800 high-speed waveform generator, (b) the user control software.

the user to reconfigure the waveform memory to the one desired. The next section discusses a popular technique for generating chirp for SAR that supports waveform reconfiguration on-the-fly, which is essential for UAV or airborne circularly polarized SAR systems.

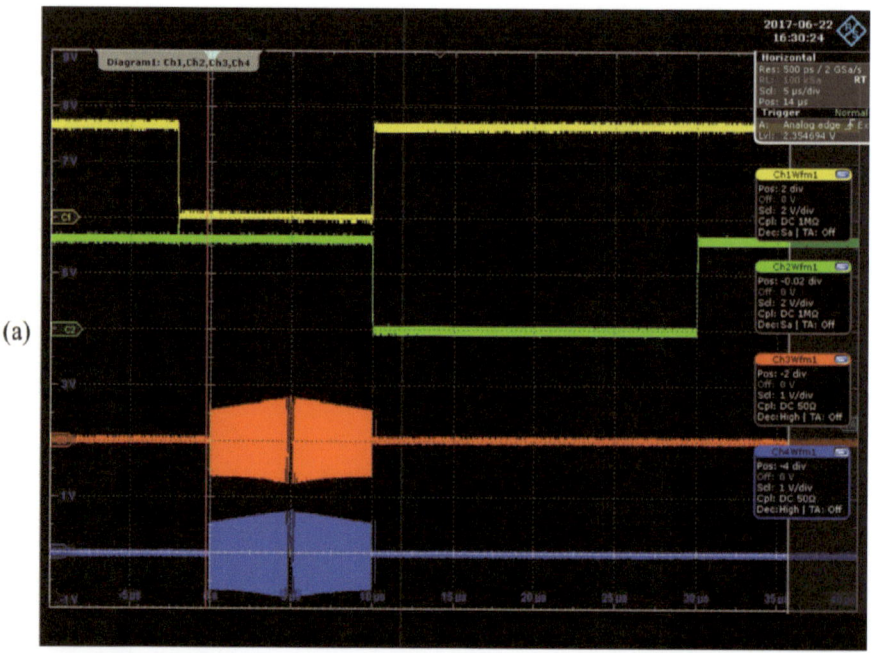

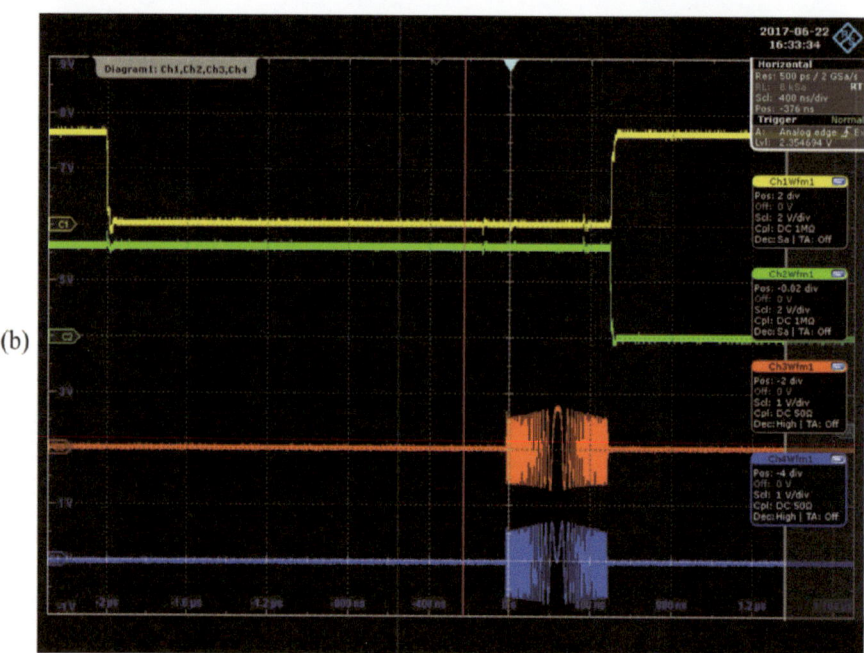

FIGURE 5.9 Chirp waveform generated using Signatec PXDAC4800 high-speed AWG; (a) 400MHz bandwidth and 10μs pulse width, (b) 400MHz bandwidth and 0.5μs pulse width.

5.2.4 CHIRP SIGNAL SYNTHESIS USING DIRECT DIGITAL SYNTHESIS (DDS) TECHNIQUE

The Direct Digital Synthesis (DDS) technique is an alternative chirp waveform synthesis technique [14]–[15]. It is a method of generating an analog periodic waveform, usually a sinusoid, by producing a time-varying signal digitally and then performing a digital-to-analog conversion. Compared to the LUT technique, DDS offers fine frequency resolution, fast switching between output frequencies, and is operable over a broad band of frequencies.

The DDS waveform synthesis technique evolved from the LUT synthesis technique was discussed earlier. As shown in Equation (5.8), the output frequency of the waveform generated using the LUT technique depends on the applied clock frequency and the size of the waveform data. A minor modification to the digital hardware is necessary if the frequency, phase, or both, are to be changed.

DDS introduces a phase accumulator block into the digital signal chain, enabling the frequency tuning capability in the system. Figure 5.10 shows the general block diagram of a DDS system for chirp waveform synthesis. As depicted in the figure, the phase accumulator replaces the address counter and takes an additional input called the (frequency) tuning word. The phase accumulator block possesses an arithmetic carry function that allows the accumulator to act as a "phase wheel."

The working principle of the DDS is dependent on the phase-frequency relationship in a sinusoidal signal. As illustrated in Figure 5.11, one can view a complete sine-wave oscillation as a rotating vector that rotates around a phase circle from $0°$ to $360°$. When the phase wheel is rotating at a constant rotation speed, the phase accumulator increments linearly and produces the linear phase information of a periodic waveform. Since the output frequency is a function of the number of steps to complete a phase wheel cycle, the step size of the phase wheel determines the number of discrete phase points in the rotation.

The phase accumulator is a modulus of 2 counter, or equivalent to n_{PA} steps counter. The counter increments its current value at every clock cycle. The additional digital input to the phase accumulator called the tuning word, M, determines the increment size and is arithmetically calculated with the current value of the counter at every

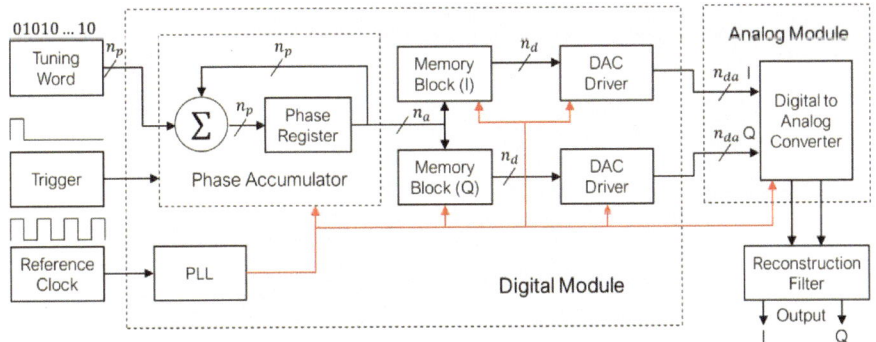

FIGURE 5.10 The block diagram of a Direct Digital Synthesizer (DDS).

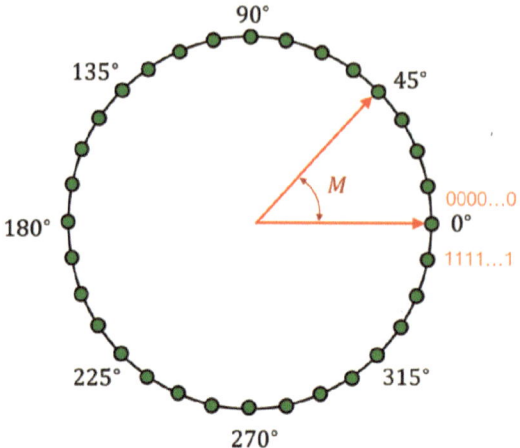

FIGURE 5.11   A complete sine-wave oscillation as a rotating vector.

clock cycle. The tuning word forms the phase step size between the subsequent reference clock updates or the immediate resolution of the phase wheel. The larger the step
size, the faster the phase accumulator overflows and completes its equivalent periodic waveform cycle. The relationship between the phase accumulator and the tuning
word gives the basic tuning equation in DDS. A change in M, results in immediate
and phase-continuous changes in the output frequency. Hence, the output frequency
of the DDS can be determined by

$$f_{out} = \frac{M \times f_s}{n_{PA}}$$ (Equation 5.9)

where

f_{out} = the frequency of the DDS's output
f_s = DDS clock frequency
M = binary tuning word
n_{PA} = accumulator length

The Nyquist sampling theorem dictates that more than two samples per cycle are
needed to reconstruct a sinusoidal waveform. Thus, setting the tuning word to a value
that is half of the length of the phase accumulator gives the maximum producible frequency by the DDS, which is

$$f_{max} = \frac{f_s}{2}$$ (Equation 5.10)

The output of the phase accumulator entails linear ramp phase information data
and cannot be directly used to generate a sine wave or any other waveform except a

ramp. Therefore, a phase-to-amplitude look-up table (memory) is needed to convert the phase accumulator's instantaneous output value into the desired periodic waveform amplitude information, which the D/A converter could then use.

Since a chirp is a frequency-modulated waveform, the DDS technique can synthesize the chirp efficiently. For this, the DDS output frequency equation has to be slightly modified. By substituting Equation (5.1) into Equation (5.9), which constitutes Equation (5.11), determining the instantaneous binary tuning word needed for a DDS hardware to generate an LFM chirp waveform can be determined.

$$M_i = \frac{\left(\alpha t - \dfrac{B}{2}\right) \times n_{PA}}{f_s}$$

(Equation 5.11)

where

M_i = the instantaneous binary tuning word w.r.t. time
α = chirp slope, B/T
B = chirp bandwidth
T = chirp pulse width

The idea of adopting DDS for synthesizing periodic waveforms emerged in the 1970s. It was eventually adopted in the industry in the 1990s and only became a well-known commercial product in the form of compact semiconductor integrated circuit (IC) packages after the 2000s. These commercial DDS ICs pack not only the essential hardware blocks – such as the phase accumulator, the waveform memory, and the DAC into the chip for a full DDS operation – but also incorporate additional hardware such as PLLs, mixers, comparators, and controllers. These DDS ICs can provide fast acquisition, highly accurate PSK and FSK modulations, or high-precision tuning resolution, among other benefits. They are an ideal solution in many applications, such as communications, test equipment, and radars, where precise or agile frequency and phase controls are required. An example of such DDS ICs is AD9174, a product from Analog Devices featuring an on-chip dual, 16-bit, 12.6 GSPS RF DAC.

Although commercial DDS ICs are high in performance, they are sometimes unsuitable for synthesizing baseband chirp for SAR and require a custom-designed DDS system. For this purpose, JMRSL developed an all-in-one, compact, and lightweight FPGA-based chirp generator named FWS0250–08, capable of synthesizing four pairs of quadrature (I&Q) chirp signals and its relevant timing and control signal for multi-band full polarimetric SAR. Figure 5.12(a) shows the hardware's PCB stack up, (b) the PCB layout in, and (c) the picture of the final assembled hardware. Meanwhile, Table 5.2 lists the hardware features of FWS0250–08.

The JMRSL FWS0250–08 hardware needs a digital system in the onboard FPGA. Figure 5.13 shows the architecture of the deployed digital system in the FPGA, which is responsible for generating the chirps, timing, and control signals. The system is comprised of: (i) four identical DDS-based reconfigurable chirp generator (CG) cores, (ii) a timing and control unit (TCU), (iii) a communication controller (CC),

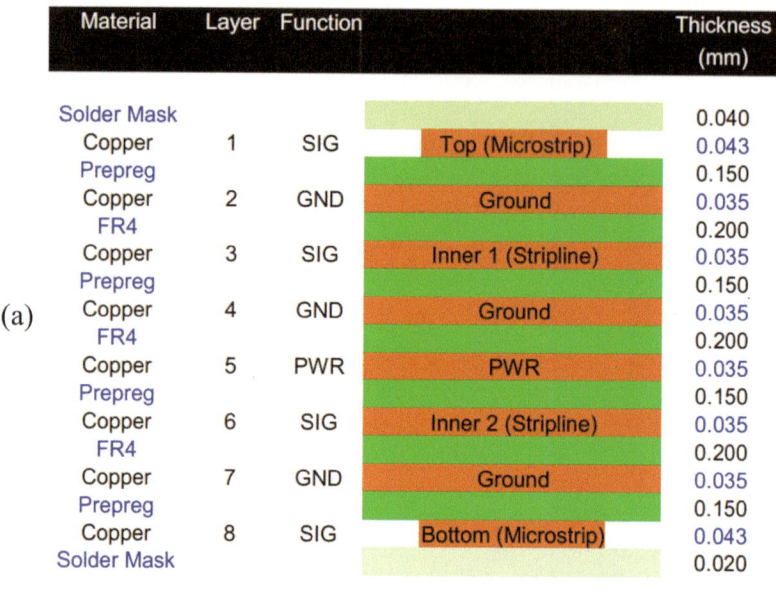

Material	Layer	Function		Thickness (mm)
Solder Mask				0.040
Copper	1	SIG	Top (Microstrip)	0.043
Prepreg				0.150
Copper	2	GND	Ground	0.035
FR4				0.200
Copper	3	SIG	Inner 1 (Stripline)	0.035
Prepreg				0.150
Copper	4	GND	Ground	0.035
FR4				0.200
Copper	5	PWR	PWR	0.035
Prepreg				0.150
Copper	6	SIG	Inner 2 (Stripline)	0.035
FR4				0.200
Copper	7	GND	Ground	0.035
Prepreg				0.150
Copper	8	SIG	Bottom (Microstrip)	0.043
Solder Mask				0.020

Total thickness: **1.556**

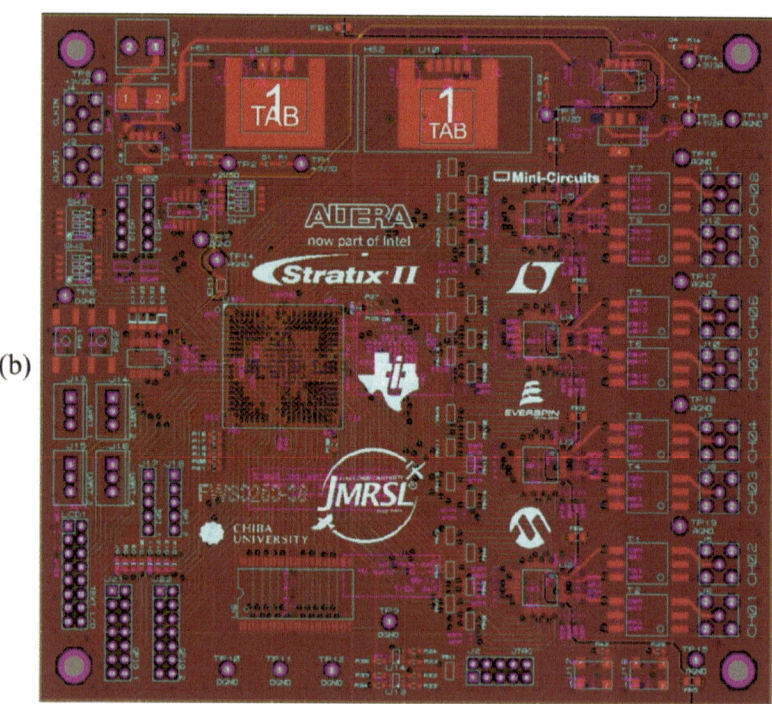

FIGURE 5.12 JMRSL FWS0250–08 reconfigurable chirp generator; (a) 8 layers PCB stackup, (b) PCB layout, (c) assembled prototype.

(c)

FIGURE 5.12 (Continued)

TABLE 5.2
Hardware Features of FWS0250–08

Onboard Module	Hardware Specifications
FPGA	• Altera Stratix II (EP2S30F484C3N) • 33,880 Les • 1,369,728 bits of on-chip memory • 128 DSP blocks • 6 PLLs • 342 I/Os
Digital-to-Analog Converter	• 275 MSPS sampling rate per channel • 14-bit resolution • 4 DACs (2 channels each) • Parallel CMOS interface
Peripherals	• Two UART interface (RS232) • Two × UART interface (3V3-TTL) • Two × SPI interface • Two × 9-ways general purpose I/O • Two × 4-ways high-speed I/O • Two × 4-ways user DIP switch • Two-ways user LED indicator • One 2-ways user push button

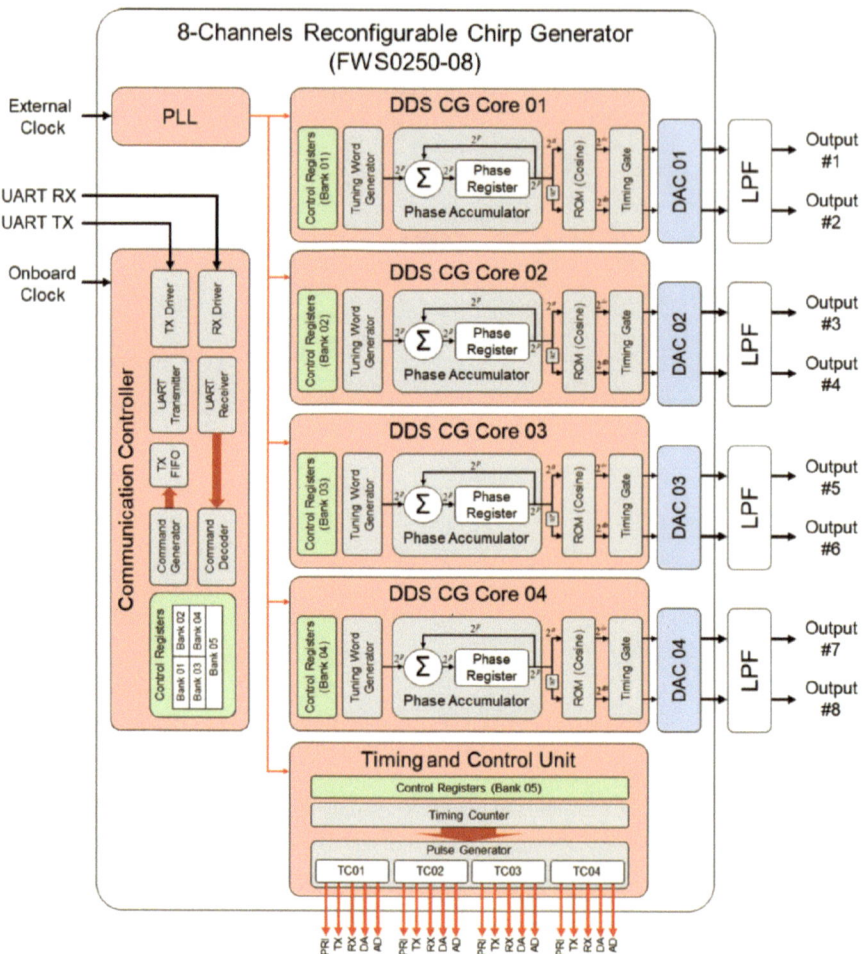

FIGURE 5.13 Digital system architecture of 8-Channels reconfigurable chirp generator.

and (iv) a phase locked loop (PLL). The filters are external devices to the board since the cut-off frequency depends on the sampling clock applied to the DACs. Table 5.3 lists the technical specifications for each submodule.

The output chirp properties such as the bandwidth, the pulse width, the PRI, and the initial phase for each channel are changeable in real-time without reconfiguring the FPGA's digital system. It can be done by connecting a computer terminal that supports the Universal Asynchronous Receiver/Transmitter (UART) interface to the system and sends the control command string to the communication controller. The system has five control register (CR) banks that manage the DDS CG cores and the timing and control unit (TCU). A received valid command through the communication controller will update the corresponding CR and change the chirp or the timing properties. The operating range and the tuning resolution of the tunable parameters are summarized in Table 5.4.

TABLE 5.3
CG Sub-Modules Technical Specifications

Module	Parameter	Value
Clock Source	Reference Clock (In)	10 MHz
	Onboard Clock #1	50 MHz
	Onboard Clock #2	27 MHz
Communication Controller	Clock Speed	50 MHz
	Baud Rate	115,200
	Data Format	8-N-1
Chirp Generator	Clock Speed	250 MHz
	No. of Output Channel	8 (4 pairs of I & Q)
	Maximum Bandwidth	250 MHz
	Pulse Width	500 ns – 2.68 s
	Maximum PRF	1 MHz
Timing and Control Unit	Clock Speed	100 MHz
	Timing Precision	10 ns
High-Speed DAC	Sampling Rate	250 MSPS
	Resolution	14-bit
	No. of Channels	8

TABLE 5.4
Chirp Generator Reconfigurable Parameters

Parameter	Operating Range	Tuning Resolution
Bandwidth, B	0.1 – 250 MHz	0.9313 Hz
Pulse width, T	100 ns – 100 μs	10 ns
Pulse Repetition Frequency, PRF	1 Hz to 10 kHz	0.3725 Hz
Initial phase, ϕ_0	0° to 359.9°	0.00000134°

Figure 5.14 shows the sample of synthesized chirp signals with 200MHz bandwidth and 10μs of pulse width recorded using a high-speed oscilloscope (Rohde & Schwarz RTO 1044 digital oscilloscope). Meanwhile, Figure 5.15 shows the timing and control signals the TCU produces that are suitable to be applied to manage the RF transceiver and the other baseband subsystem. The system can produce four pairs of quadrature chirp (I & Q) signals for multiband SAR applications by configuring Channel 1 and 2 as the first pair, Channel 3 and 4 as the second pair, Channel 5 and 6 as the third pair, and Channel 7 and 8 as the last pair. Figure 5.16 plots the spectrum for one of the chirp signals. The spectrum of the chirp spans from -100MHz to 100MHz. The spurs after $f_s/2$ at 125MHz are due to the quantization effect of the DAC and are removable using a low-pass filter.

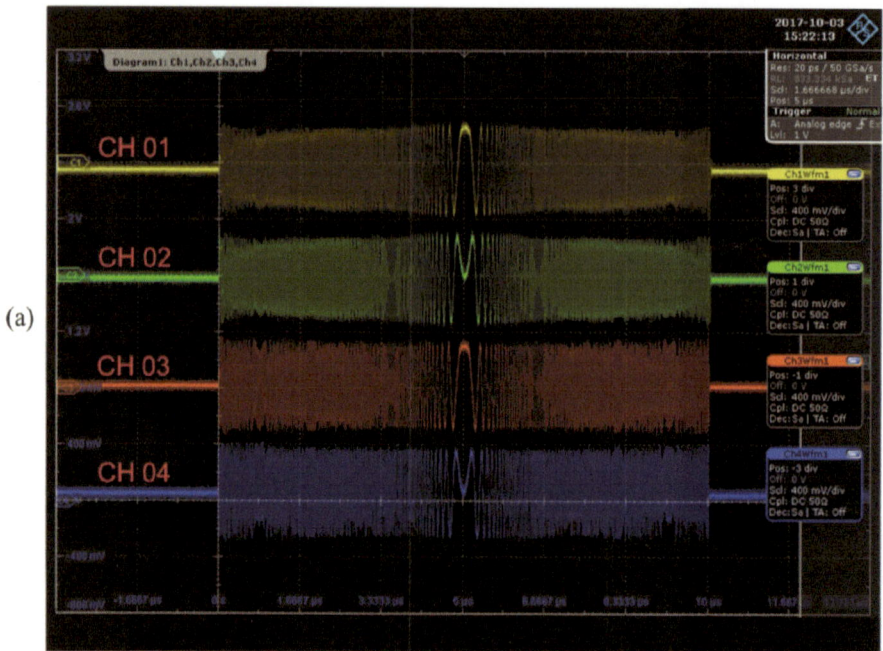

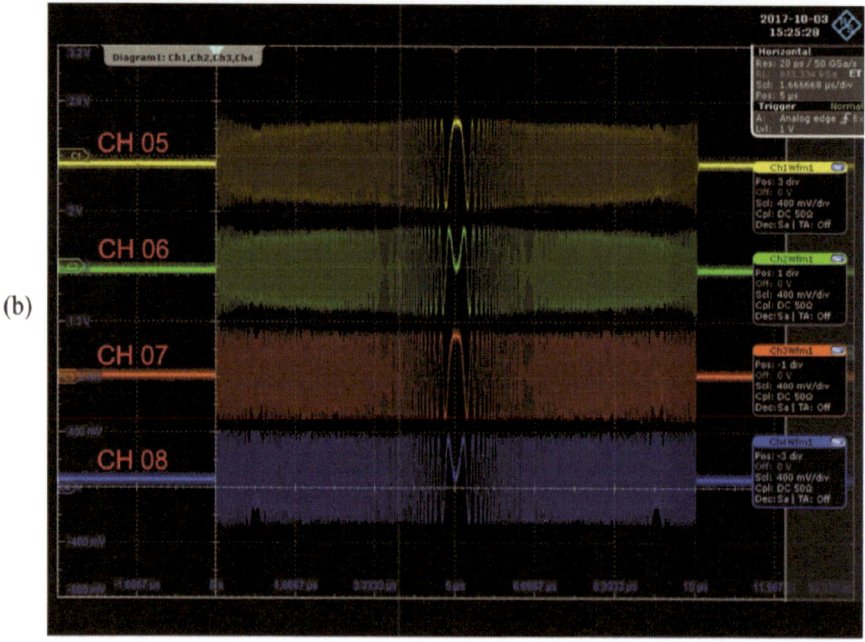

FIGURE 5.14 Synthesized chirp signals (200MHz bandwidth, 10μs pulse width); (a) channel #1 (yellow), #2 (green), #3 (orange), and #4 (blue), (b) channel #5 (yellow), #6 (green), #7 (orange), and #8 (blue).

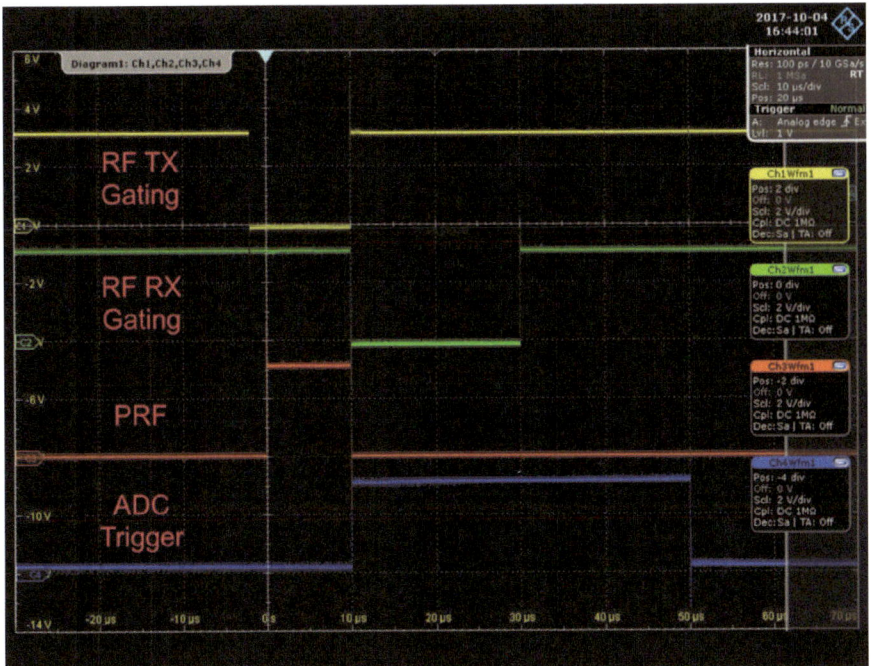

FIGURE 5.15 The generated timing and control signals; RF transmitter gating signal (yellow), RF receiver gating signal (green), PRF pulse (orange), ADC trigger signal (blue).

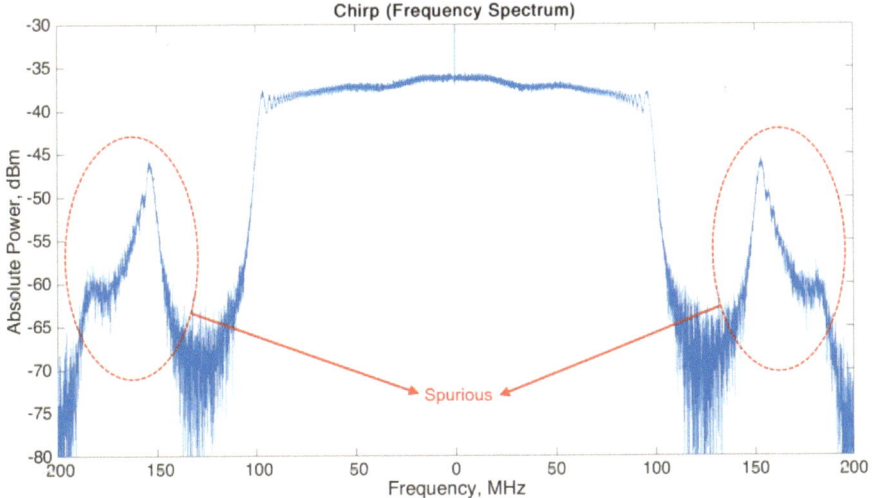

FIGURE 5.16 The spectrum of the chirp signal (200MHz bandwidth, 10µs pulse width).

5.3   DATA ACQUISITION UNIT (DAU)

SAR systems need a high-speed data acquisition unit (DAU) to capture all the echoes for processing. Most satellite-borne SAR systems have their own custom-designed compact and lightweight embedded data acquisition units due to the limited space in the spaceborne SAR payload. However, to cut down the development time, the Hinotori-C circularly polarized SAR system employed a PC-based data acquisition system using Dell Precision R7910 rack mount workstation with two Intel® Xeon® processors, 16 giga-bytes of 2,400MHz DDR RDIMM ECC random access memory (RAM), and two TB V-NAND solid-state disks (SSD) as PC-based controller, and Gage Applied CSE16502 high-speed digitizer as the analog-to-digital converter (ADC).

Since the chirp bandwidth is usually high, the component selection of the data acquisition system for SAR systems should consider the following factors:

a)   A high sampling rate digitizer that meets the Nyquist sampling theorem of at least $f_{max}/2$, where f_{max} is the highest frequency in the chirp.

b)   The digitizer has sufficient onboard memory to temporarily hold the digitized data before being recorded into the data storage system.

c)   The digitizer has a high-response trigger system to respond to the trigger-timing signal immediately.

d)   The digitizer and the PC system have an appropriate high-speed data interface to transfer the digitized data from the digitizer's onboard memory to the PC's data storage system. The estimated data rate, r, is

$$r = f_s \times \tau_{RX} \times b \times c \times PRF \times L \qquad \text{(Equation 5.12)}$$

where

f_s = the sampling rate of the digitizer
τ_{RX} = the receiving window time
b = the number of bits per sample
c = the number of acquisition channel
PRF = the pulse repetition frequency
L = link factor

e)   The data storage system has sufficient storage space to save the data captured by the digitizer during the flight mission.

Many commercial digitizers that support high-speed computer to peripheral interface (e.g., PCIe) are suitable for SAR data acquisition units [16]–[17]. Figure 5.17 illustrates two commercial digitizers used by Hinotori-C circularly polarized SAR and the user control software developed to control the SAR echo acquisition for these high-speed digitizers using the software development kit (SDK) provided by these digitizers. The GUI supports four acquisition modes used for data acquisition in the flight mission,

(a)

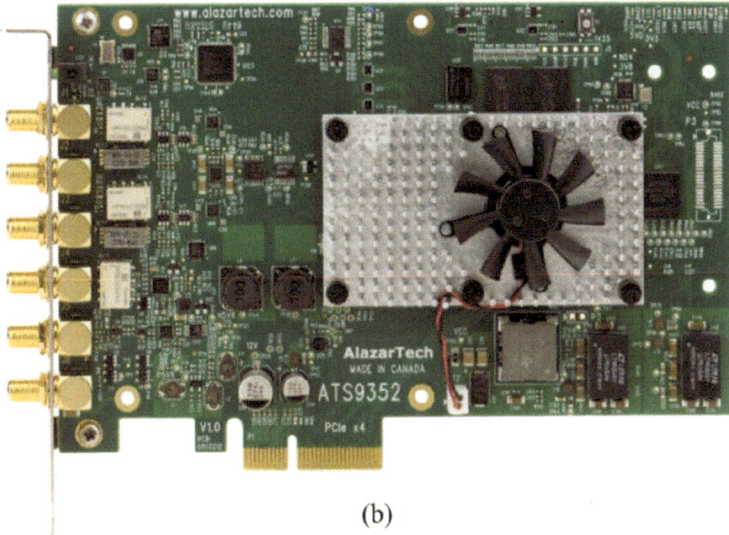

(b)

FIGURE 5.17 High-speed digitizer for in Hinotori-C Circularly Polarized SAR; (a) Gage Applied CSE16502, (b) Alazartech ATS9352, and (c) the user software panel.

ground test, and laboratory test. Table 5.5 tabulates the digitizer's configuration for each mode, and Figure 5.18 plots a sample of data captured using the system.

5.4 TIMING AND CONTROL UNIT (TCU)

The timing and control unit (TCU) is one of the essential components of an SAR system. With the SAR geometry of a flight mission, a set of timing information can be derived and used to generate accurate and precise signals to control the operating

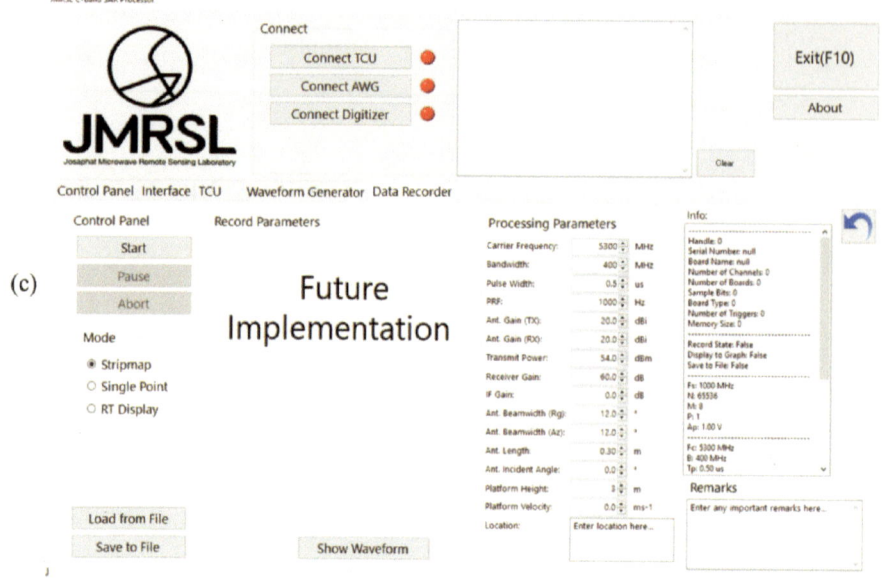

FIGURE 5.17   (Continued)

TABLE 5.5
Hinotori-C Circularly Polarized SAR Data Acquisition Modes

Mode	Configuration	Value
Stripmap (for flight mission)	Sampling Rate	500 MSPS
	Sample Length	16384 samples
	Segment Number	65536 segments
	Acquisition Mode	Two channels, external reference clock
	Channel Mode	1000 mV$_{pp}$, DC Coupling, 50Ω for both channels
	Trigger	External, Rising Edge, 20%
	File Type	Binary file
Stop-N-Go (for ground point target test) & Single Acquisition (for system test)	Sampling Rate	1000 MSPS
	Sample Length	32768 samples
	Segment Number	16 segments
	Acquisition Mode	Two channels, external reference clock
	Channel Mode	1000 mV$_{pp}$, DC Coupling, 50Ω for both channels
	Trigger	External, Rising Edge, 20%
	File Type	Binary file
Real-time Display (for waveform real-time viewing)	Sampling Rate	1000 MSPS
	Sample Length	32768 samples
	Segment Number	16 segments
	Acquisition Mode	Two channels, external reference clock
	Channel Mode	1000 mV$_{pp}$, DC Coupling, 50Ω for both channels
	Trigger	External, Rising Edge, 20%
	File Type	Data will not be recorded

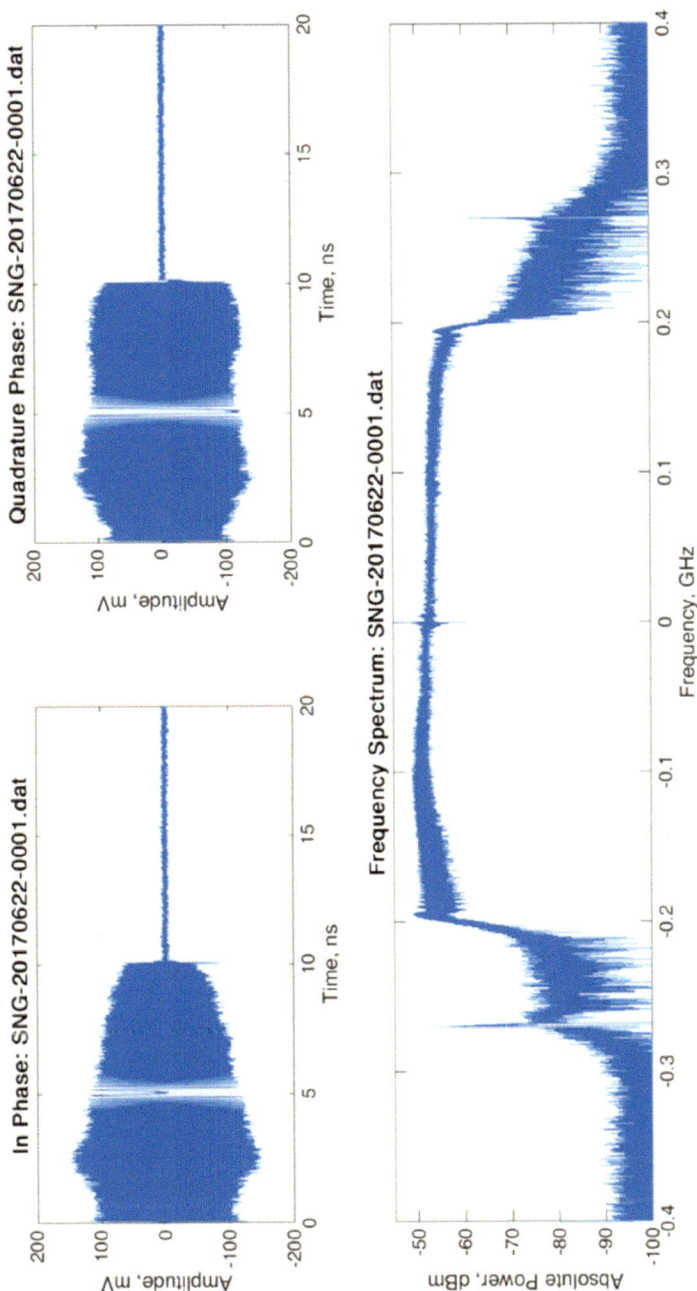

FIGURE 5.18 Example of SAR loopback echo (bandwidth = 400MHz, pulse width = 10µs) recorded by the data acquisition system.

sequence of the rest of the modules in the SAR system. Besides this, the TCU also synchronizes the timing information among these modules to maintain/retain the clock coherency of the SAR system so that the signal processor can properly focus echoes on the azimuth domain during the image formation process.

TCU usually is a standalone electronic module in a SAR system. Since the control signals must be accurate and precise, a custom-designed digital system using FPGA is a more viable solution compared to a microcontroller. The TCU receives the master reference clock (usually 10MHz) from the LO unit and uses it to derive the necessary timing signals. Depending on the system architecture of the particular SAR, each may need a different amount and type of timing and control signals. The Hinotori-C circularly polarized SAR needs four important control signals, each for controlling: (i) the RF transmitter, (ii) the RF receiver, (iii) the chirp generator, and (iv) the data acquisition system. Figure 5.19(a) illustrates the TCU hardware design and the user control software in Figure 5.19(b), while Figure 5.19(c) illustrates the architecture of the TCU digital system deployed in the FPGA. Furthermore, Table 5.6 lists the summary of the hardware features.

The TCU system contains submodules, such as the UART controller module, the timing generator module, the PLL module, and the clock indicator module. The timing generator (TG) module functions to generate the control signals (the TCU00 – TCU09 outputs). It contains a control register (CR) group that determines the timing for each generated control signal. Any compatible external devices can connect to the system and update the value of these CRs through the UART controller. The timing counter (TC) is a large counter in the TG that counts every clock cycle from its initial value (usually 0) until its overflow value. The PRI control register sets the counter's overflow value which determines the pulse repetition interval of the control signals. The pulse generators in the module read and compare the value of the timing counter and set or clear the control output logic (pulse ON or pulse OFF), accordingly. Meanwhile, the PLL in the TCU generates the internal clocks for TG using the reference clock from the local oscillator unit (LOC). The UART controller uses the onboard clock so that the TCU will remain reachable even though the reference clock is unavailable. Figures 5.20–5.22 show some samples of control signals with different timing settings recorded using an oscilloscope.

5.5   INERTIAL NAVIGATION SYSTEM

An inertial navigation system (INS) uses the outputs of sensors such as the global positioning system (GPS), accelerometers, and gyroscopes to predict an autonomous indication of aircraft position, velocity, and altitude. Airborne circularly polarized SAR, especially lightweight UAVs with unstable flight dynamics, requires an INS system to record and predict the flight attitude when the SAR system collects data during the mission. The predicted flight attitude information is necessary during SAR image formation, especially when applying the image formation technique with a motion compensation algorithm for data collected during unstable flight attitudes.

The Hinotori-C circularly polarized SAR INS system integrates MEMSIC AHRS440CA-200 6-DOF inertial measurement unit (IMU) and u-blox EVK-6 GPS

(a)

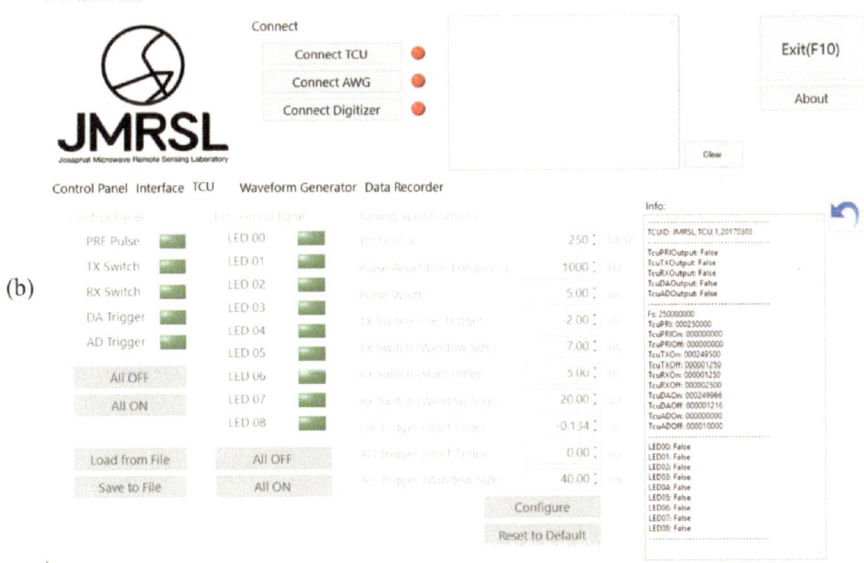

(b)

FIGURE 5.19 Hinotori-C SAR TCU system; (a) hardware, (b) the user control software, and (c) TCU digital system.

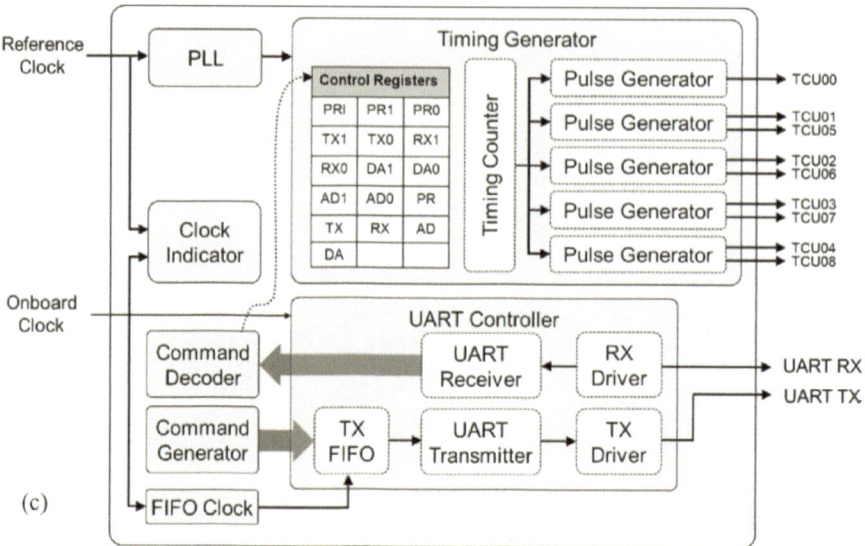

FIGURE 5.19   (Continued)

TABLE 5.6
Hinotori-C Circularly Polarized SAR TCU Hardware Features

Hardware Features	Value
Output channel number	9
Interface	Two × RS232
Analog input	Four
External clock input	Two
Output I/O standard	+3.3V LVTTL
Onboard FPGA	Altera DE0-Nano (EP4CE22F17C6N)

receiver to collect the X-Y-Z acceleration data, roll-pitch-yaw rotation data, GPS location, and aircraft's speed. Figure 5.23(a) shows the MEMSIC AHRS440CA-200 IMU and the u-blox EVK-6 GPS module in Figure 5.23(b), while Figure 5.23(c) shows the user interface to control the data collection. Figure 5.24 plots one of the datasets collected during the maiden flight of Hinotori-C Circularly Polarized SAR.

5.6   POWER UNIT

All electrical power comes from the power unit (PU). The PU in Hinotori-C circularly polarized SAR is composed of: (i) APC Smart-UPS uninterrupted power supply (UPS) for providing a stable AC source, (ii) Meanwell RSP-500 series AC/DC converter for converting the AC source to +28V DC supply, and (iii) a custom-designed

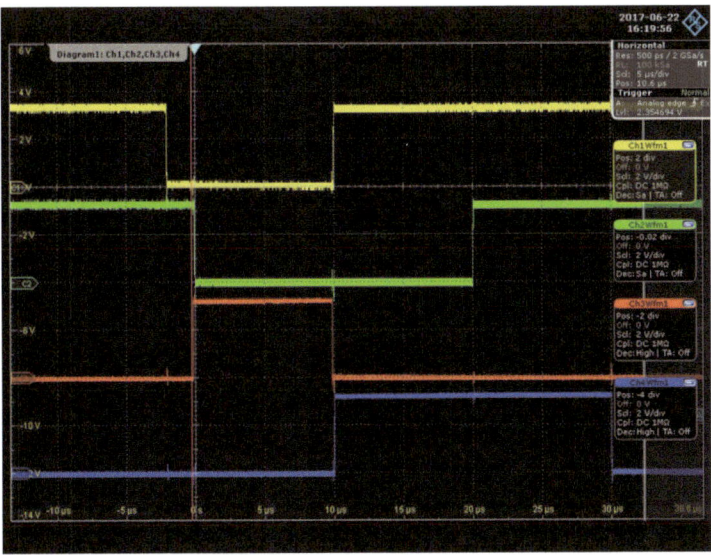

FIGURE 5.20 Timing and control signals (PRF = 1000Hz, pulse width = 10μs, TX pre-trigger = -2μs, TX window size = 12μs, RX start time = 0μs, RX window size = 20μs, DA start time = -0.134μs, AD start time = 10μs, AD window size = 20μs).

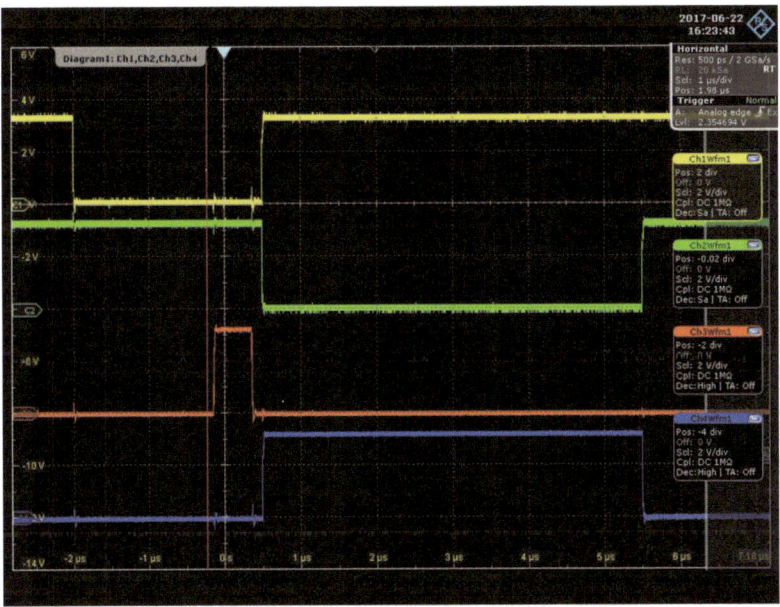

FIGURE 5.21 Timing and control signals (PRF = 1000Hz, pulse width = 0.5μs, TX pre-trigger = -2μs, TX window size = 2.5μs, RX start time = 0.5μs, RX window size = 10μs, DA start time = -0.134μs, AD start time = 0.5μs, AD window size = 10μs).

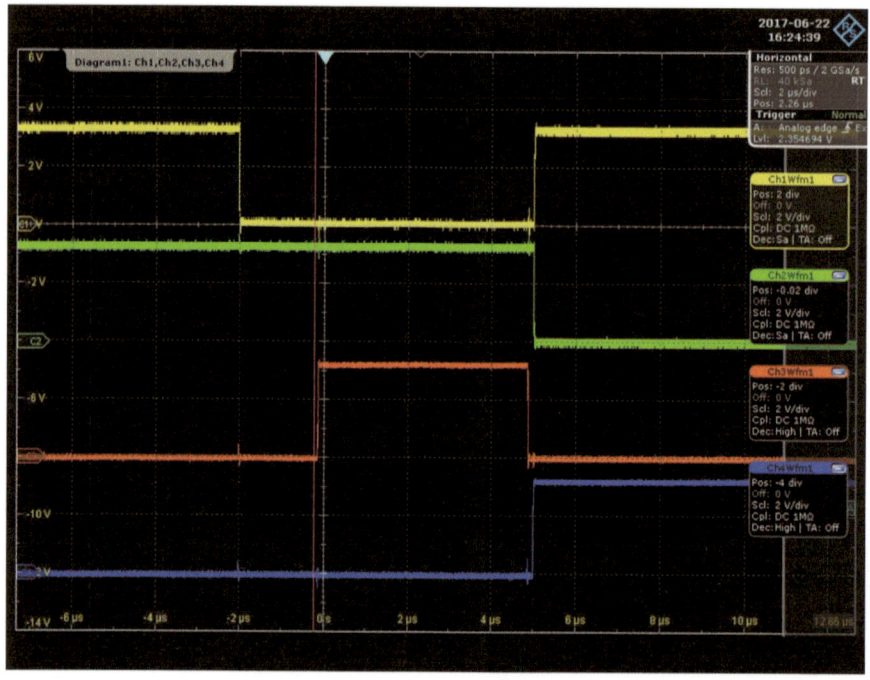

FIGURE 5.22 Timing and control signals (PRF = 1000Hz, pulse width = 5μs, TX pre-trigger = -2μs, TX window size = 7μs, RX start time = 5μs, RX window size = 10μs, DA start time = -0.134μs, AD start time = 5μs, AD window size = 10μs).

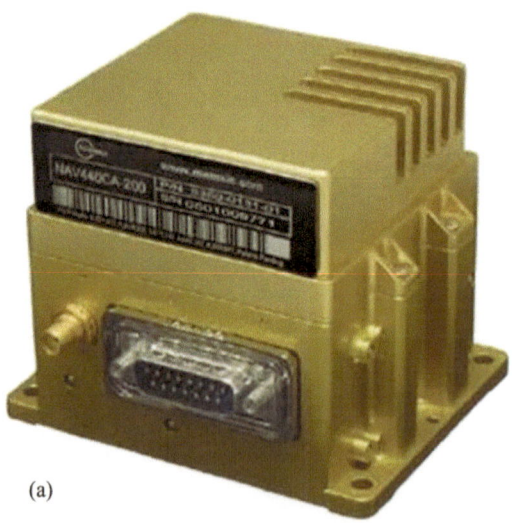

(a)

FIGURE 5.23 Hinotori-C INS system; (a) MEMSIC AHRS440CA-200 IMU, (b) u-blox EVK-6 GPS module, and (c) the user interface software.

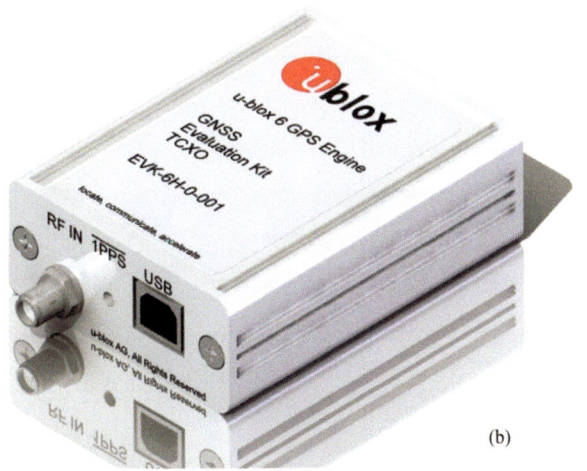

(b)

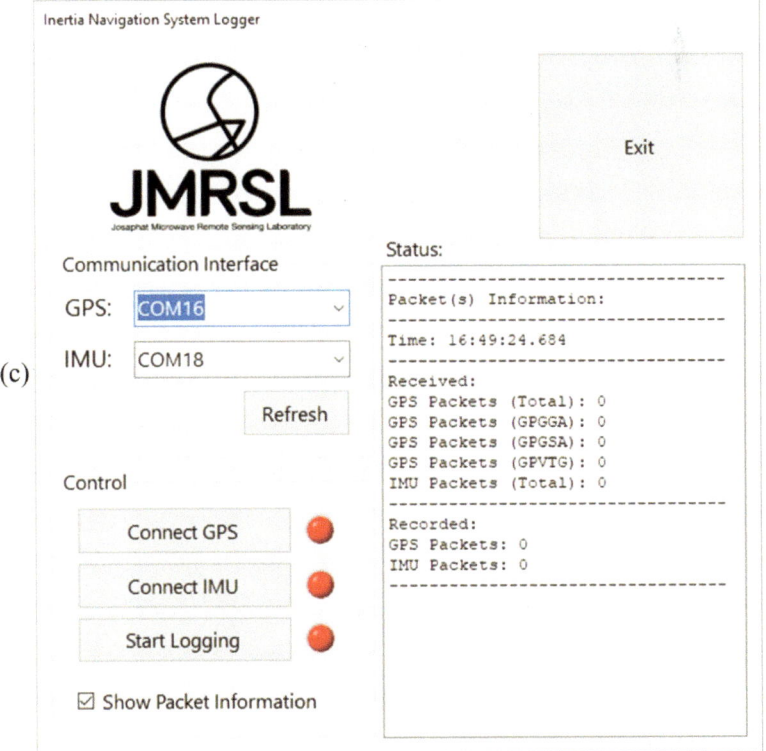

(c)

FIGURE 5.23   (Continued)

FIGURE 5.24   The plots of the recorded INS data: (a) GPS data, (b) INS data.

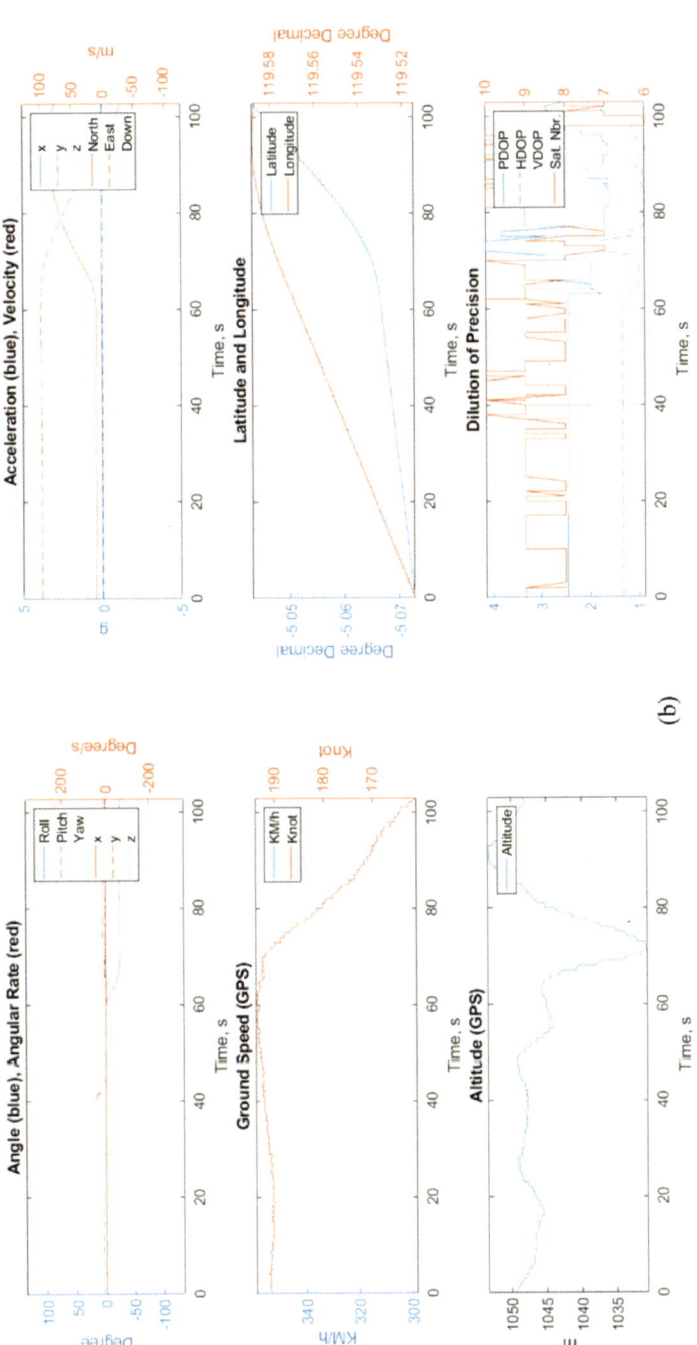

FIGURE 5.24 (Continued)

DC/DC converter module for creating a few DC sources with various voltages from the +28V DC supply. Figure 5.25 shows the snapshot of the power supplies.

The subsystem described in Chapter 4 and 5 was an example of the practical realization of the major parts needed in developing a SAR system. Each is replaceable

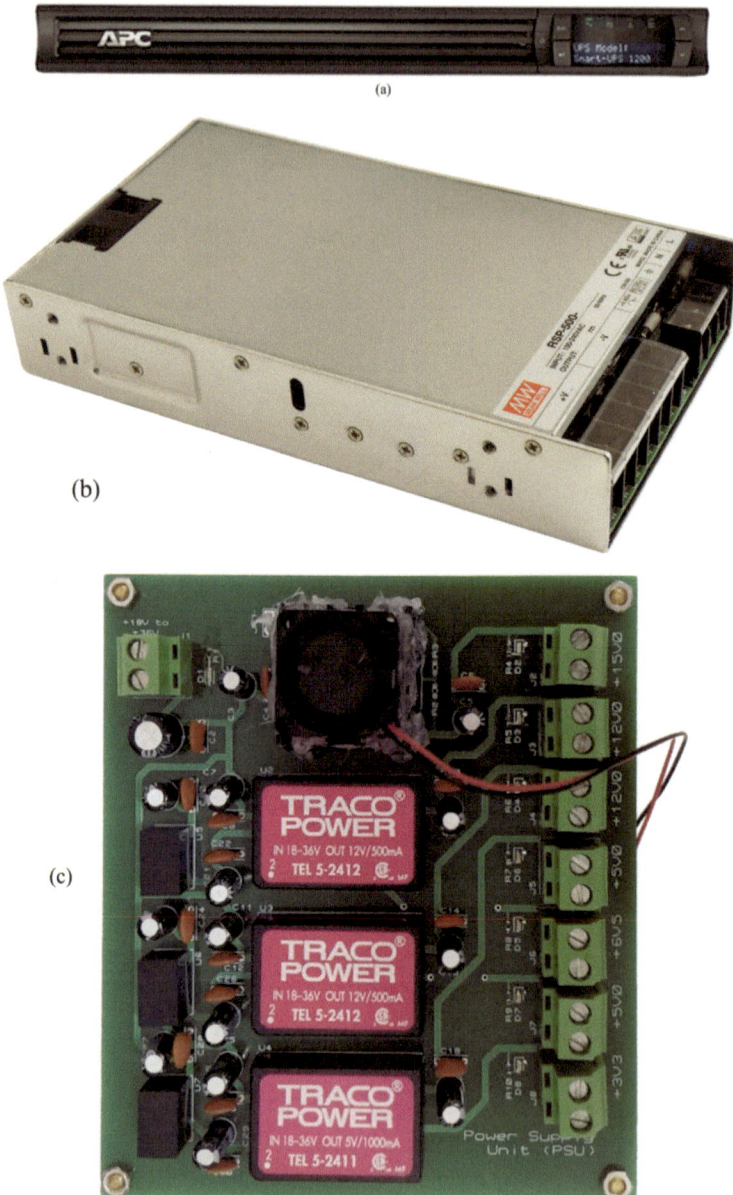

FIGURE 5.25 The Hinotori-C Circularly Polarized SAR power distribution unit; (a) uninterrupted power supply, (b) AC-DC converter, (c) in-System DC-DC converter.

with other solutions as long as it meets its requirements. The test setups and techniques used to validate the functionality of each subsystem's functionality and the entire integrated system are discussed in Chapter 7.

REFERENCES

[1] JT Sri Sumantyo, MY Chua, CE Santosa, GF Panggabean, T Watanabe, B Setiadi, FD Sri Sumantyo, K Tsushima, K Sasmita, A Mardiyanto, E Supartono, ET Rahardjo, G Wibisono, MA Marfai, RH Jatmiko, Sudaryatno, TH Purwanto, BS Widartono, M Kamal, D Perissin, S Gao, and Ki Ito, "Airborne Circularly Polarized Synthetic Aperture Radar," IEEE Selected Topics in Applied Earth Observations and Remote Sensing (JSTARS), Vol.14, pp.1676–1692, January 2021, DOI:10.1109/JSTARS.2020.3045032.

[2] MY Chua, JT Sri Sumantyo, CE Santosa, GF Panggabean, FD Sri Sumantyo, T Watanabe, YQ Ji, Peberlin, P Sitompul, M Nasucha, F Kurniawan, A Babag B Purbantoro, A Awaludin,, K Sasmita, ET Rahardjo, W Gunawan, RH Jatmiko, S Sudaryatno, TH Purwanto, BS Widartono, and M Kamal, "The Maiden Flight of Hinotori-C: The First C Band Full Polarimetric Circularly Polarized Synthetic Aperture Radar in the World", IEEE Aerospace and Electronic Systems Magazine, Vol. 34, Issue 2, pp. 24–35, June 2019..

[3] JT Sri Sumantyo, MY Chua, CE Santosa, GF Panggabean, T Watanabe, B Setiadi, K Tsushima, FD Sri Sumantyo, K Sasmita, A Mardiyanto, E Supartono, ET Rahardjo, G Wibisono, RH Jatmiko, TH Purwanto, BS Widartono, M Kamal, RH Triharjanto, S Gao, and K Ito, "Hinotori-C2 mission: CN235MPA Aircraft Onboard Circularly Polarized Synthetic Aperture Radar (CP-SAR)", 2019 IEEE International Geoscience and Remote Sensing Symposium (IGARSS2019), pp. 8538–8541, July 2019.

[4] JT Sri Sumantvo, MY Chua, CE Santosa, GF Pariguabean, K Taushima, T Watanabe, K Sasmita, A Mardiyanto, FD Sri Sumantyo, Eko Tjipto Rahardjo, Gunawan Wibisono, E Supartono, S Gao, SP Parulian, M Nasucha, F Kurniawan, A Awaludin, B Purbantoro, YQ Ji, N Imura, "Hinotori-C: A Full Polarimetric C Band Airborne Circularly Polarized Synthetic Aperture Radar for Disaster Monitoring", 2018 Progress in Electromagnetics Research Symposium (PIERS-Toyama), pp. 1466–1473, August 2018.

[5] JT Sri Sumantyo, MY Chua, CE Santosa, A Takahashi, and K Ito, "Aircraft and High Altitude Platform System Onboard Circularly Polarized Synthetic Aperture Radar (CP-SAR)", 2021 IEEE International Geoscience and Remote Sensing Symposium (IGARSS2021), pp. 8515–8518, July 2021.

[6] MY Chua, VC Koo, "FPGA-based Chirp Generator for High Resolution UAV SAR", Progress in Electromagnetics Research, Vol. 99, pp. 71–88, 2009.

[7] MY Chua, VC Koo, HS Lim, HS Lim, JT Sri Sumantyo, "Phase-coded Stepped Frequency Linear Frequency Modulated Waveform Synthesis Technique for Low Altitude Ultra-Wideband Synthetic Aperture Radar", IEEE Access, Vol. 5, pp. 11391–11403, May 2017.

[8] YK Chan, MY Chua, VC Koo, "Sidelobes Reduction Using Simple Two and Tri-stages Non Linear Frequency Modulation (NLFM)", Progress in Electromagnetics Research, Vol. 98, pp. 33–52, 2009.

[9] MY Chua, HS Boey, CH Lim, VC Koo, HS Lim, YK Chan, TS Lim, "A Miniature Real-time Reconfigurable Radar Waveform Synthesizer for UAV based Radar", Progress in Electromagnetics Research C, Vol, 31, pp. 169–183, 2012.

[10] MY Chua, JT Sri Sumantyo, YQ Ji, "An 8-Channels FPGA-Based Reconfigurable Chirp Generator for Multi-Band Full Polarimetric Airborne/Spaceborne CP-SAR", 2018 Progress in Electromagnetics Research Symposium (PIERS-Toyama), pp. 876–881, Aug 2018.

[11] MY Chua, VC Koo, HS Lim, JT Sri Sumantyo, "FPGA-based Reconfigurable Chirp Generator for L-band UAV CP-SAR", International Symposium on Remote Sensing 2017, 2017.

[12] MY Chua, VC Koot, HS Lim, YK Chan, "An FPGA based Real-Time Multi-Target Synthetic Aperture Radar Echoes Synthesizer", Journal of Engineering Technology and Applied Physics, Vol. 2, Issue 2, pp. 10–16, Dec 2020.

[13] VC Koo, MY Chua, HS Lim, HS Boey, YK Chan, CH Lim, TS Lim, HT Chuah, "High Speed AD DA for Synthetic Aperture Radar", Proceedings of PIERS 2013, pp. 994–997, 2013.

[14] K Gentile, R Cushing, A Technical Tutorial on Digital Signal Synthesis, Analog Devices Education Library, 1999.

[15] J Tierney, C Rader, and B Gold, "A Digital Frequency Synthesizer", IEEE Transactions on Audio and Electroacoustics, Vol. 19, Issue 1, pp. 48–57, Mar 1971.

[16] MY Chua, JT Sri Sumantyo, CE Santosa, GF Panggabean, YQ Ji, SP Parulian, M Nasucha, "An PC-based Airborne SAR Baseband System", 2018 Progress in Electromagnetics Research Symposium (PIERS-Toyama), pp. 882–888, Aug 2018.

[17] CH Lim, CS Lim, MY Chua, YK Chan, TS Lim, VC Koo, "A New Data Acquisition and Processing System for UAVSAR", IEICE Electronics Express, Vol. 8, Issue 20, pp. 1716–1722, 2011.

[18] K Suto, JT Sri Sumantyo, VC Koo, MY Chua, WG Cheaw, "Development of SAR Base-band Signal Processor using FPGA and Onboard PC", 2014 IEEE Geoscience and Remote Sensing Symposium, pp. 672–675, Jul 2014.

APPENDIX I
DATA FORMAT OF THE DATA ACQUISITION FILE

Field No.	Data Length	Data Format	Description
1	6	string	Sensor ID (AAABBB)
2	9	string	Sensor Name (CCCCCCCCC)
3	8	string	Date (YYYYMMDD)
4	6	string	Time (HHMMSS), in 24-hours format
5	36	string	Filename: No specific format
6	1	string	File type, T: 0 = TCU, 1 = AWG, 3 = Digitizer, 4 = Processor, 5 = SAR Design, 6 = IMU
7	1	string	Author initial
8 – 16	1	string	reserved for header
17	1	double	Sampling frequency, f_s
18	1	int	Record length, N
19	1	int	Record number, M
20	1	int	Page number, P
21	1	double	Maximum amplitude, A_p
22	1	string	Mode: 0 = Stripmap, 1 = Stop&Go, 2 = SinglePoint

Field No.	Data Length	Data Format	Description
23 – 30	1	double	reserved for header
31	$N \times M$	float	Data: In phase
32	$N \times M$	float	Data: Quadrature phase
33	1	double	Carrier Frequency, f_c
34	1	double	Bandwidth, B
35	1	double	Pulse width, T_p
36	1	double	Pulse repetition frequency, PRF
37	1	double	Transmitter Antenna Gain, G_{at}
38	1	double	Receiver Antenna Gain, G_{art}
39	1	double	Transmit Power, P_t
40	1	double	Receiver Gain, G_r
41	1	double	IF Gain, G_{if}
42	1	double	Antenna Beamwidth (Range), μ_R
43	1	double	Antenna Beamwidth (Azimuth), μ_{AZ}
44	1	double	Antenna Length, L_{sa}
45	1	double	Antenna Incident Angle, ϕ
46	1	double	Platform Height, h
47	1	double	Platform Velocity, v
48	1	string	Location
49	1	string	Remarks
50	M	double	Time Stamp
51 – 64	1	double	reserved for header

6 Circularly Polarized Antenna

Cahya Edi Santosa and Josaphat Tetuko
Sri Sumantyo
With the collaboration of Dr. Steven Gao and
Dr. Koichi Ito

6.1 INTRODUCTION

One of the most significant parts of an SAR system is the antenna. The antenna acts as the "eyes" to observe objects on the land surface. It can convert guided radio waves into electromagnetic waves in free space, and vice versa. As an active imaging radar, the SAR system transmits electromagnetic waves through a transmitter antenna (Tx). Then the receiver antenna (Rx) receives the scattering signal from illuminated objects in the swath area. The antenna's characteristics determine the SAR radar image's quality and resolution. Therefore, research on an SAR antenna with high gain and efficiency, broad bandwidth, compact physical form, simple design, easy to manufacture at low cost, and other advantages, is continuous.

Based on experience, the SAR antenna specification is defined by SAR mission requirements and platform types that carry out flights using the SAR system. Applications of SAR systems onboard airborne and spaceborne platforms require different electrical, technical, and material specifications. However, both applications have weight, volume dimensions, and radio power limitations. Therefore, the SAR system specification is sometimes adapted to the carrying platform and has to trade off another advantage. Airborne SAR antennas are more straightforward than the requirements of spaceborne SAR antennas, such as in materials and construction used. Airborne SAR is commonly used in approximately the same environment in the atmosphere, although operated at different altitudes. In comparison, spaceborne SAR must pay attention to environmental differences, such as extreme temperature and pressure, gravity influence, space radiation, and others.

Recently, various antenna designs and technologies have been multiplying rapidly. Conventional and smart array antenna technologies are commonly adopted for SAR antenna applications to reach the SAR system requirements, including microstrip arrays, slot waveguides, and reflector antennas. Microstrip array antennas are more suitable for airborne SAR applications that operate in low frequencies or lower than 10GHz, such as L-band and C-band. Although the level of power efficiency is slightly lower, the high dielectric constant of the substrate material allows the SAR antenna dimensions that operate in low frequencies to be compact. This advantage aligns with the physical requirements for an airborne SAR platform, which requires antennas that can be installed in a limited space and have a simple aerodynamic form. The narrow bandwidth and losses issue can be overcome using several techniques in advanced design. At the same time, the waveguide slot antenna technology has excellent power efficiency performance. However, a SAR system that employs a waveguide slot array antenna and operates in low frequencies probably results in a large and heavy array antenna. A waveguide slot array antenna with an air dielectric constant is more suitable for airborne SAR applications that operate at high frequencies, such as X-band or higher.

On the other hand, refectory antenna technology is more suitable for spaceborne SAR applications. The spaceborne SAR antenna requires a very high gain, narrow beam, lightweight, and quality components. The current spaceborne SAR antenna uses a fabric or mesh reflector to reduce its weight, while a horn antenna supports it as the feeder. The dimensions do not matter as long as they are lightweight, can be packaged compactly on the launcher vehicle, and can be deployed into space quickly.

6.2 BASIC ANTENNA PARAMETERS

In the design and analysis of antennas, several factors must be defined to meet the SAR antenna's requirements, including electrical, technical, and costs. The following antenna parameters should be considered to produce a high-performance SAR antenna.

6.2.1 RETURN-LOSS AND VOLTAGE STANDING WAVE RATIO (VSWR)

The input impedance (Z_{in}) on the input port of the antenna is expressed in Equation (6.1)[1]. The input impedance is declared as a complex number dependent on a certain frequency. The real value or resistance component (R_{in}) includes the loss resistance and radiation resistance. The imaginer part, or reactance component (X_{in}), consists of capacitance and inductance values that also depend on the operational frequency. In the feeding line design, the width of a microstrip line determines the resistance part of the impedance. Meanwhile, the length of a microstrip line controls the reactance part of the impedance.

$$Z_{in} = R_{in} + jX_{in}$$
(Equation 6.1)

The input port of the antenna is connected to a signal source module using a short transmission line with a characteristic impedance. The reflection coefficient is dependent on impedance at the input port antenna and transmission line characteristic impedance as expressed in Equation (6.2). The characteristic impedance of the transmission line is adjusted equally to a nominal impedance (Z_o) with a value of 50Ω. The impedance mismatch between the antenna input port and transmission line will generate a reflection wave that leads back to the signal source. If the intensity of the reflection wave exceeds the limit, the electronic components of the signal source module will be damaged.

$$\Gamma = \left(Z_{in} - Z_0\right)/\left(Z_{in} + Z_0\right)$$
(Equation 6.2)

Return loss (*RL*) is the reflection coefficient expressed in dB, as defined in Equation (6.3). The antenna's impedance bandwidth (IBW) is the spreading frequency with at least 10dB of return loss. Sometimes the reflection coefficient is also expressed in S-parameter (S_{11}) with a value of -dB, as illustrated in Figure 6.1(a). The lower and upper frequencies achieved in the IBW are denoted by f_L and f_H.

$$RL = -20log\,|\Gamma| \qquad \text{(Equation 6.3)}$$

The antenna's efficiency in transmitting a radiation wave from the power source is defined as a voltage standing wave ratio (*VSWR*) determined by the voltage ratio between maximum and minimum values on the transmission line, expressed in Equation (6.4). The *VSWR* values are between 1 to 2, as illustrated in Figure 6.1(b). If the *VSWR* value equals 1.0, the reflection coefficient $|\Gamma| = 0$. This means the input impedance ideally matches, and no wave is reflected. In contrast, if the reflection coefficient $|\Gamma| = 1$, the *VSWR* value is infinite. This means the input impedance significantly does not match, and all waves are reflected. The *VSWR* value of 2.0 means the return loss is near to the limit of 10dB.

$$VSWR = \left|V_{max}\right|/\left|V_{min}\right| = \left(1+|\Gamma|\right)/\left(1-|\Gamma|\right) \qquad \text{(Equation 6.4)}$$

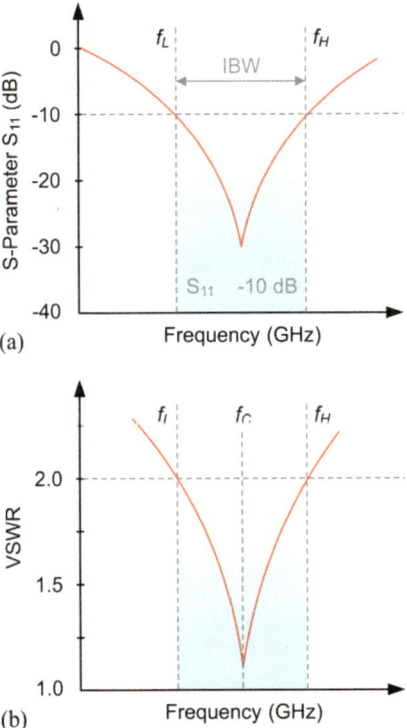

FIGURE 6.1 The IBW is defined in the spreading frequency. (a) S-parameters. (b) The VSWR.

6.2.2 RADIATION PATTERN

Antennas do not radiate electric fields equally in all directions but radiate more power in some directions and form a particular pattern depending on antenna design, dimensions, environment, and other factors. A radiation pattern is a graphical representation of a radiated electric field of an antenna as a function of spherical coordinates at all angle directions. Figure 6.2(a) illustrates a radiation pattern of an antenna in a symmetrical three-dimensional plot in a spherical coordinate with a linear scale or a logarithmic scale (in dB). The main lobe is a radiated electric field with a maximum magnitude, and the direction of the main lobe is measured from theta $\theta = 0°$ in a spherical coordinate. Half power beamwidth (HPBW), or 3dB beamwidth, is the angular separation between two directions at the main lobe, which has a half-power radiation intensity (at 3dB) compared to the maximum power of the main lobe. All the lobes except the main lobe in any direction are called sidelobes. Figure 6.2(b) depicts the radiation pattern of an antenna in a two-dimensional plot in a normalized field with a linear scale or a logarithmic scale (in dB). A sidelobe level (SLL) is the power intensity difference between the maximum power of the main lobe and the nearest sidelobe. Sometimes sidelobes are undesired and suppressed to -20dB or smaller. A radiated electric field in the opposite direction of the main lobe is called a back lobe. The first null beamwidth (FNBW) is the angular separation between the first null of the main lobe.

6.2.3 DIRECTIVITY

A general definition of directivity $D (\theta, \varphi)$ of the antenna is the radiation intensity in a given direction, observed in the far-field region compared to the average radiation intensity overall isotropic directions. It can be estimated by Equation (6.5) [2], where r, E, H, and P_{rad} denote the distance between the source and test point in the far-field region, the peak value of the electric field, the peak value of the magnetic field, and the radiated power of the antenna. The far-field region (R) is written as Equation (6.6), where D is the maximum dimension of the antenna.

$$D(\theta, \varphi) = \frac{r^2 R_e \left[E \times H^* \right]}{P_{rad} / 4\pi} \qquad \text{(Equation 6.5)}$$

$$R > \frac{2D^2}{\lambda} \qquad \text{(Equation 6.6)}$$

6.2.4 GAIN AND EFFICIENCY

Gain of the antenna $G (\theta, \varphi)$ is the ratio of the radiation intensity in a given direction observed in the far-field region to the radiation intensity if the input power of the antenna is radiated isotropically. The antenna gain includes the radiation efficiency (η) of the antenna and is expressed in Equation (6.7).

$$G(\theta, \varphi) = \eta D(\theta, \varphi) \qquad \text{(Equation 6.7)}$$

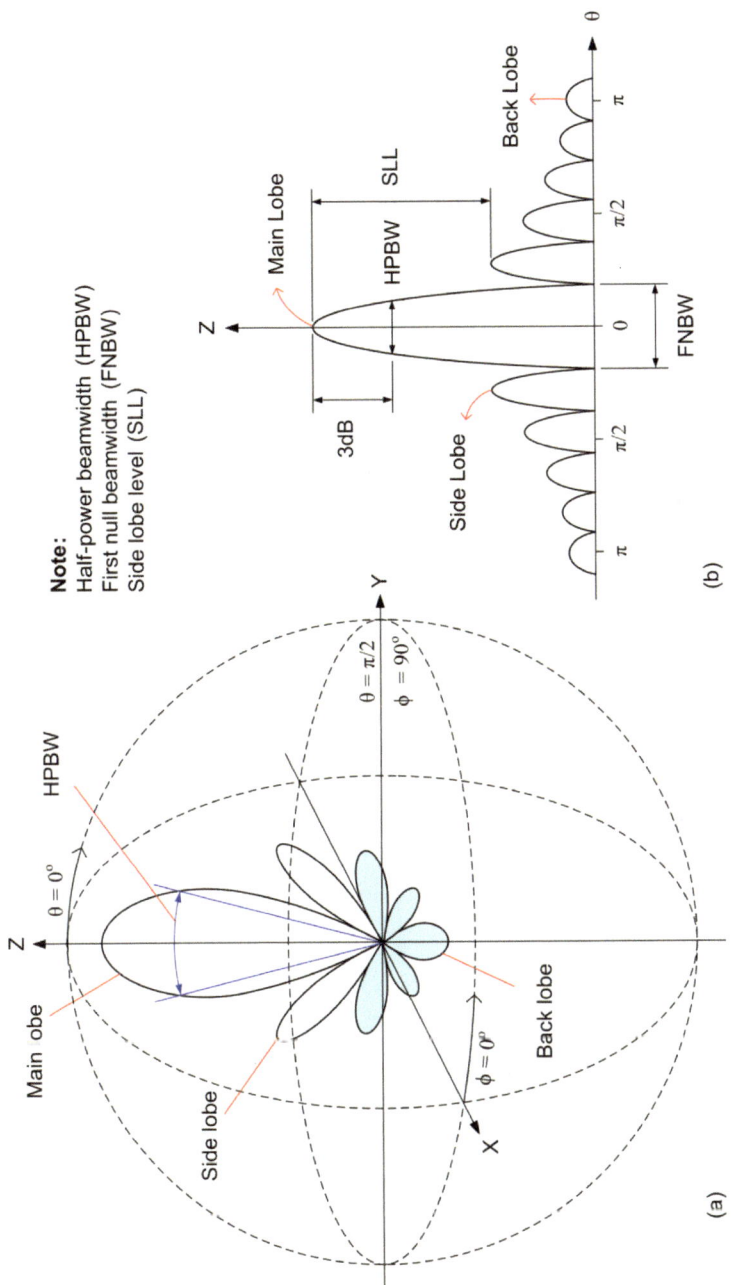

FIGURE 6.2 A radiation pattern of an antenna in a symmetrical. (a) Three-dimensional plot. (b) Two-dimensional plot.

The radiation efficiency η describes how much the input power of the antenna P_{in} is radiated by the antenna. It can be expressed using Equation (6.8), where P_{rad} is the radiated power by the antenna.

$$\eta = \frac{P_{rad}}{P_{in}}$$ (Equation 6.8)

Another description of the radiation efficiency η is the ratio of the effective aperture area (A_{eff}) to the actual physical dimension of the antenna A, as written in Equation (6.9), where the effective aperture area A_{eff} can be calculated by Equation (6.10).

$$\eta = \frac{A_{eff}}{A}$$ (Equation 6.9)

$$A_{eff} = \frac{\lambda^2}{4\pi} G$$ (Equation 6.10)

However, the antenna has some losses, such as impedance mismatching, conduction, and the material's dielectric constant. The overall efficiency of the antenna can be expressed using Equation (6.11), where η_r, η_c, η_d denote the reflection (mismatch) efficiency, conduction efficiency, and dielectric efficiency.

$$\eta = \eta_r \eta_c \eta_d$$ (Equation 6.11)

6.2.5 POLARIZATION

Polarization is the orientation of the electric field oscillation while an electromagnetic wave propagates in the medium. There are three types of wave polarization: linear, elliptical, and circular – see Figure 6.3. Linear polarization (LP) occurs in an electromagnetic wave that oscillates over time in one fixed direction. In synthetic aperture radar applications, the LP wave is classified into vertical polarization (V) and horizontal polarization (H). A circular polarization occurs in the electromagnetic wave that has two linear electric fields (E-fields) with the same amplitude, a perpendicular orientation to each other, and 90° phase differences. The resultant E-field produces a circle rotation wave in its propagation direction. Elliptical polarization (EP) occurs in the electromagnetic wave with two linear electric fields with different amplitudes. The EP and circular polarization are generally classified into right-handed polarization (RH) and left-handed polarization (LH). The agreed direction of polarization is defined from the source of the radiated wave to the propagation direction. The right-handed circular polarization (RHCP) explains electrics field rotation is clockwise, and the left-handed circular polarization (LHCP) when the electrics field rotation is counter clockwise, as shown in Figure 6.3 (c).

Axial Ratio (AR) is a quantity of circularly polarized radiated waves measured in dB units. The AR values can be used to distinguish between elliptical and circular

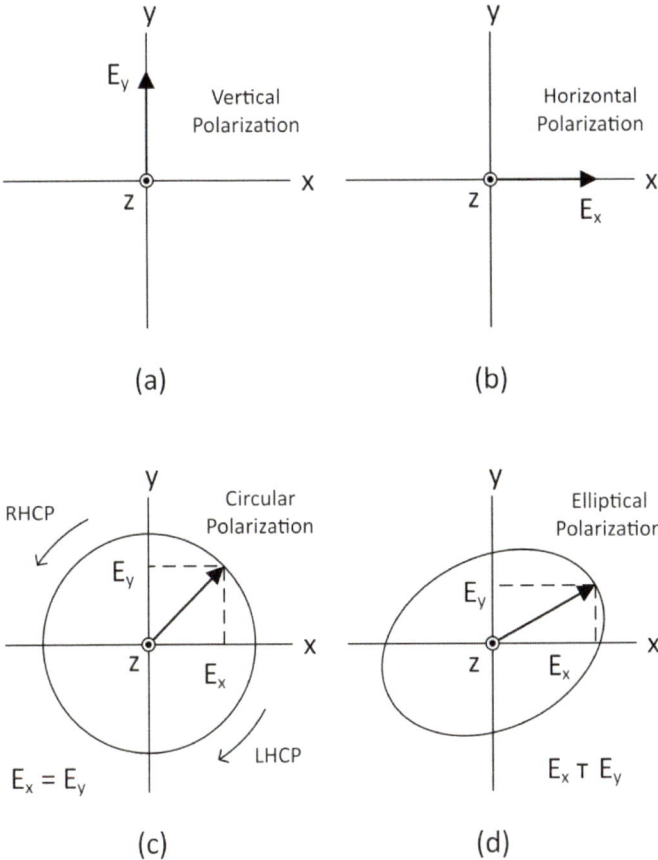

FIGURE 6.3 Visualization of some polarization types. (a) Vertical polarization (V). (b) Horizontal polarization (H). (c) Circular polarization. (d) Elliptical polarization (EP).

polarization. A perfectly circularly polarized microwave has an AR value of 0dB, but in reality, this is very difficult to achieve. In practical applications, a circularly polarized wave has an AR value that varies between 0dB to 3dB and an elliptical polarization with an AR value greater than 3dB. Polarization with AR values greater than 3dB are thin ellipses and considered to be close to linear polarization. The AR value expresses perfectly linear polarization on a value >40dB.

6.3   THE SAR ANTENNA REQUIREMENTS

A sidelooking imaging radar has the disadvantage of low resolution in the azimuth direction. Two ways to increase the azimuth resolution are by using an antenna with a narrow beamwidth or increasing the radar signal's wavelength [3]. Obtaining an antenna with a narrow beamwidth will require a very long antenna on the flying platform, which is difficult to realize because radar platforms have weight and size limitations. Conversely, using a radar signal with a shorter wavelength reduces the antenna's dimensions. The use of high frequencies for radar should pay attention to the atmospheric attenuation of signal and RF power requirements of the radar system. Another way to increase the azimuth resolution is to use signal processing techniques, such as those implemented in the SAR system.

The SAR antenna transmits chirp signals and receives echo signals from target objects located at certain distances. In the single pulse radar equation, the antenna subsystem should contribute to the signal-to-noise ratio (SNR) performance according to the signal processing system requirements to display the received signal above the background noise. SNR is the ratio between the power of the echo signal received by the radar system compared to the noise characteristics of the radar system. It is expressed in Equation (6.12) [4] and uses units of dB.

$$SNR = \left(\frac{P_t G^2 \lambda_o^2}{(4\pi)^3 R^4 F} \right) \left(\frac{\sigma}{kTB} \right) \left(\frac{c\tau_p}{2\sin\theta_i} \right) \left(\frac{\lambda_o R}{L_a} \right) \qquad \text{(Equation 6.12)}$$

where P_t is the average transmitted power, G is antenna's gain, λ_o is wavelength in free space, R is a range, F is noise factor, σ is the object's cross-section coefficient for backscattering signal, k is Boltzmann's constant, 1.38 x 10–23 J/K, T is temperature, B is bandwidth, c is the speed of light, τ_p is radar pulse duration, θ_i is the incident angle, and L_a is the length of the antenna in azimuth.

The following section will describe some of the antenna parameters that significantly control the SNR performance of the SAR system directly and indirectly.

6.3.1 Antenna's Aperture and Beamwidth

The sidelooking SAR geometry is shown in Figure 6.4(a)[4]. The SAR system is carried on a moving platform, such as an airborne or spaceborne vehicle. The platform is assumed to fly at a constant altitude, h, and speed, v, but achieving this practically is not easy. Therefore, corrections are needed in the SAR data processing system by adding moving compensation calculations. The airborne vehicle flies in a straight trajectory to synthesize a very long antenna aperture in the attempt to achieve high azimuth resolution of the SAR images. The SAR antenna has an aperture width dimension, W_a, and length, L_a. The look angle of the SAR antenna is defined by the direction of the antenna's main beam, θ. The 3dB main beam antenna has a range beamwidth of θ_V and an azimuth beamwidth of θ_H. The SAR antenna radiates chirp signals with pulse length, τ_p, pulse-repetition-interval (PRI), and pulse repetition frequency (PRF) properties. The 3dB main beam antenna illuminates the target on the ground surface and produces a footprint area with a swath width, W_g.

Generally, SAR antennas use rectangular antennas, considering the possibility of optimally controlling the 3dB antenna's beamwidth according to the SAR system's design of the illumination target area – see Figure 6.4(b). The use of a parabolic antenna will produce a symmetrical pencil beam directivity. The height dimensions of the aperture antenna, W_a, controls the 3dB range beamwidth, θ_V, the swath width, W_g, range ambiguity, and range resolution, ρ_g. At the same time, the length dimension, L_a, of the aperture antenna will manage the 3dB azimuth beamwidth, θ_H, azimuth ambiguity, azimuth resolution, ρ_a, and PRF selection on the SAR system design. The width of the SAR antenna, W_a, controls the 3dB range beamwidth, θ_V, in the range direction, as expressed in Equation (6.13). Then, the length of the ground swath, Wg, depends on the 3dB antenna's beamwidth in the range direction, θ_V, established by the antenna width, W_a, as defined in Equation (6.14), where R_m is the slant range.

$$\theta_V = \frac{\lambda_o}{W_a} \qquad \text{(Equation 6.13)}$$

$$W_g = \frac{\lambda_o R_m}{W_a \cos\theta} \qquad \text{(Equation 6.14)}$$

Figure 6.4(c) illustrates the SAR geometry from the side view with the look angle, θ, as the angle between the SAR main beam and the Earth's surface, which equals the

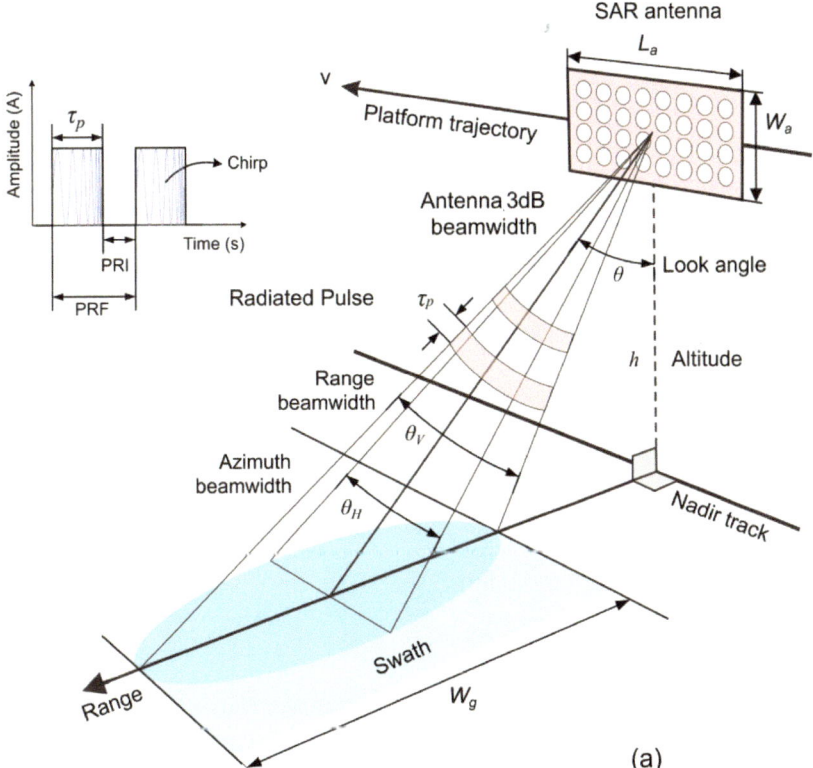

(a)

FIGURE 6.4 The geometry of sidelooking SAR is described from some viewpoints. (a) Illustration on three-dimensional view. (b) Illustration on the top view of the SAR antenna's illumination. (c) Illustration on the side view.

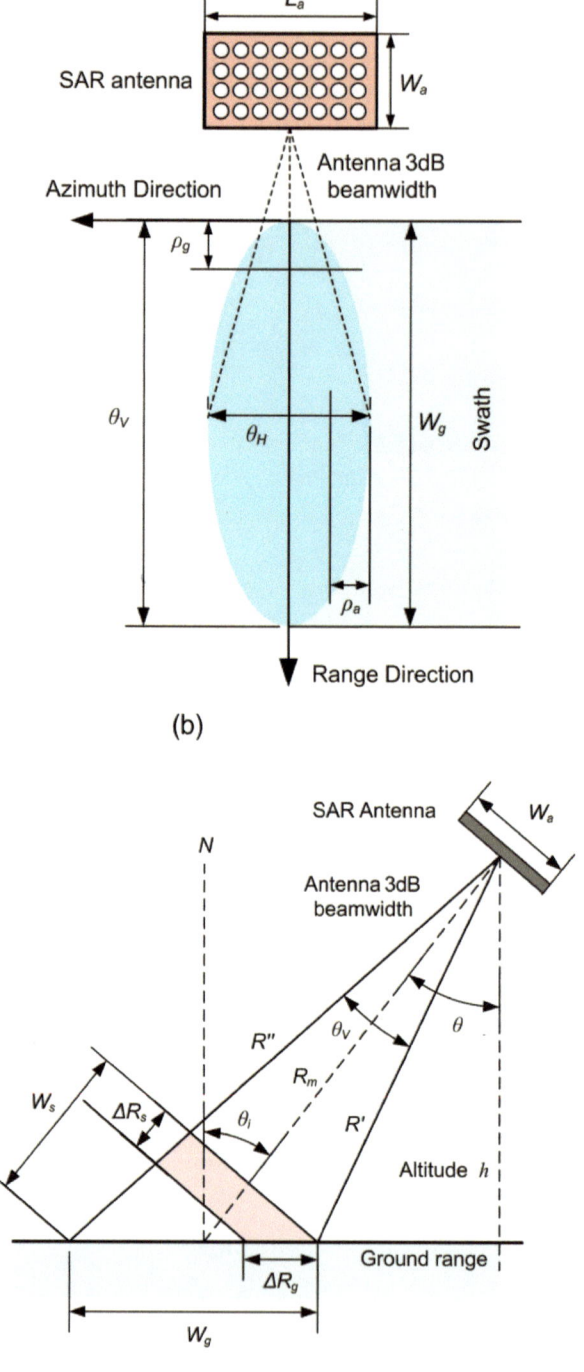

(b)

(c)

FIGURE 6.4 (Continued)

incident angle, η, ($\eta = \theta$). The range resolution of range aperture radar, $\Delta R_g = \rho_g$, is given in Equation (6.15), where ΔR_s is the slant range resolution expressed in Equation (6.16). Range resolution is the minimum distance between two objects on the ground, which the SAR system can recognize and plot as separate images in the range direction. The range resolution is the function of look angle θ and pulse width τ_p.

$$\Delta R_g = \frac{\Delta R_s}{sin\theta} = \frac{c\tau_p}{2sin\theta} = \rho_g \qquad \text{(Equation 6.15)}$$

$$\Delta R_s = c\tau_p \qquad \text{(Equation 6.16)}$$

The length of the SAR antenna, L_a, controls the 3dB range beamwidth, θ_H, in the azimuth direction, as expressed in Equation (6.17). Similar to the range section, azimuth resolution is the minimum distance between two objects on the ground that the SAR system can recognize and plot as separate images in the azimuth direction or parallel to the platform trajectory. The azimuth resolution, ρ_a, depends on aperture length, L_a, and can be written as Equation (6.18). The maximum resolution, ρ_a, in the azimuth direction, θ_H, achievable equals half of the aperture length, L_a, as expressed in Equation (6.19).

$$\theta_H = \frac{\lambda_o}{L_a} \qquad \text{(Equation 6.17)}$$

$$\rho_a = R_m\theta_H = \frac{\lambda_o R_m}{L_a} \qquad \text{(Equation 6.18)}$$

$$\rho_a \geq \frac{L_a}{2} \qquad \text{(Equation 6.19)}$$

The SAR antenna beamwidth determines the swath width and the range resolution of the SAR images. In the array antenna design, the antenna's beamwidth is controlled by the configuration of the number of constituent patches and the distance between the patches. More patches configured on an array formation with a broader patch separation will produce a narrower beamwidth with higher directivity gain. In other words, an SAR antenna with a wide range of beamwidth creates a broad swath area, but will reduce the antenna gain.

6.3.2 Antenna's Gain

Antenna gain will significantly affect the performance of the SAR system. An SAR antenna with a high gain will have the advantage of increasing the SNR value of the SAR system. In addition, SAR systems with high-gain antennas operate using less RF power for the same scanning height. The array design should avoid high-intensity grating and sidelobe levels (SLL) because it can increase the possibility of decreasing the antenna gain, double-scanning in the SAR data, and producing add-itional problems related to SAR ambiguity. The grating lobes and sidelobes level of the array antenna can be suppressed by arranging the configuration, orientation, and spacing of each patch aligned with the radiation pattern characteristics of the single patch from the constituent elements. The wider antenna aperture increases the antenna gain, G, and narrowing beamwidth. However, widening the antenna aperture must consider the limitation of the antenna's installation area on the SAR platform.

At the same time, the power gain efficiency is influenced by the matching input impedance and feeding network on the antenna array. Implementing a circularly polarized SAR antenna has also proven to increase the intensity of the echo signal from the backscattering object, especially in the combination of cross-polarization mode, RL and LR, compared to SAR's antennas with linear polarization, VH and HV.

6.3.3 ANTENNA'S BANDWIDTH

As shown in Equation (6.12), a high-resolution SAR image in range direction is obtainable with a shorter pulse duration, τ_p. The pulse duration becomes too short to deliver sufficient energy per pulse to obtain a fair echo signal-to-noise ratio (SNR) for reliable detection. A longer pulse duration is needed for high resolution and a high SNR of the SAR signal. Usually, a pulse compression technique is employed to get a pulse length with enough energy, and a high resolution of SAR images is obtained. The reasonable ground range resolution can be expressed in Equation (6.20), where B_R is the frequency bandwidth of the transmitted pulse and $\tau_p = 1/B_R$. This means that broadening the pulse bandwidth of the SAR signal would increase the ground range resolution. On the other hand, broadening operational bandwidth usually comes at the cost of decreasing antenna gain.

$$\delta R_g = \frac{c}{2B_R sin\eta} \qquad \text{(Equation 6.20)}$$

6.4 CIRCULARLY POLARIZED ANTENNA DESIGN

A circularly-polarized antenna radiates two linear E-field electromagnetic waves with the same amplitude in a perpendicular orientation to each other, and has 90° phase differences. The resultant E-field produces a circle rotation wave in its propa-gation direction. The features of the circularly polarized antenna provide several advantages compared to LP antennas. In spaceborne SAR applications, such as sat-ellite platforms, the transmitted signal from the antenna and scattering signals of the object propagates pass through the ionosphere layer. Unfortunately, LP microwave is sensitive to the Faraday rotation effect, which is considerable in an SAR system that operates at low frequency. The orientation angle of the LP microwave may change, causing power degradation due to a polarization mismatch. The circularly polarized antenna is expected to overcome the power losses due to mismatched polarization. In airborne SAR applications, the Faraday rotation effect on signal propagation has not been considered. The airborne SAR platform usually flies closer to the Earth's surface, escaping the Faraday effect at the ionosphere layers.

Another advantage of a circularly polarized antenna in wireless telecommunication is that it has less chance of cross-polarization interference. Polarization from the reflected signal is an inversion of the incident signal. An incident signal of an RHCP is reflected from the ground to an LHCP signal, and vice versa. These RHCP antennas reject reflected LHCP signals.

The next section will demonstrate the design of a circularly polarized microstrip array antenna for an airborne circularly polarized SAR application on the Hinotori mission. The discussion includes how to solve and compromise between existing limitations with some antenna parameters.

6.4.1 HINOTORI C2 AIRBORNE CIRCULARLY POLARIZED SAR

The Hinotori C1 mission entailed a flight test of the C-band airborne circularly polarized SAR system for land surface observation and disaster monitoring onboard the Cessna 172 aircraft. The Hinotori system operated at a center frequency of 5.3GHz with a maximum operating bandwidth of 400MHz. This mission was postponed due to a problem with the radome antenna installation that affected the aircraft's flight stability. Then the flight test was continued by the Hinotori C2 mission on the CN235 MPA aircraft as the platform, see Figure 6.5(a) [5]. Figure 6.5(b) illustrates the conceptual construction of the Hinotori C2 circularly polarized SAR antenna inside the nosecone of the CN235 MPA aircraft. The circularly polarized SAR system designed works in stripmap scanning mode with an off-nadir angle of 55°. The fully polarimetric circularly polarized SAR system on Hinotori C2 was equipped with four-unit antennas with transmitter antennas (Tx) and receiver antennas (Rx) using the same design; these are RHCP-Tx, LHCP-Tx, RHCP-Rx, and LHCP-Rx, with the configuration shown in Figure 6.5(c). Tx antennas are mounted on the upper side of the antenna installation and the Rx antennas are on the bottom side. The combination of Tx and Rx was expected to produce fully polarimetric, circularly polarized SAR images . The Hinotori C2 mission was conducted in Makassar, Indonesia, in 2018 and successfully captured several high-resolution circularly polarized SAR images. The design specifications of the circularly polarized SAR antenna are briefly summarized in Table 6.1.[3][6] The total weight of the Hinotori C2 antenna installation should be less than 20kg.

6.4.2 MICROSTRIP ARRAY ANTENNA

A microstrip antenna is a printed antenna on a thin copper layer on a dielectric substrate sheet with a ground plane layer on the reverse side. Commonly, microstrip antennas are fed by the pin feeding or strip-line feeding method. It was a reasonable candidate for developing an excellent array of antenna with several attractive features, such as its compact size for various applications with low operating frequency, lightweight, easy to fabricate, and low cost to manufacture. The dielectric substrate provides an advantage in compromising and adjusting the microstrip antenna's dimension and requirements. However, dielectric constant and substrate thickness degrade antenna parameters, such as low gain, low efficiency, and narrow bandwidth [7].

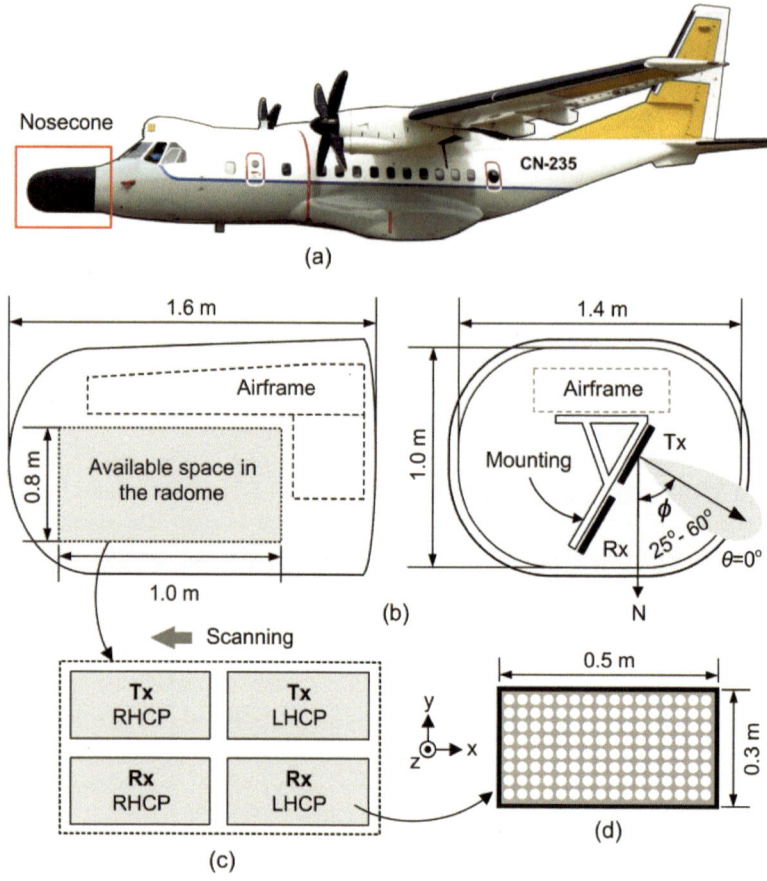

FIGURE 6.5 The airborne circularly polarized SAR antenna's system. (a) The CN-235-MPA aircraft as the SAR platform. (b) The position of the circularly polarized SAR antenna system installed inside the radome. (c) Configuration of four unit circularly polarized SAR antennas for full-polarimetric images generation. (d) The maximum allowed size of each antenna panel.

TABLE 6.1

Design Requirements of the Circularly Polarized SAR Antenna

Parameters	Value	Unit
Frequency (f_c)	5.3	GHz
Impedance bandwidth (IBW)	400	MHz
Axial-ratio bandwidth (ARBW)	400	MHz
Range beam-width (β_r)	10°	Degree
Azimuth beam-width (β_a)	5°	Degree
Total gain (G)	20	dBic
Dimension ($x \times y \times z$)	500 × 300 × 3	mm

The first step in microstrip antenna design is choosing suitable dielectric substrates for the commercial market to realize the circularly polarized SAR antenna requirements and application. Perhaps it would be wise to consider saving costs in looking for a suitable substrate to fulfill all antenna parameters for excellent performance. It has been proven that the impedance bandwidth, gain, and efficiency of the microstrip antenna depend on the thickness and the dielectric substrate characteristics. Sometimes we have to trade off certain advantages. The impedance bandwidth of an antenna increases when the substrate thickness increases and the dielectric constant hen decreases. Decreasing the dielectric constant of substrates will affect the scope of an antenna change more significantly. The efficiency of an antenna decreases proportionally with increasing the dielectric constant and thickness.

6.4.3 Unit Element of the Circularly Polarized Antenna

The circularly polarized SAR antenna on the Hinotori C2 system applied a planar microstrip array antenna. The 3D model and detailed geometry of the unit element of the Hinotori C2 antenna are shown in Figure 6.6.[8]. A thick dielectric substrate with a low dielectric constant was selected to broaden IBW and raise the gain. The antenna design is printed on a polytetrafluoroethylene (PTFE) substrate with a thickness of 1.6mm, dielectric constant (ε_r) of 2.17, a copper layer of 0.035mm, and dielectric dissipation factor (tan-δ) of 0.0005. The antenna structure consists of two stacked substrates with identical specifications – see Figure 6.6(a). The circularly polarized wave radiator and proximity-coupled feeding network are printed on the top of the lower substrate. In contrast, a circle-slotted parasitic patch was printed on the top of

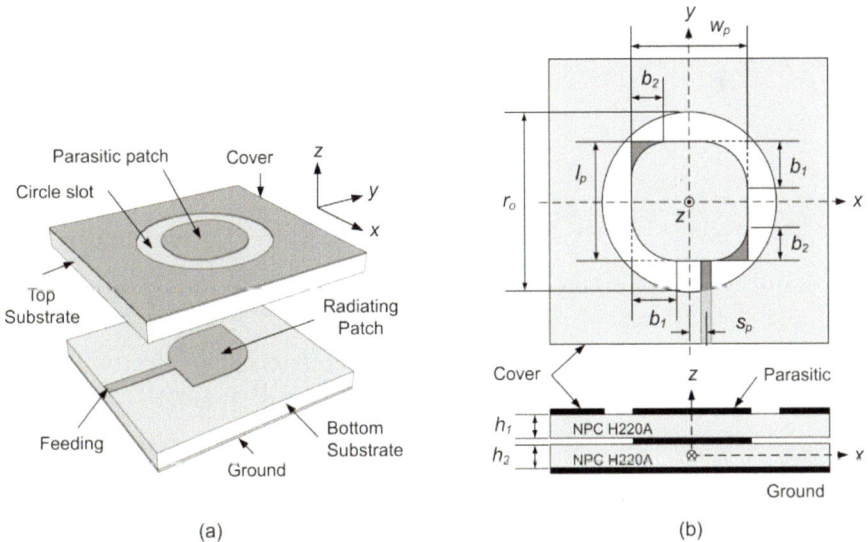

(a)   (b)

FIGURE 6.6 The single patch with circular polarization. (a) Geometry in a three-dimension view. (b) The detailed geometry.

TABLE 6.2
Optimized Dimensions of the Single Patch

Variable	Unit (mm)	Variable	Unit (mm)	Variable	Unit (mm)	Variable	Unit (mm)
w_p	17.8	l_p	17.8	r_o	26.5	s_p	2.5
b_1	7.0	b_2	5.5	h_1	1.6	h_2	1.6

the upper substrate to attempt to improve gain. A square patch with curved truncation on a diagonal corner was adopted as the main radiator to generate circularly polarized waves in a broad bandwidth. Curve truncation, b_1 and b_2, drive the quality of the axial ratio of radiation waves – see Figure 6.6(b). The axial ratio bandwidth is optimized by adjusting the shifting value, s_p, of the feed point. The dimension of the square patch, l_p and w_p, was designed resonant to the operating frequency of 5.3GHz and estimated by referring to Equations (6.21)–(6.24)[2]. Furthermore, the optimum dimensions of the unit element result of simulations and iteration are summarized in Table 6.2.

$$w_p = \frac{v_o}{2f_c}\sqrt{\frac{2}{\varepsilon_r + 1}} \qquad \text{(Equation 6.21)}$$

$$l_p = \frac{v_o}{2f_c\sqrt{\varepsilon_{reff}}} - 2\Delta l_p \qquad \text{(Equation 6.22)}$$

$$\varepsilon_{reff} = \frac{\varepsilon_r + 1}{2} + \frac{\varepsilon_r - 1}{2}\left[1 + 12\frac{h}{w_p}\right]^{-\frac{1}{2}} \qquad \text{(Equation 6.23)}$$

$$\frac{\Delta l_p}{h} = 0.412\frac{\left(\varepsilon_{reff} + 0.3\right)}{\left(\varepsilon_{reff} - 0,258\right)}\frac{\left[\frac{w_p}{h} + 0.264\right]}{\left[\frac{w_p}{h} + 0.8\right]} \qquad \text{(Equation 6.24)}$$

6.4.4 BROADENING CIRCULARLY POLARIZED ANTENNAS BANDWIDTH

In microstrip circularly polarized antenna technology, researchers have investigated several methods to broaden the impedance bandwidth (IBW) and axial ratio bandwidth (ARBW), and improve the achieved gain. A modification of antenna structure in the multi-layer substrate using various thicknesses and dielectric constants broadened IBW. A proximity-coupled feeding network on stacked substrate layers has also demonstrated bandwidth expansion. A serial sequential rotation technique on a feeding network that applies a 90° phase delay in four array elements with circular configuration has been proven to broaden ARBW. Meanwhile, additional parasitic patches around the antenna's radiator effectively improve the gain.

The evolution of unit elements to broaden IBW and ARBW is illustrated in Figure 6.7, with simulated characteristics of the S-parameter (S_{11}) and the axial ratio (AR) depicted in Figure 6.8. The shadowed bar on the frequency axis indicates the operating bandwidth of the circularly polarized SAR antenna spanned from 5.1GHz to 5.5GHz. At first, the patch design is a square patch with conventional straight-line truncation (a_1) on the diagonal corner and printed on a single layer substrate (see A1 in Figure 6.7). Modifying curve-truncation (b_1) and applying a shift on the feed point (sp) affect the minimum axial ratio. The S_{11} of the A2 antenna becomes aligned at the center frequency of 5.3GHz. An additional parasitic patch and double-layer stacked substrates in the A3 antenna significantly broaden ARBW and IBW. On the A4 antenna, the ARBW and IBW get wider after implementing a circle slotted around

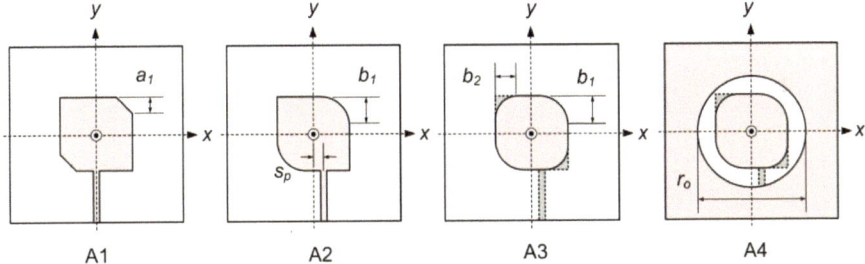

A1 = Antenna with single layer substrate and conventional corner truncation .
A2 = Antenna with single layer substrate , curved truncation, and shift feeding.
A3 = Antenna with double layer substrates , curved truncation, shift feeding, and parasitic patch.
A4 = Antenna with double layer substrates , curved truncation, shift feeding, and circle slotted parasitic patch .

FIGURE 6.7   The evolution of the single patch to broaden ARBW and IBW.

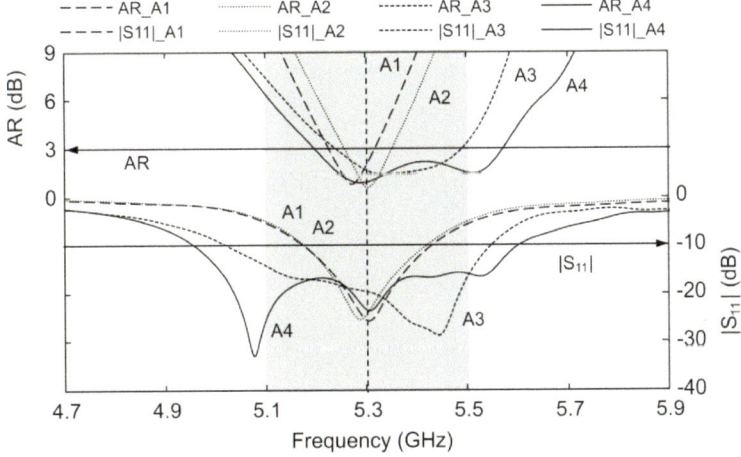

FIGURE 6.8   Simulated characteristics of the single patch during shape evolution for bandwidth optimization.

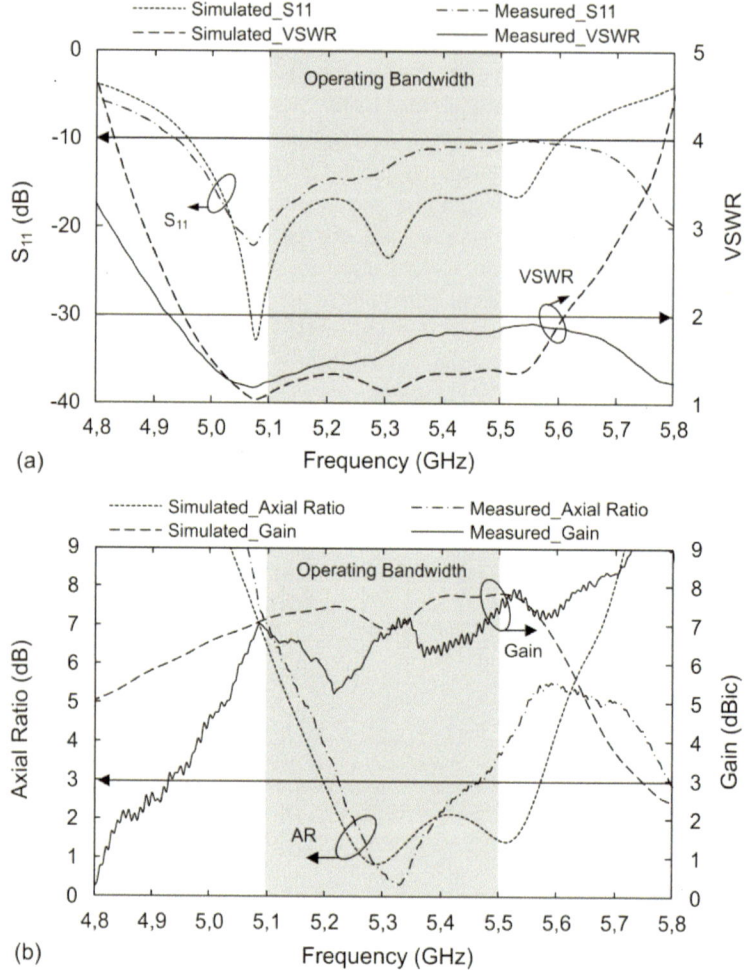

FIGURE 6.9 Simulated and measured performance of the single patch. (a) S_{11} and VSWR. (b) Axial Ratio and Gain.

the parasitic patch (r_o). A comparison between the A1 and A4 antenna shows that the A4 antenna has a broader ARBW at 422% and IBW at 256%. The realized A4 antenna produces 0.25GHz ARBW, 1.11GHz IBW, and 7dBic gain measured at the center frequency of 5.3GHz. The optimum performance of the single patch antenna is plotted in Figure 6.9.

6.4.5 THE 2 × 2 SUBARRAY ARRANGEMENT

A single patch antenna can achieve optimum performance. However, the performance of some antenna parameters do not meet the target requirements. A number of single patches need to be arranged into a more extensive antenna array to achieve suitable performance. The arraying technique is commonly used to improve the accumulation of gain, set the desired beamwidth, and broaden bandwidth. The dimension of arrays sometimes needs be compromised to improve array performance and installation limitations, while being limited by available materials on the market to realize it. In the case example of the Hinotori C2 circularly polarized SAR arrays, huge array dimensions gave higher gain, better aperture, and narrower beamwidth. However, it requires a vast installation space, heavyweight, strong mounting structure, a larger platform vehicle, and high cost in implementation.

6.4.5.1 Element Separation

Critical parameters in arraying design are the configuration and distance between array elements. In the Hinotori C2 antenna, the first step of arraying is configuring four patches into a subarray with a 2×2 matrix formation [9]. The separation between array elements is chosen between $0.5–1.0\lambda_0$ (λ_0 is wavelength in free space) to avoid grating lobes appearing on the array. Figure 6.10 provides the study overview of the radiation pattern characteristics of arrays with various configurations and element separations. In the computer simulation, each array element is supplied with a power source of the same intensity to ignore the effect of different power distributions on the feeding network variation of arrays. Simulation results show that the grating lobe appears at the element separation greater than $1.0\lambda_0$. Longer array spacing results in a higher gain and narrower beamwidth in the range of $0.5–1.0\lambda_0$. The Hinotori C2 system used the uniform element separation of $0.5\lambda_0$ to suppress the antenna dimension and avoid exceeding the available space inside the aircraft's nosecone – see Figure 6.5 (c). The $0.5\lambda_0$ separation is not the best array element separation associated with the electrical performance of this array antenna but offers a suitable compromise for the requirements and limitations in the Hinotori C2 system.

6.4.5.2 The Serial Sequential Rotation Configuration

The subarray adopts a serial sequential rotation (SSR) technique in the effort to broaden ARBW [10][11]. Four patches were combined using a serial feeding network to perform a sequential rotation with a phase current difference of 90° between one patch and the next patch in the subarray. The SSR formation in the subarray starts on the patch marked A, then patch B, patch C, and patch D, as illustrated in Figure 6.11(a). The subarray was arranged with uniform element separation $d_x = d_y = 0.5\lambda_0$ in a circle fan formation of the RHCP mode. A counter-clockwise circular arrow indicates the RHCP rotation. The feeding line of each patch is directed to the center coordinate of the subarray (O) by rotating parallel to the x'-axis and y'-axis by 45° rotation from the original x-axis and y-axis. This formation was the most straightforward and shortest feeding network to improve the transmission power efficiency in the SSR subarray.

The disadvantage of an array configuration with a serial feeding network is that the closest patch element to the input power will receive a stronger power distribution than the farthest patch element from the input port. The received power in the first patch and the last patch must be balanced so that the subarray beam is close to boresight at 0°. A subarray beam with a boresight close to 0° will be easier in the next wider array arrangement. Modifying the impedance ratio of the subarray feeding

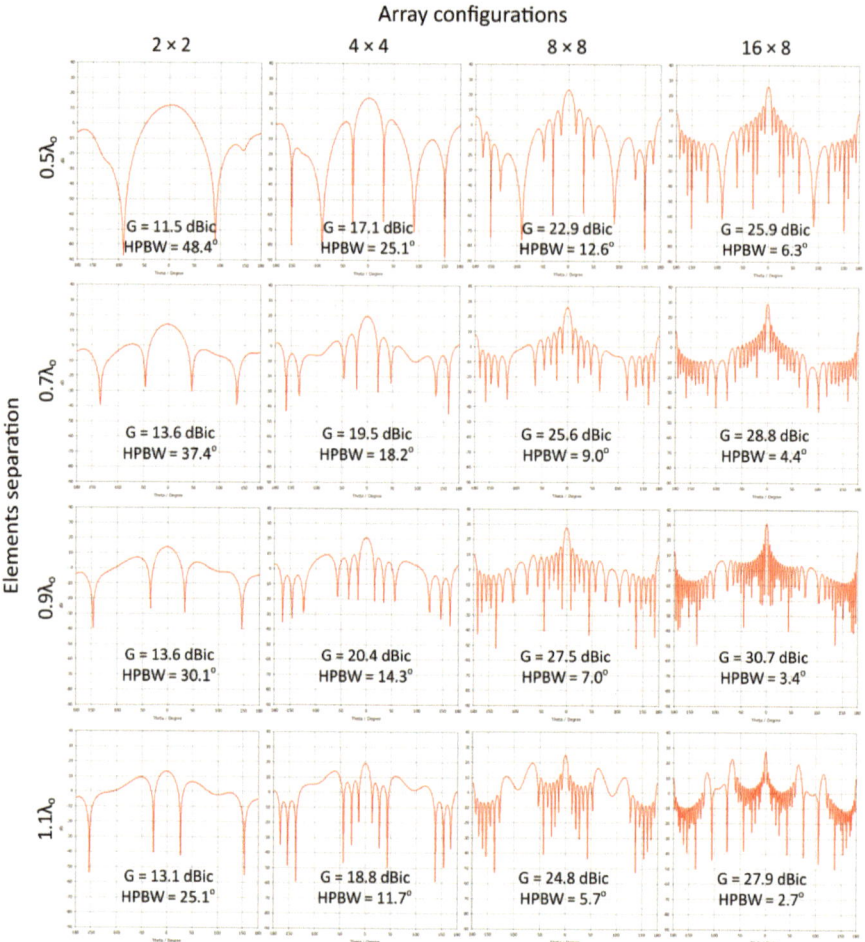

FIGURE 6.10 Investigation of array characteristics with various configurations and element separation.

network can balance the power distribution in a serial circuit. This method is more common and does not waste power. Another way is to connect the ends of the feeding network to a load or the ground plane with a particular proportion. However, this technique causes power losses and is rarely implemented.

6.4.5.3 Impedance Matching of the Feeding Network

The detailed design of the 2×2 subarray feeding network is depicted in Figure 6.11(b). The phase shifter uses a microstrip line with the length of a quarter lambda $(\lambda_g/4)$. The phase shifter affects the phase lagging $90°$ between one patch to the next in the SSR subarray. The equivalent circuit model is shown in Figure. 6.11(c) [12]. The input power goes to the subarray, depicted as P. The input power supplies to each

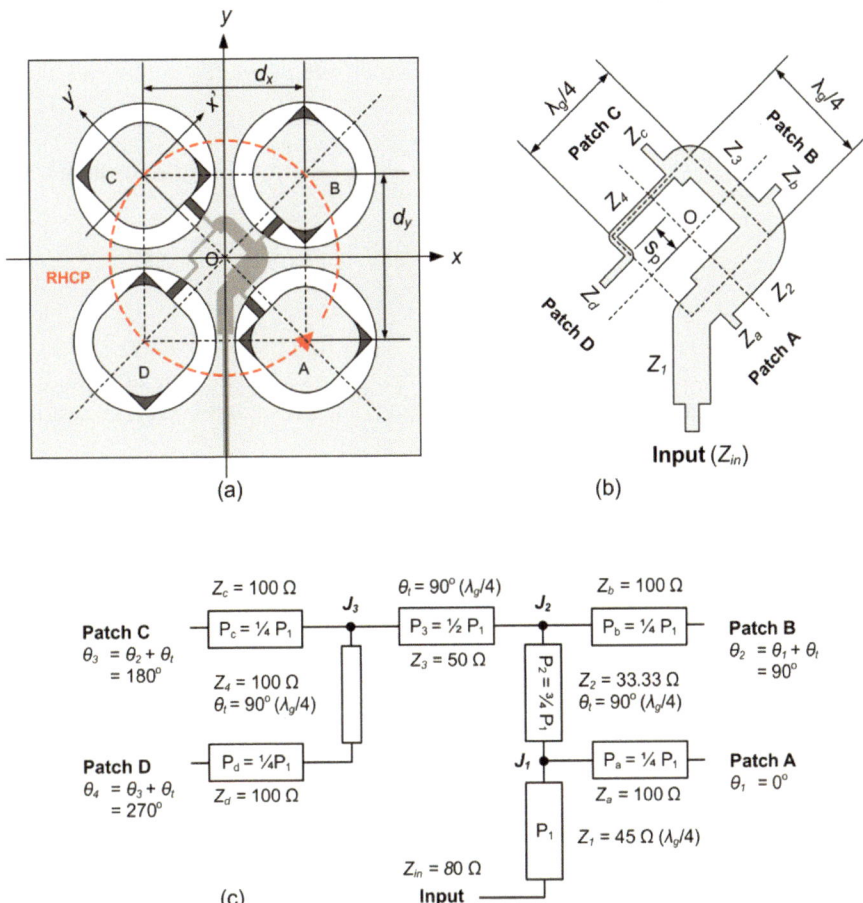

FIGURE 6.11 Configuration of the 2 × 2 subarray. (a) Implementation geometry of serial-sequential rotation technique with RHCP mode and patch separation of d_x and d_y. (b) The feeding network with impedance matching. (c) The equivalent circuit of the subarray feeding network.

subarray element (patch A to patch D) are equally distributed at 0.25P. Each patch is distributed parallel with a T-junction power divider of J_3, J_2, and J_1. The impedance of the feeding line is defined as $Z_a = Z_b = Z_c = Z_d = 100\Omega$ considering the feasibility to realize the minimum width of the strip-line feeding the subarray. The ratio of the power divider at each T-junction of J_3, J_2, and J_1 is expressed in Equations (6.25) to (6.28), respectively.

$$J_3 \rightarrow \frac{P_d}{P_c} = \frac{0.25P_1}{0.25P_1} = \frac{Z_c}{Z_4} \qquad \text{(Equation 6.25)}$$

$$J_2 \rightarrow \frac{P_3}{P_b} = \frac{0.50P_1}{0.25P_1} = \frac{Z_b}{Z_3} \qquad \text{(Equation 6.26)}$$

$$J_1 \rightarrow \frac{P_2}{P_a} = \frac{0.75P_1}{0.25P_1} = \frac{Z_a}{Z_2} \qquad \text{(Equation 6.27)}$$

$$Z_1 = \sqrt{Z_{in} \cdot J_1} \qquad \text{(Equation 6.28)}$$

The calculation results of Equations (6.25) to (6.27) were $Z_4=100\Omega$, $Z_3=50\Omega$, and $Z_2=33.33\Omega$. The impedance of Z_{in} is set to 80Ω. And Equation (6.28) defines the impedance of the quarter-lambda transformer Z_1, where J_1 is the parallel resultant between Z_2 and Z_{P1}. We got $Z_1=45\Omega$ and $J_1=25\Omega$.

Following the SSR technique, patch A assumed to have a phase of $\theta_1 = 0°$, and patch B set on a degree of $\theta_2 = 90°$ by delayed $\theta_t = 90°$ using a $\lambda_s/4$ transformer. The step between one patch to the next patch is sequentially delayed $\theta_t = 90°$ on RHCP rotation from patch A to patch D. Flipping the RHCP antenna with the centerline on the y-axis can perform the LHCP subarray antenna with SSR in the reverse direction.

6.4.5.4 Subarray Evaluation

The 2 × 2 subarrays with SRR feeding network configuration in various uniform element separations have been simulated at the center frequency of 5.3GHz. Simulated antenna parameters are plotted in Figure 6.12. The shadowed bar on the graph indicates the operational bandwidth of the circularly polarized SAR antenna in the Hinotori-C2 system. The SSR implementation on the subarray configuration significantly broadened the IBW and ARBW. Total gain increase proportional to increasing the number of subarray elements on the extensive array configuration. The detailed characteristics of the 2 × 2 subarray antenna with SSR configuration in various patch separations are summarized in Table 6.3. The subarray with an element separation of $0.7\lambda_0$ gives excellent performance on the overall antenna parameters and achieves the highest gain. Unfortunately, the estimation dimension for the 16 × 8 array configuration with the element separation of $0.7\lambda_0$ provides a length dimension of 641mm, exceeding the maximum length dimensions of 500mm.

6.4.6 LARGE ARRAY EXPANSION

The position and formation of each element in the array will determine the mutual coupling, characteristics of the radiation pattern, beamwidth, and the complexity of the feeding network. The extensive array of circularly polarized antennas design should maintain the efficiency of power distribution, synchronous phase, impedance bandwidth, and axial-ratio bandwidth. An excellent power distribution arrangement entails the shortest possible Path to the feeding network. Also, it matches impedance in each junction power divider to reduce undesired reflection waves.

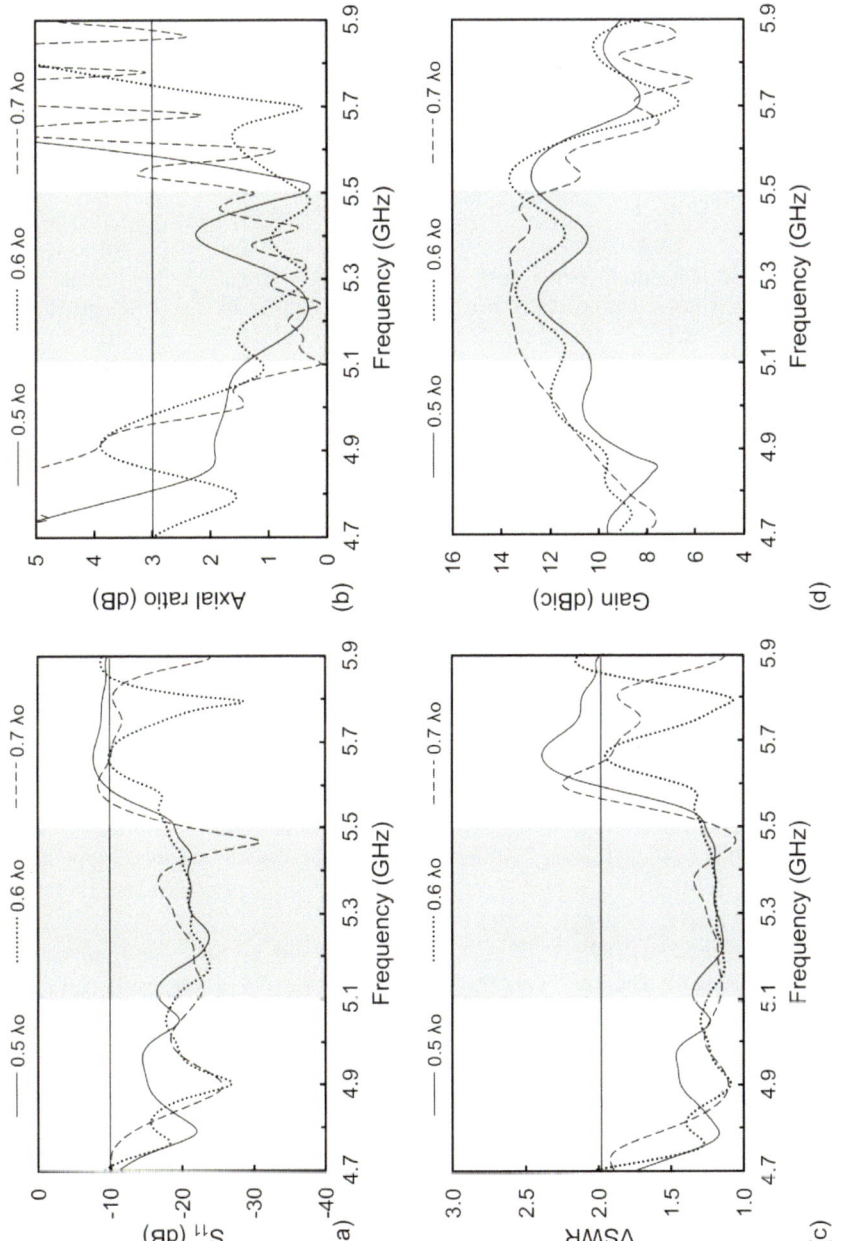

FIGURE 6.12 Simulated characteristics of the 2 × 2 subarray in various element separation of $0.5\lambda_o$, $0.6\lambda_o$, and $0.7\lambda_o$. (a) S-parameter. (b) Axial ratio. (c) VSWR. (d) Gain.

TABLE 6.3

Performance of the 2×2 Subarray Simulated at the Center Frequency of 5.3 GHz

Parameters		$d_x=d_y=0.5\lambda_0$	$d_x=d_y=0.6\lambda_0$	$d_x=d_y=0.7\lambda_0$	Unit
S-Parameter (S_{11})	f_L	4.68	4.71	4.67	GHz
	f_H	5.59	5.65	5.56	GHz
	IBW	910 (17.17)	940 (17.74)	890 (16.79)	MHz (%)
Axial-Ratio	f_L	4.81	4.98	4.96	GHz
(AR)	f_H	5.58	5.75	5.54	GHz
	ARBW	770 (14.53)	770 (14.53)	580 (10.94)	MHz (%)
Radiation pattern	Direction	6.0	3.0	2.0	Degree
H-plane ($\varphi=0°$)	BM	44.20	38.30	33.40	Degree
	SLL	-18.20	-16.00	-11.90	dB
Radiation pattern	Direction	0.0	0.0	0.0	Degree
E-plane ($\varphi=90°$)	BM	43.50	38.40	33.40	Degree
	SLL	-28.80	-20.40	-11.8	dB
Gain		12.00	13.23	13.54	dBic
VSWR (1:2)		1.2	1.2	1.2	N/A
Input Impedance (Z_{11})		49.80+j2.70	49.82+j0.68	51.20-j0.02	Ohm
Array Dimension ($x \times y$)		471.5 × 245.2	336.3 × 284.8	641 × 324.4	mm

6.4.6.1 The 4 × 4 Subarray

The conceptual design of the large array with the shortest possible route to an SSR feeding path is illustrated in Figure 6.13(a). The 4 × 4 subarrays consist of four sets of 2 × 2 subarrays arranged in different orientations and positions. Subarray C is the optimum design for the 2 × 2 subarray with SSR configuration, then set as the original position and orientation, indicated by the shadowed subarray. It does not need to rotate. Subarray B, A, and D are duplicates of subarray C. The direction changed to the right rotation of 90°, 180°, and 270°. Each group of the 2 × 2 subarrays is separated with a uniform separation of $d_x = d_y = 0.5\lambda_0$ and combined using an SSR feeding network in RHCP mode, see Figure 6.13(b).

The detailed feeding network of the 4 × 4 subarray with an SRR configuration is shown in Figure 6.14. The 4 × 4 subarrays consist of four sets of 2 × 2 subarrays, A, B, C, and D. The first patch of each 2 × 2 subarray on the serial feeding network is A_1, B_1, C_1, and D_1. These first patches are set on the inside part of the subarray with the aim to concentrate the radiation power on the center of the subarray for directivity gain improvement and performance of the SSR in RHCP mode – indicated by the yellow background. The feeding network of the 4 × 4 subarray has the exact impedance matching with the 2 × 2 subarray. This is different in configuration path design only, as illustrated in Figure 6.14(b).

6.4.6.2 The 8 × 8 Subarray

The arrangement of the 8 × 8 subarray antenna uses the same technique as applied in the 4 × 4 subarray arrangement. The 8 × 8 subarrays consist of four sets of 4 × 4

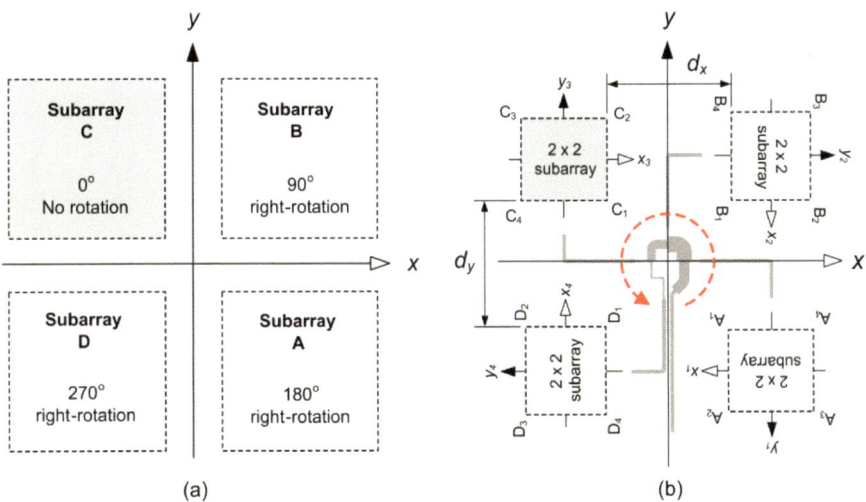

FIGURE 6.13 (a) The scenario for enlarge array antenna. (b) Duplicating and rotating subarray to implement the SSR technique in RHCP mode.

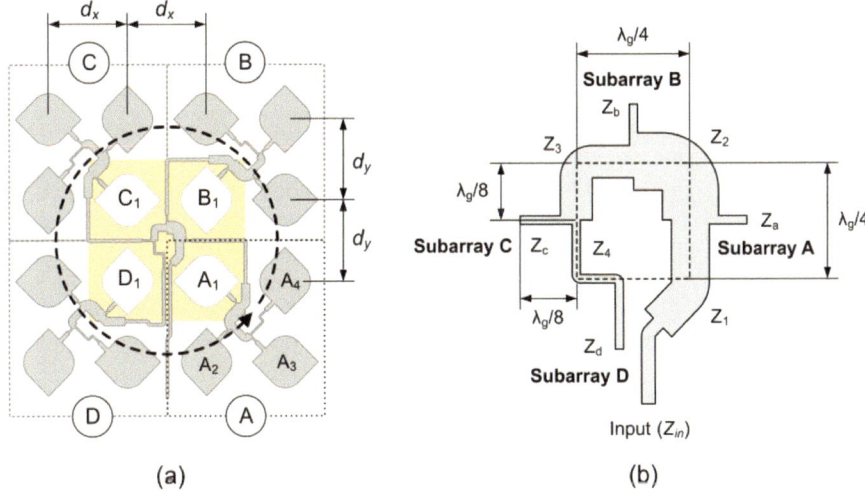

FIGURE 6.14 (a) The detailed feeding network of the 4 × 4 subarray with the SRR configuration in the RHCP mode. (b) Geometry of the power divider with $\lambda_g/4$ phase shifter in the SSR formation.

subarrays arranged in different orientations and positions, these are P, Q, R, and S – see Figure 6.15(a). Following the SSR technique, the first 4 × 4 subarray in the 8 × 8 subarray is P, assumed to have a phase of $\theta_1 = 0°$ Between the first 4 × 4 subarray to the next subarray, the phase delayed on $\theta_t = 90°$ as the result of a quarter lambda phase shifter. The next 4 × 4 subarray is Q, R, and S with phases 90°, 180°, and 270°, respectively.

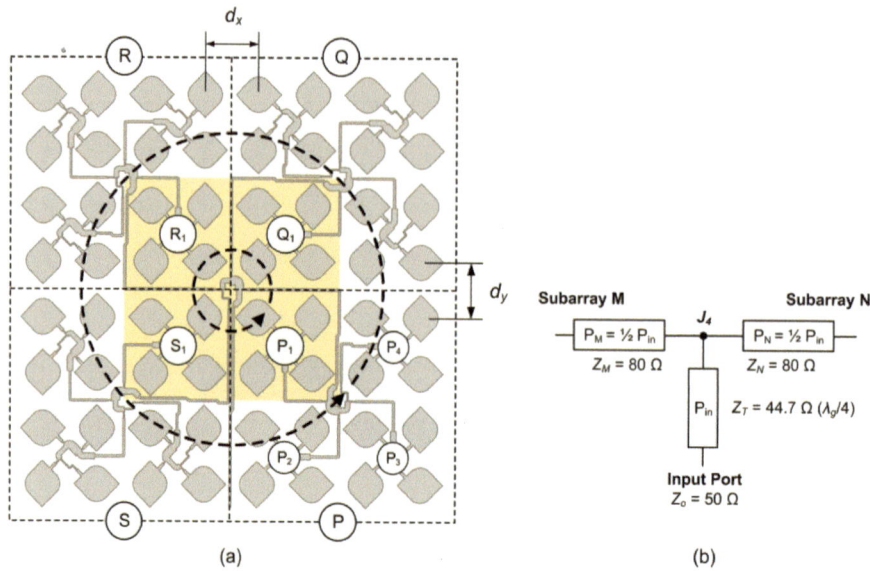

FIGURE 6.15   (a) The detailed feeding network of the 8 × 8 subarray with the SRR configuration in the RHCP mode. (b) The equivalent circuit of the power divider combined two 8 × 8 subarrays into a 16 × 8 array.

The first 2 × 2 subarray of each 4 × 4 subarray on the serial feeding network is defined as P_1, Q_1, R_1, and S_1, indicated by the shadowed yellow background. These first 2 × 2 subarrays set on the inside part of the 8 × 8 subarray aim to accumulate the radiation power on the center of the 8 × 8 subarray for directivity gain improvement and performance of the SSR during co-polarization of RHCP mode. The feeding network of the 8 × 8 subarray has both exact impedance matching and configuration with the 4 × 4 subarray. Four 4 × 4 arrays integrated into an 8 × 8 array use a simple power divider, whose equivalent circuit is shown in Figure 6.11(c).

6.4.6.3 The 16 × 8 Subarray

The conceptual design of the 16 × 8 array with an SSR configuration is illustrated in Figure 6.16. The 16 × 8 array configuration consists of two sets of 8 × 8 subarrays arranged in the same orientations dan positions. These are M and N. This configuration is intended to keep the mutual coupling of the radiation pattern synchronous during co-polarization on RHCP mode. The radiation pattern of a subarray that has a boresight does not precisely fit at 0°, therefore the subarray should be arranged into a uniform position and orientation to maintain the narrow beamwidth of the main lobe. The feeding network configuration of the 16 × 8 array, which is composed of the gradual composition of 2 × 2 subarrays, 4 × 4 subarrays, and 8 × 8 subarrays, as illustrated in Figure 6.17. At the same time, simulated characteristics of axial-ratio and gain of some antennas in arraying stages from the single patch to the 16 × 8 subarray dimension are plotted in Figure 6.18. Implementing the SSR principle in

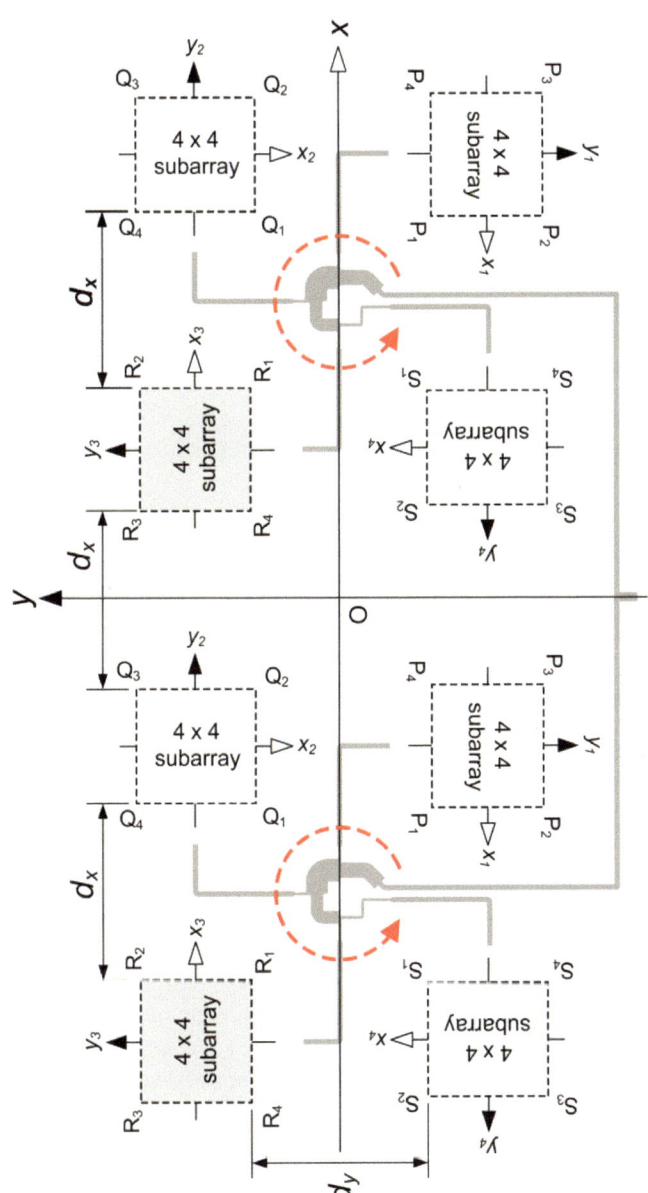

FIGURE 6.16 Conceptual design to arrange the 16 × 8 array antenna.

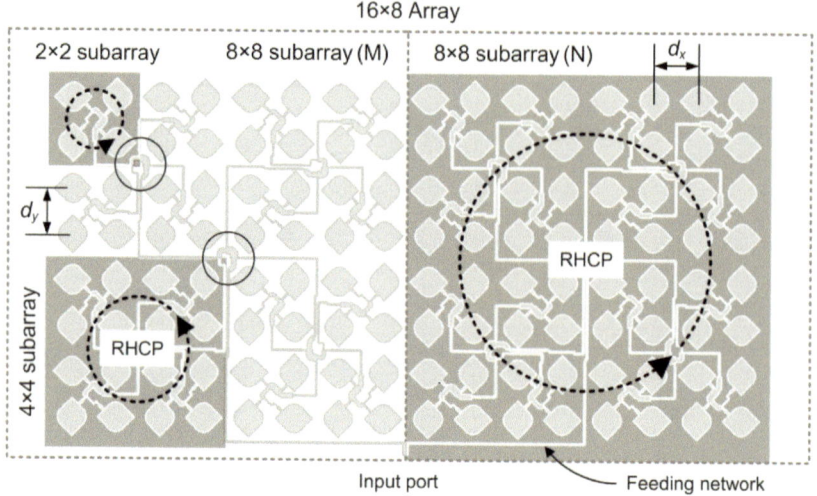

FIGURE 6.17 The gradual composition of 2 × 2 subarrays, 4 × 4 subarrays, and 8 × 8 with uniform separation $d_x = d_y$ and the RHCP rotation in the 16 × 8 array configuration.

the feeding network configuration significantly broadens the axial-ratio bandwidth. Increasing the number of array elements with a uniform separation of $0.5\lambda_o$ provides the gain improvement with an acceptable value.

Two 8 × 8 arrays integrated into a 16 × 8 array use a simple T-junction power divider whose equivalent circuit is shown in Figure 6.15(b). The value T-junction is $J_4 = 40\Omega$. It is obtained from calculating the parallel impedance between Z_M and Z_N. At the same time, Z_T is a quarter lambda transformer for impedance matching between J_4 and a normalized input port of $Z_0 = 50\Omega$.

6.4.7 THE PROTOTYPE ANTENNA

The designed structure of the 16 × 8 array is depicted in Figure 6.19. The 16 × 8 array with two dielectric substrates is stacked and fixed by 2mm plastic screws at multiple locations to avoid misaligning. Then it is put on the 3mm aluminum plate as a ground plane and makes the antenna's construction more robust during implementation. The prototype is framed by a square hollow aluminum bar with a diameter of 1.2cm. The prototype dimension of the 16 × 8 array antenna has a rectangular shape with 30cm length, 27cm width, and 3.3cm thickness. The N-type port with 50 Ω impedance is used to deliver the maximum RF power of 200W.

The manufacturing of the 16 × 8 array in RHCP mode is shown in Figure 6.20. The feeding network is printed at the top layer of the bottom substrate and the parasitic patch is printed at the top layer of the top substrate. The prototype has a total weight of 2.3kg. Four prototypes of the 16 × 8 array antenna were manufactured at Chiba University Japan – two prototypes of RHCP antenna and two prototypes of LHCP antenna – to generate a fully polarimetric circularly polarized SAR image.

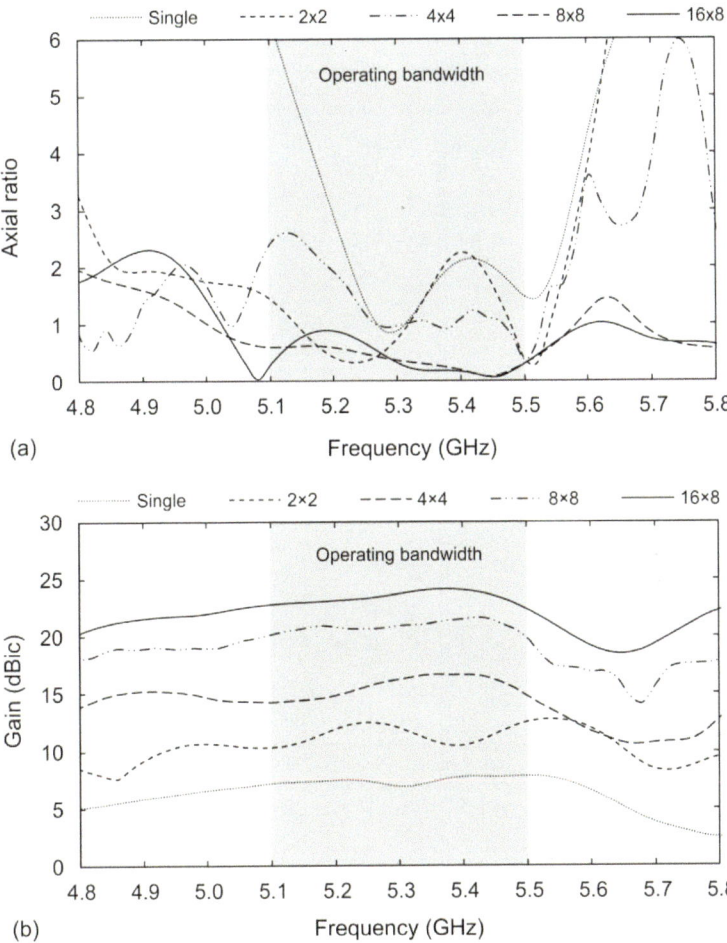

FIGURE 6.18 Simulated characteristics of antenna parameters in arraying evolution from the single patch to the 16 × 8 array. (a) Broadening of axial-ratio. (b) Gain improvement.

6.4.8 PERFORMANCE OF THE HINOTORI C2 ANTENNA

The performance characteristics of the 16 × 8 array antennas have been measured in an anechoic chamber with E8364C PNA at Chiba University Japan. Measured parameters include the S-parameter (S_{11}), VSWR, axial ratio, gain, and radiation pattern. Comparison performances of the 16 × 8 antenna between design simulation and measurement in RHCP mode are depicted in Figures 6.21 to Figure 6.23 and summarized in Table 6.4. The shadowed bar in the frequency axis of the graph indicates the operational bandwidth of the circularly polarized SAR antenna. The mismatching between simulated and measured characteristics of the designed array contributed by accumulation errors include the prototype manufacturing process, assembling, and measurement tolerances between the designed antenna and the prototype antenna.

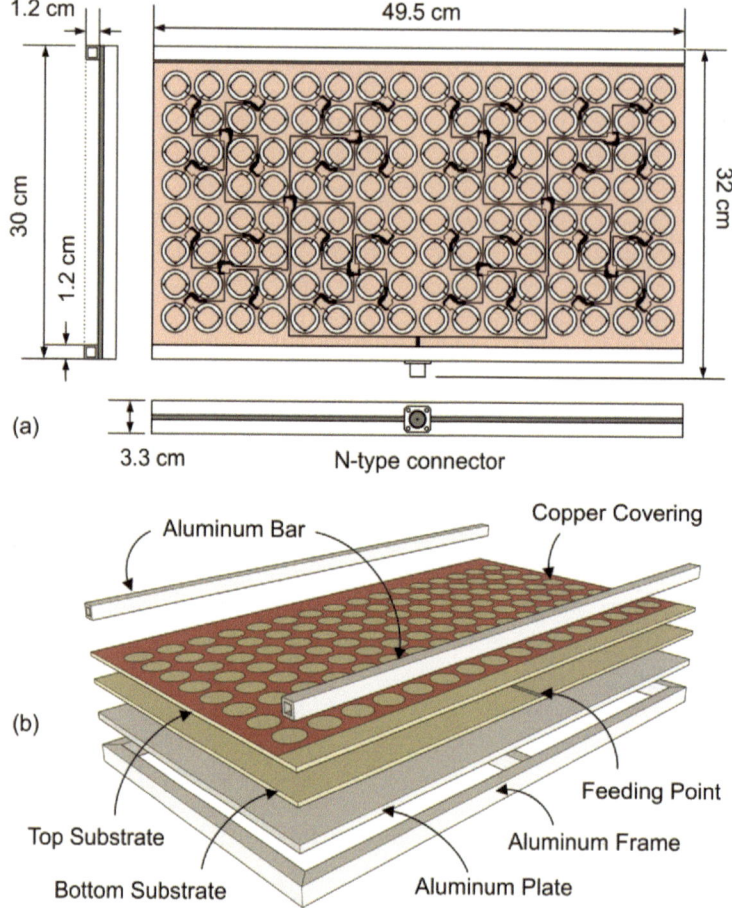

FIGURE 6.19 Structure of the prototype antennas for the Hinotori C2 mission. (a) Detailed dimension. (b) Detailed antenna construction.

Figure 6.21(a) shows the impedance bandwidth and axial ratio bandwidth have met the requirements of 400MHz. As plotted in Figure 6.21(b), the antenna gain has been higher than 20dBic as the target requirement. Meanwhile, the axial ratio and gain of the 3dB beamwidth in the main beam's range field and azimuth field met the required target – see Figure 6.22. Validation of the radiation pattern at three frequencies – 5.1 GHz, 5.3 GHz, and 5.5 GHz – is shown in Figure 6.23. The grating lobe does not appear even though the sidelobe level is relatively high in the H-plane or azimuth field. A sidelobe level should be as low as possible to avoid double scanning when the circularly polarized SAR operates.

Overall, the measured result of the 16 × 8 array antenna fulfilled the target requirement and was acceptable to integrate into the airborne circularly polarized SAR system. The antenna installation layout for full polarimetric image generation onboard CN235MPA in the Hinotori-C2 mission is illustrated in Figure 6.24. Measured characteristics of all prototype antennas are plotted in Figure 6.25 and summarized in Table 6.5.

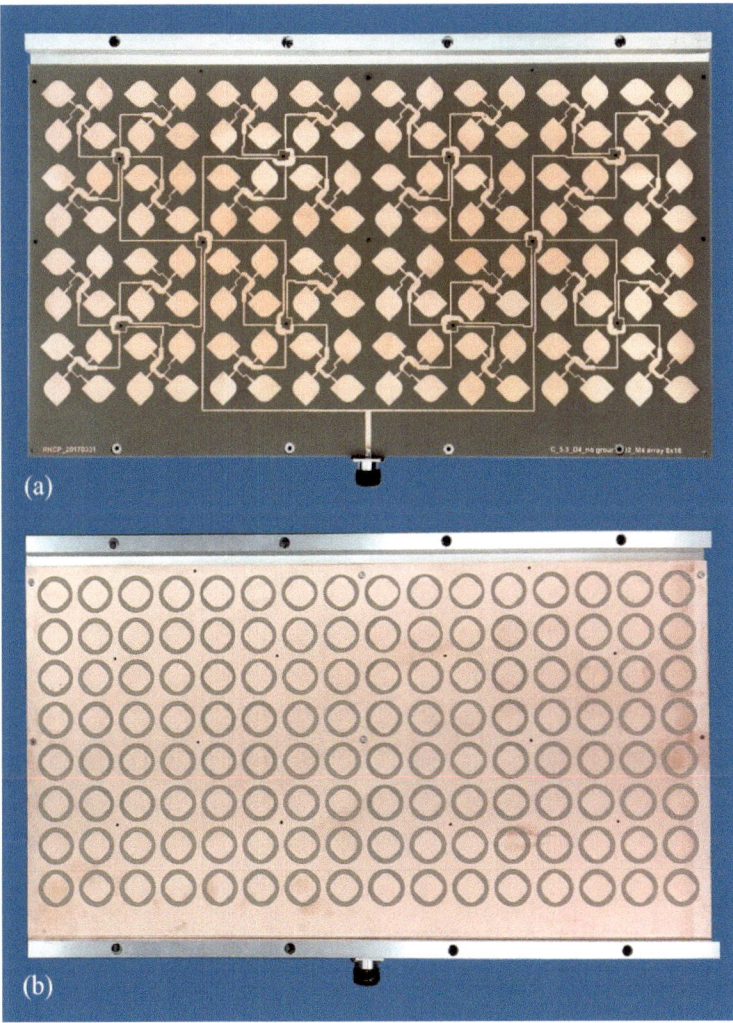

FIGURE 6.20 The realized of the 16 × 8 array antenna prototype. (a) Printed feeding network at the top layer of the bottom antenna substrate. (b) Printed parasitic patch at the upper layer of the top antenna substrate.

TABLE 6.4

Summarized Performances of the 16 × 8 Array Antenna (RHCP)

Parameters		The 16 × 8 Array Performance		Unit
		Simulation	Measurement	
S-parameter (S_{11})	Frequency (f_c= 5.3GHz)	-15.9	-19.5	dB
	Lower frequency (f_L)	4.54	4.94	GHz
	Upper frequency (f_H)	6.12	5.8	GHz
	Impedance bandwidth (IBW)	1580	860	MHz
	Fractional bandwidth (FBW)	29.74	16.23	%
Axial-ratio (AR)	Frequency (f_c= 5.3GHz)	0.3	2.3	dB
	Lower frequency (f_L)	4.62	4.95	GHz
	Upper frequency (f_H)	6.64	6.27	GHz
	Axial ratio bandwidth (ARBW)	2020	1330	MHz
	Fractional bandwidth (FBW)	38.11	25.00	%
H-plane ($\phi = 0°$)	Direction	0.0	1.0	Degree
	Beamwidth (BM)	6.0	6.0	Degree
	Side lobe level (SLL)	-10.0	-8.57	dBi
E-plane ($\phi = 90°$)	Direction	0.0	-1.0	Degree
	Beamwidth (BM)	13	13	Degree
	Side lobe level (SLL)	-15.8	-13.5	dBi
Gain (G)		23.7	21.3	dBic
VSWR (1:2)		1.39	1.22	-
Input impedance (Z_{11})		51.20 + j1.72	55.16 − j6.76	Ohm
Dimension (x × y)		495 × 300 × 33	495 × 300 × 35	mm

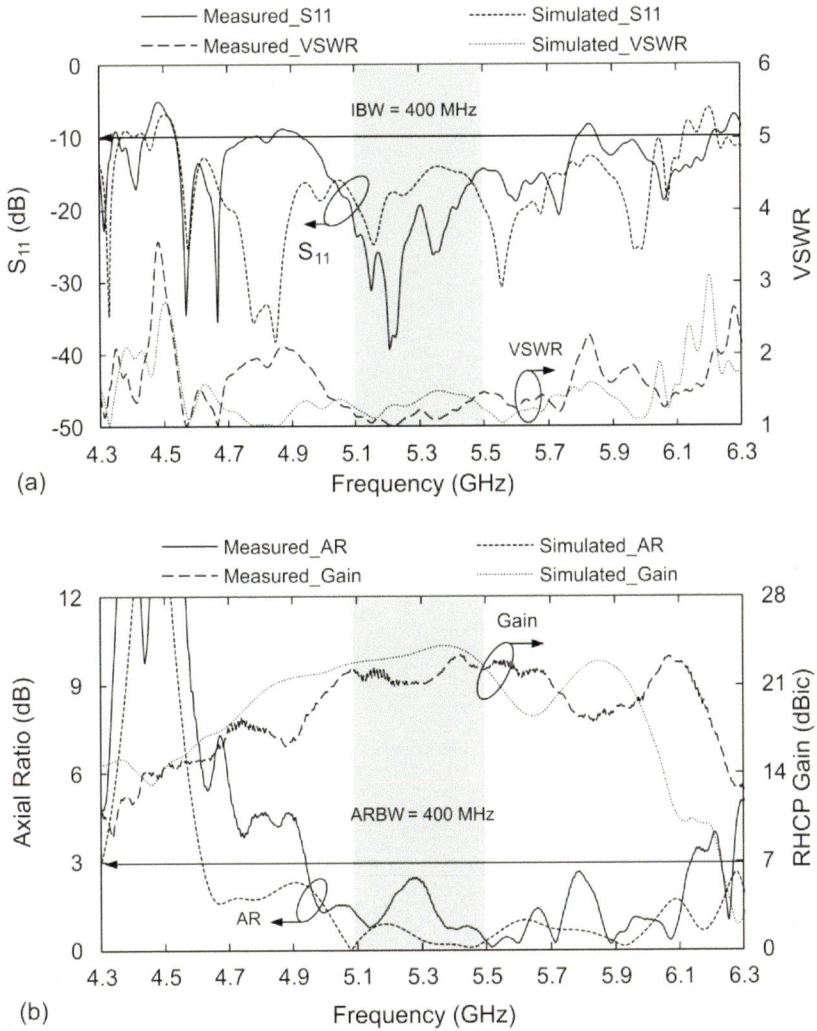

FIGURE 6.21 Measured and simulated characteristics of the 16 × 8 array antenna plotted in spreading frequency. (a) S_{11} and VSWR. (b) Axial Ratio and Gain.

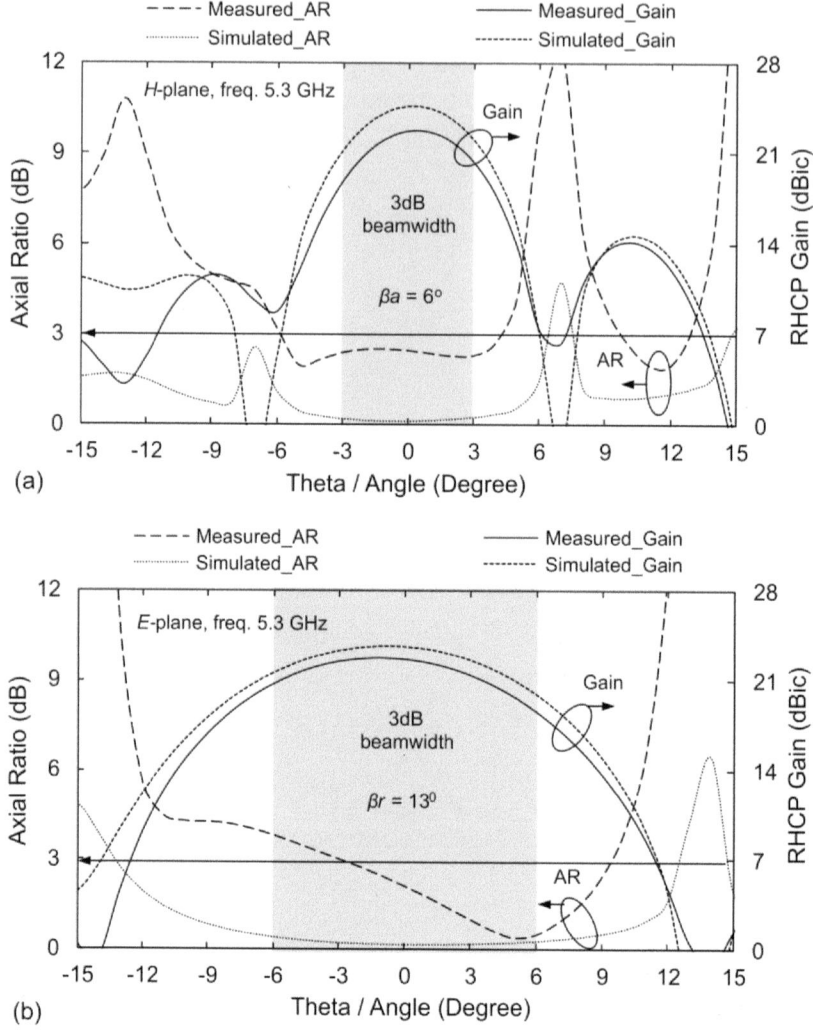

FIGURE 6.22 Measured and simulated axial ratio and gain on the mainbeam of the 16 × 8 array antenna plotted in spreading angle. (a) Azimuth direction (*H*-plane). (b) Range direction (*E*-plane).

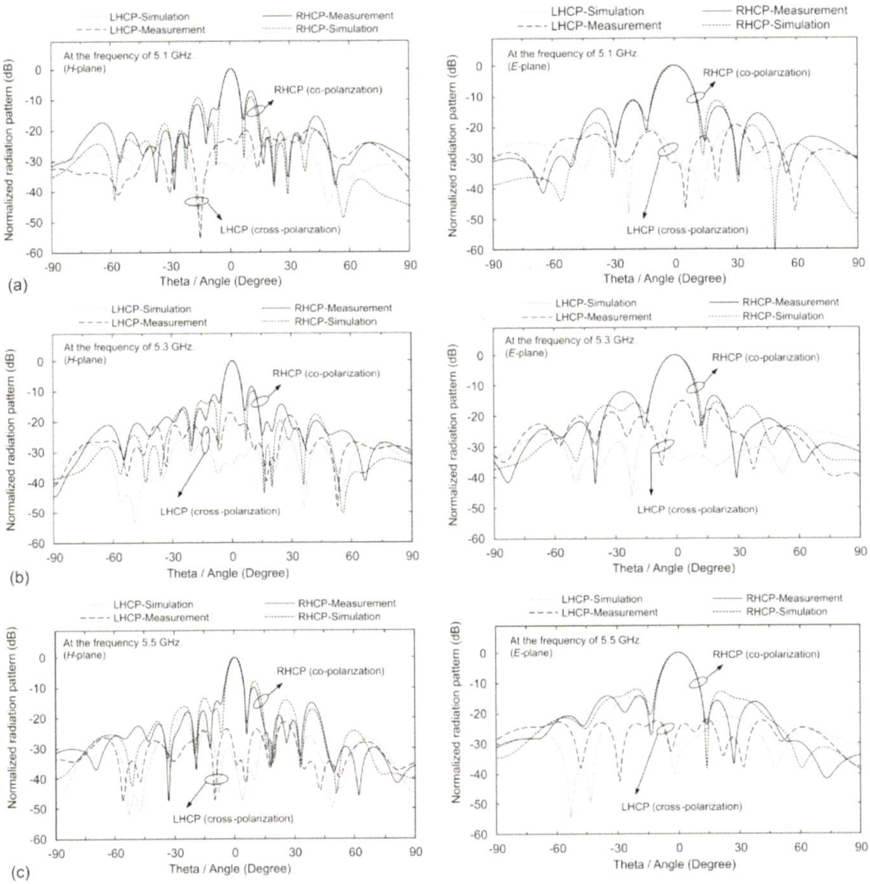

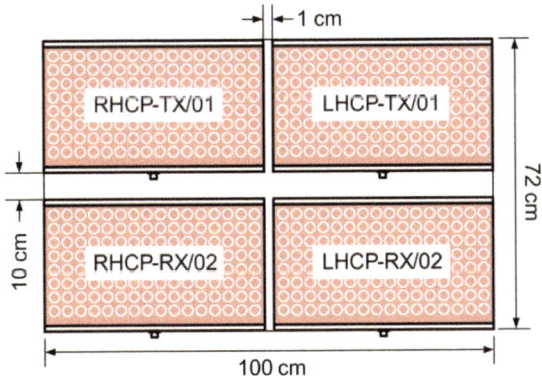

FIGURE 6.23 The measured and simulated radiation pattern of the 16 × 8 array antenna was investigated in several planes and frequencies. (a) 5.1 GHz. (b) 5.3 GHz. (c) 5.5 GHz.

FIGURE 6.24 The circularly polarized SAR antenna installation layout for full polarimetric image generation onboard CN-235 for the Hinotori C2 mission.

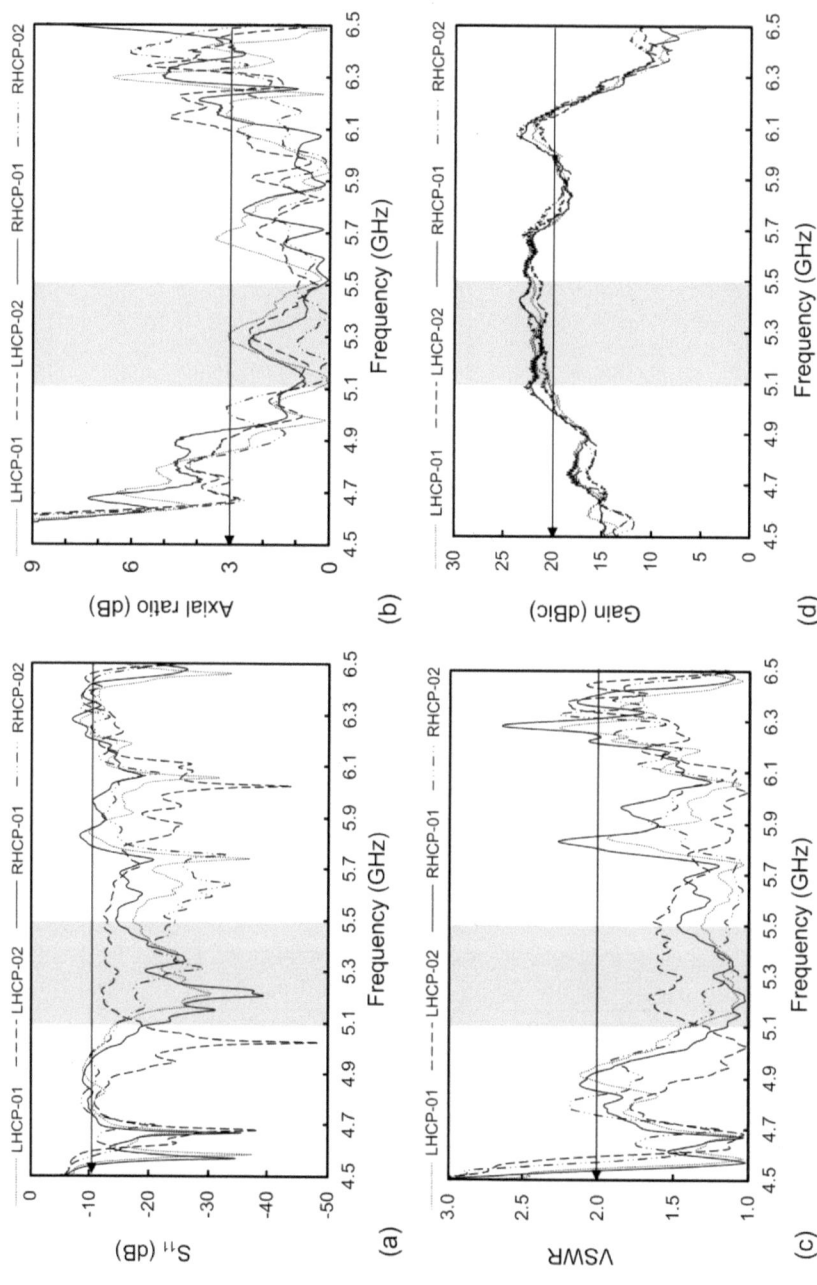

FIGURE 6.25 Measured characteristics of four prototype antennas on the Hinotori C2 mission (LHCP-01, LHCP-02, RHCP-01, RHCP-02) plotted in spreading frequency. (a) S_{11}. (b) Axial Ratio. (c) VSWR. (d) Gain.

TABLE 6.5
Summarized Performances of all Circularly Polarized SAR Antenna Series for the Hinotori C-2 Mission

Parameters		RHCP-01	RHCP-02	LHCP-01	LHCP-02	Unit
		The antenna series				
S_{11}	$fc = 5.3\text{GHz}$	-19.9	-25.1	-23.0	-14.8	dB
	f_L	4940	4840	4960	4600	MHz
	f_H	5800	6310	6250	6330	MHz
	IBW	860	1470	1290	1730	MHz
	FBW (%)	16.23	27.7	24.3	32.6	%
AR	$fc = 5.3\text{GHz}$	2.3	0.8	3.0	2.3	dB
	f_L	4950.0	5040	4860	4920	MHz
	f_H	6270.0	6320.0	6150	6100	MHz
	ARBW	1330	1280	1290	1980	MHz
	FBW (%)	25.0	24.2	24.3	37.4	%
Azimuth	Cross-Polarization	-18.0	-29.6	-15.3	-17.1	dB
	Beamwidth	6.0	6.0	6.0	6.0	Degree
	SLL	-8.57	-9.4	-8.4	-10.0	dB
Range	Cross-Polarization	-17.0	-26.5	-24.7	-25.0	
	Beamwidth	13.0	13.0	13.0	13.0	Degree
	SLL	-13.5	-11.3	-11.4	-11.6	dB
Gain		21.3	22.0	22.0	21.4	dBic
VSWR		1.22	1.1	1.1	1.4	

REFERENCES

[1] C. A. Balanis, *Antenna Theory Analysis and Design, 3rd Edition*, John Wiley and Sons, 2005.

[2] Q. Luo, S. Gao, W. Liu, and C. Gu, *Low-cost Smart Antennas*, John Wiley and Sons, 2019.

[3] Y. K. Chan and V. C. Koo, "An Introduction to Synthetic Aperture Radar (SAR)", *Progress in Electromagnetics Research B*, vol. 2, pp. 27–60, 2008.

[4] J. C. Curlander and R. N. McDonough, *Synthetic Aperture Radar System and Signal Processing*, John Wiley and Sons, 1991.

[5] J. T. Sri Sumantyo, et al., "Airborne Circularly Polarized Synthetic Aperture Radar", *IEEE Journal of Selected Topics in Applied Earth Observations and Remote Sensing*, vol.14, pp. 1676–1692, Dex 2020.

[6] M. Y. Chua et al., "The Maiden Flight of Hinotori-C: The First C Band Full Polarimetric Circularly Polarized Synthetic Aperture Radar in the World", *IEEE Aerospace and Electronic Systems Magazine*, vol. 34, no. 2, pp. 24–35, Feb. 2019.

[7] J. R. James and P. S. Hall, *Handbook of Microstrip Antennas*, Peter Peregrinus, 1989.

[8] C. E. Santosa, J. T. Sri Sumantyo, K. Urata, C. M. Yam, K. Ito, and S. Gao, "Development of a Low Profile Wide-bandwidth Circularly Polarized Microstrip Antenna for C-band Airborne CP-SAR Sensor", *Progress in Electromagnetics Research*, vol. 81, pp. 77–88, Feb. 2018.

[9] C. E. Santosa, J. T. Sri Sumantyo, C. M. Yam, K. Urata, K. Ito, and S. Gao, "Subarray Design for C-band Circularly Polarized Synthetic Aperture Radar Antenna Onboard Airborne", *Progress in Electromagnetics Research*, vol. 163, pp. 107–117, Aug. 2018.

[10] S. Gao, Y. Qin, and A. Sambell, "Low-Cost Broadband Circularly Polarized Printed Antennas and Array", *IEEE Antennas and Propagation Magazine*, vol. 49 issue 4, pp. 57–64, Aug 2007.

[11] S. Gao, Q. Luo, and F. Zhu, "Circularly Polarized Antenna", *Willey-IEEE Press*, 2014.

[12] C. E. Santosa, J.T. Sri Sumantyo, S. Gao, and K. Ito, "Broadband Circularly Polarized Microstrip Array Antenna With Curved-Truncation and Circle-Slotted Parasitic", *IEEE Transactions on Antennas and Propagation*, vol. 69 issue 9, pp. 5524–5533, Sep 2021.

7 Ground Test, Airborne Installation, and Flight Mission

Ming Yam Chua, Cahya Edi Santosa, and Josaphat Tetuko Sri Sumantyo

7.1 INTRODUCTION

The Hinotori-C SAR hardware comprises a number of subsystems, as depicted in Figure 7.1. This chapter discusses testing the SAR sensor for performance validation and integrating the SAR system with the aircraft for flight tests.

7.2 HINOTORI-C SAR SUBSYSTEMS TEST

The following subsections detail the test setup, procedure, results, and analysis for the various hardware subsystems of the Hinotori-C SAR system [1]–[2].

7.2.1 CLOCK

All subsystems in the Hinotori-C SAR use the common clock from the local oscillator unit (LOU) to retain the subsystem's coherency. The two clock variants generated in LOU are STALO and 10MHz reference clocks. These clocks must be exact and stable (low jitter), and have low noise. These specifications can be measured using a spectrum analyzer or an oscilloscope (for low-frequency clocks only), as depicted in Figure 7.2.

Figures 7.3(a) and 7.3(b) show the measured 10MHz sinusoidal reference clock in the time domain and frequency domain, respectively. The measured time-domain waveform shows a consistent period (low jitter) across the cycles and has an almost identical shape with every cycle (low noise). It also can be validated by looking at its spectrum, in which a low jitter clock exhibits low-phase noise (sharper peak at its fundamental frequency component), and a low noise clock will have low amplitude spurious effects (> 35dBc). Figures 7.4(a) and 7.3(b) show the measured 10MHz rectangular reference clock. Similarly, it shows the clock is low in jitter and noise. The high amplitude odd-order harmonics in the measured spectrum are the harmonics that shape the rectangular waveform.

On the other hand, Figure 7.5 shows the measured spectrum of STALOs, which were tuned to 1,550MHz (LO1) and 3,750MHz (LO2) after the maiden flight. Both

DOI: 10.1201/9781003282693-7

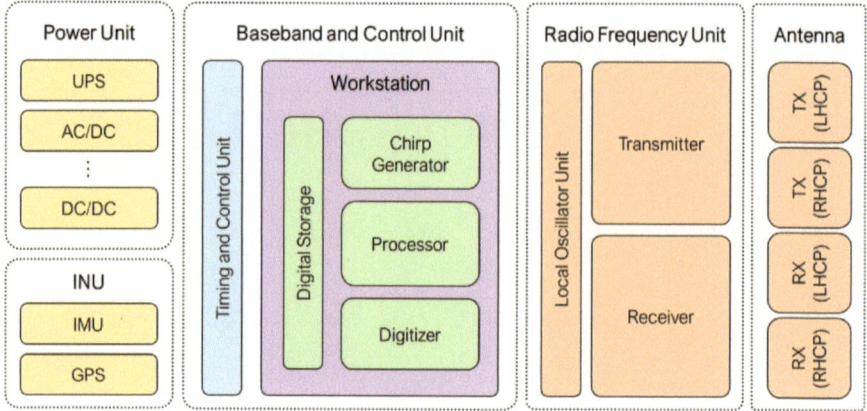

FIGURE 7.1   Subsystems in the Hinotori-C SAR.

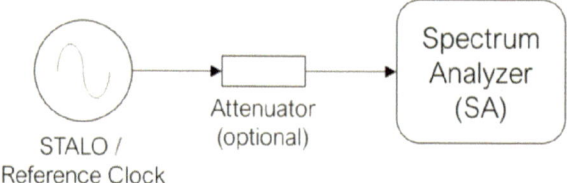

FIGURE 7.2   Clock measurement setup.

spectrums show low phase noise and low jitter. Some spurious effects appeared in the spectrum of LO1. However, its relative magnitude to the carrier is greater than 55dBc and would not affect the operation of the subsystems.

7.2.2   RF TRANSMITTER

The primary function of the RF transmitter is to generate the required microwave signal. The transmitter modulates and upconverts the baseband waveform to a suitable frequency and then amplifies it to an appropriate power level for transmission. Hence, the following tests and measurements are necessary.

7.2.2.1   Output Power Measurement

SAR has a stringent requirement on the power level of its transmitting signal. Considering the SAR geometry of operation specifications, the minimum required power level is determined during the SAR system-level design, and the SAR transmitter will be built based on it. The output power level of the transmitter must be validated every time before the flight test to ensure it is within the desired level.

The transmitting signal's power level can be measured using either a spectrum analyzer or a power meter. Since the transmitting power is usually high, ranging

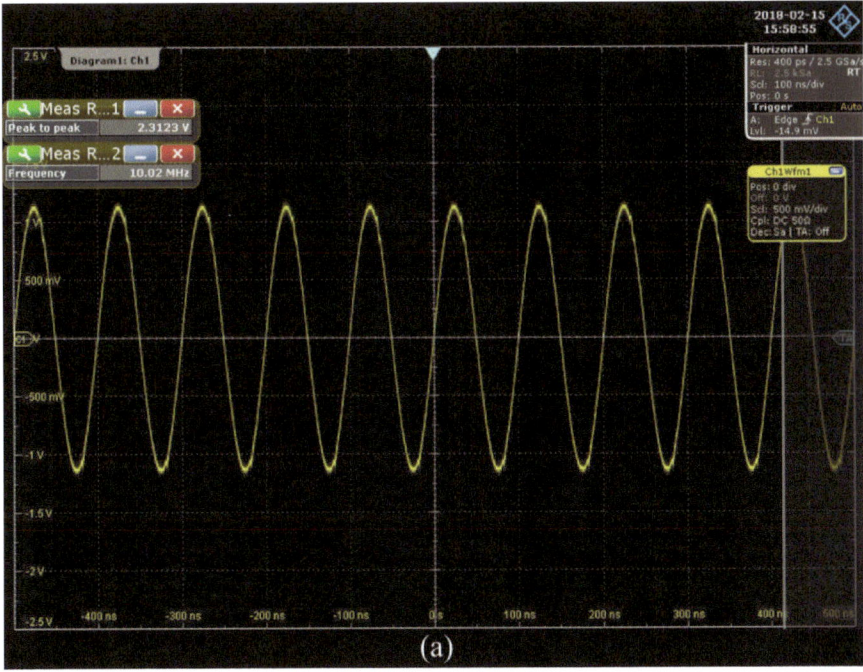

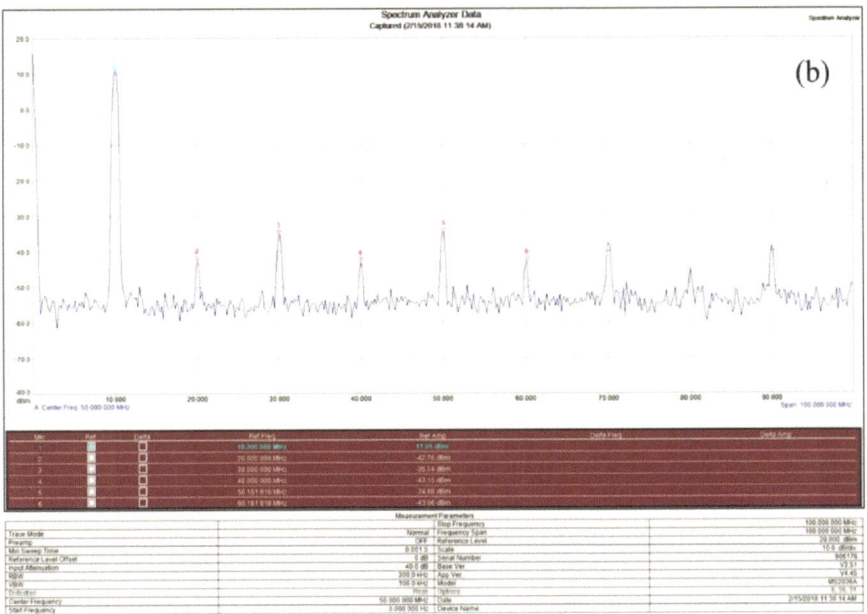

FIGURE 7.3 Measured 10MHz reference clock (sinusoidal shape); (a) time domain, (b) spectrum.

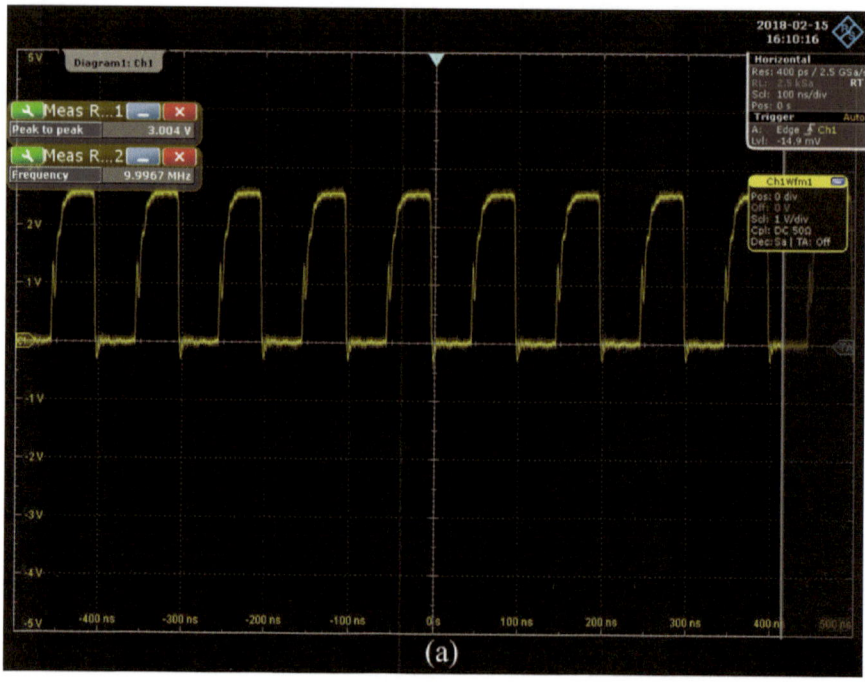

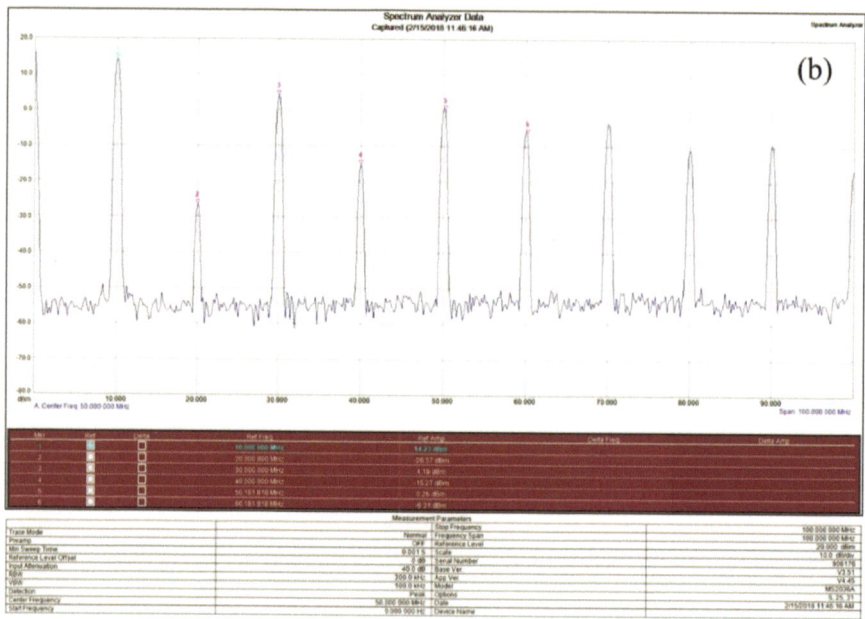

FIGURE 7.4 Measured 10MHz reference clock (rectangular shape); (a) time domain, (b) spectrum.

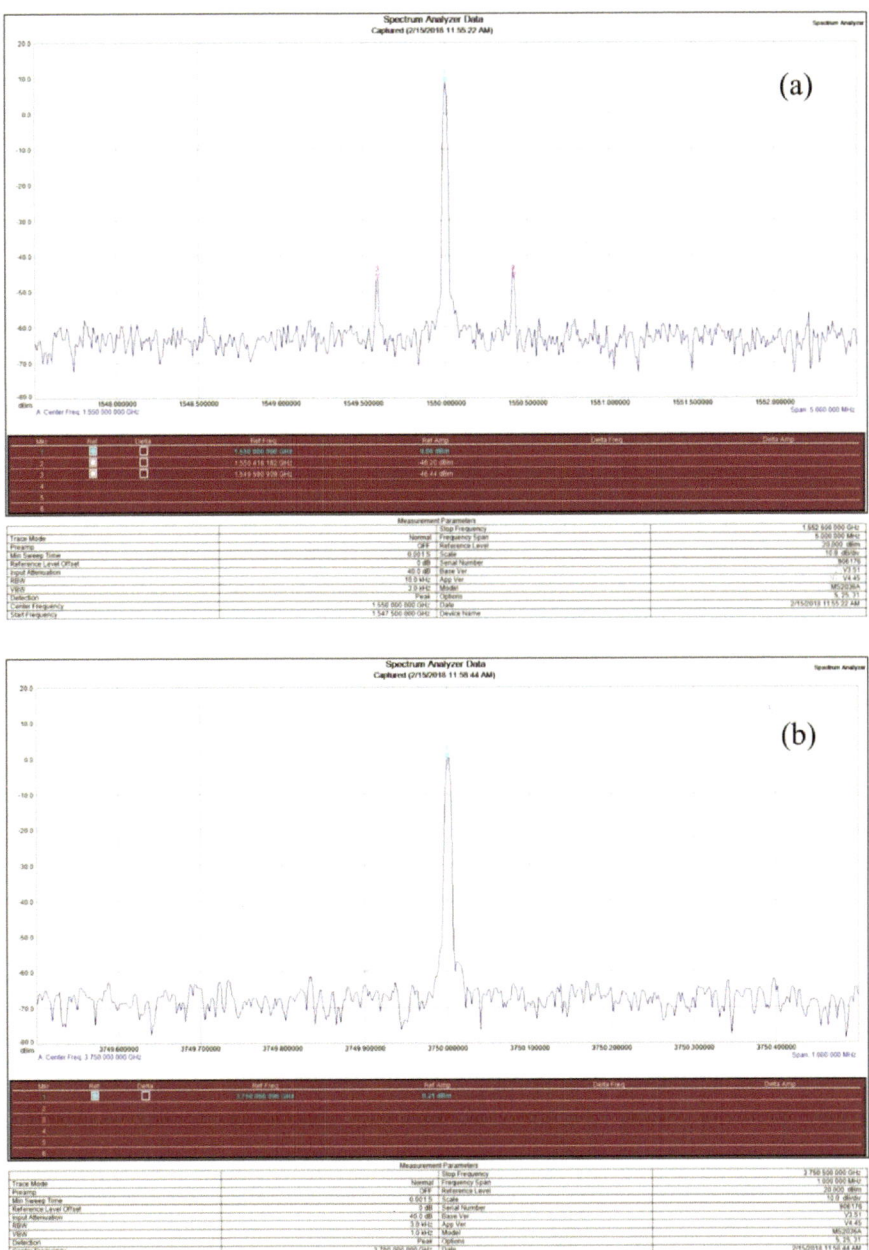

FIGURE 7.5   Measured STALO outputs; (a) LO1 (adjusted to 1,550 MHz), (b) LO2 (adjusted to 3,750MHz).

from 100W to 2,000W, it cannot be measured directly, and its power level must be attenuated below the equipment's maximum RF input signal level. The required attenuation can be obtained by inserting some suitable attenuators at the transmitter's output, as depicted in Figure 7.6(a). However, care must be taken when selecting these attenuators to ensure the attenuator not only gives the required amount of attenuation but can also handle the input power level without damaging itself.

In one of the output power measurements for the RF transmitter of the Hinotori-C SAR, three 50W attenuators (i.e., 40dB, 30dB, and 20dB) were inserted into the signal chain to attenuate the 400W (~56dBm) output signal to around -30dBm, which is safe to measure by the spectrum analyzer. When calculating the transmitter's output power, the losses of the connecting cables and adaptors must be measured beforehand and considered. Figure 7.6(b) shows the peak power measured by the spectrum analyzer, and Table 7.1 shows the output power calculation estimating the output power of 509.3W (+57.07dBm).

When a spectrum analyzer is used to measure the peak power of RF pulses, the envelope detector in the spectrum analyzer detects the peak amplitude of a

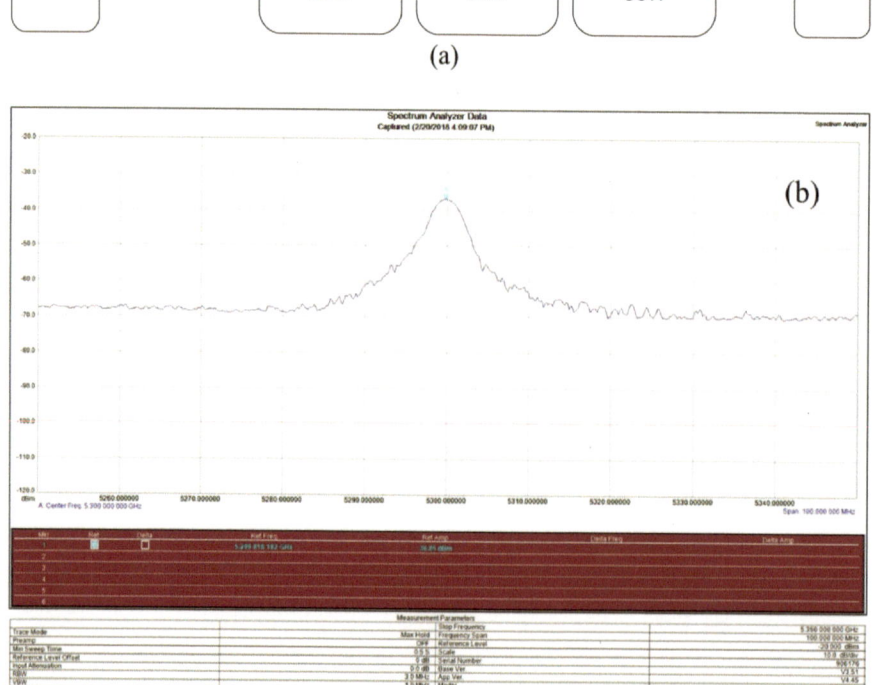

FIGURE 7.6 Transmitter's output power measurement; (a) measurement setup, (b) measured power level.

TABLE 7.1

Transmitter's Output Power Calculation (Measured Using Spectrum Analyzer)

Component		Value
Chain Loss	Cable A	-2.30 dB
	Cable B	
	Adaptor A	
	Attenuator A	-40.32 dB
	Attenuator B	-30.58 dB
	Attenuator C	-20.72 dB
	Total	-93.92 dB
Measured output power		-36.85 dBm
Calculated output power		+57.07 dBm
		509.33 Watt

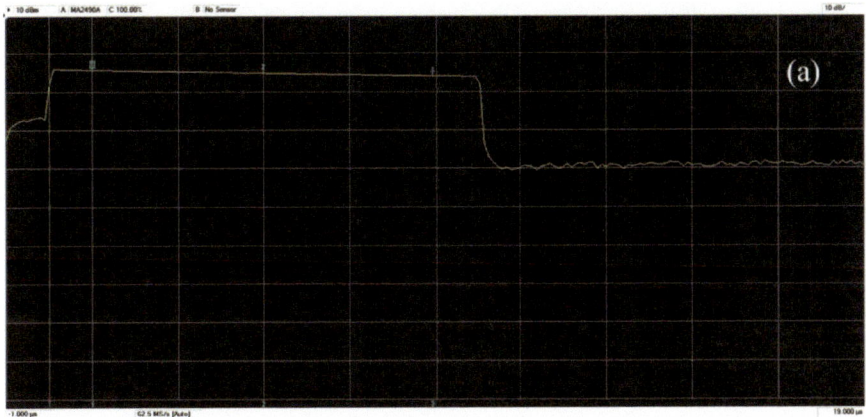

FIGURE 7.7 RF pulse power measured using power meter and power sensor; (a) chirp pulse of 100MHz bandwidth and 10µs pulse width, (b) chirp pulse of 200 MHz bandwidth and 10µs pulse width, (c) chirp pulse of 400MHz bandwidth and 10µs pulse width.

bucket of downconverted IF signals, and it will not show any timing information of the RF pulses. Some RF power meters can measure the RF pulses' peak power and display their timing information. Figure 7.7 shows the snapshots of the chirp pulse with various chirp bandwidths measured using the Anritsu ML2438A Power Meter. The calculation of the transmitter's output power is summarized in Table 7.2.

7.2.2.2 Bandwidth Measurement

In SAR, the slant range resolution is proportional to the bandwidth of its transmitting chirp pulse. A spectrum analyzer can efficiently measure the bandwidth of the chirp pulse (at carrier frequency). Similarly, since the transmitting power is too large compared to the power level the spectrum analyzer can take in, the signal must

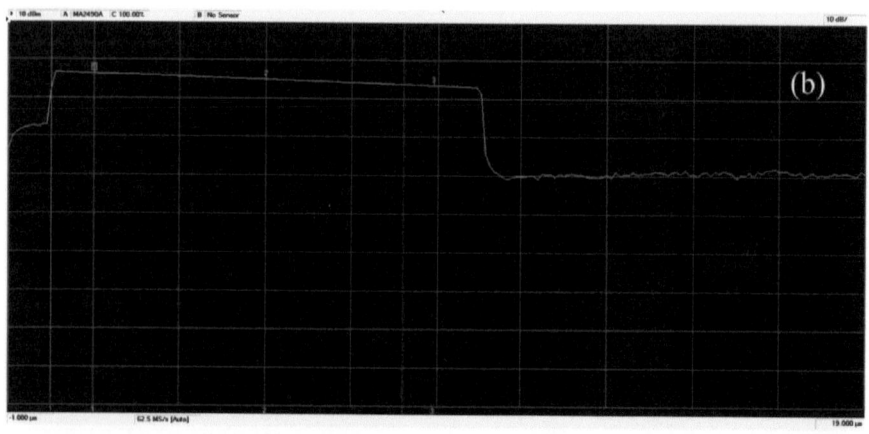

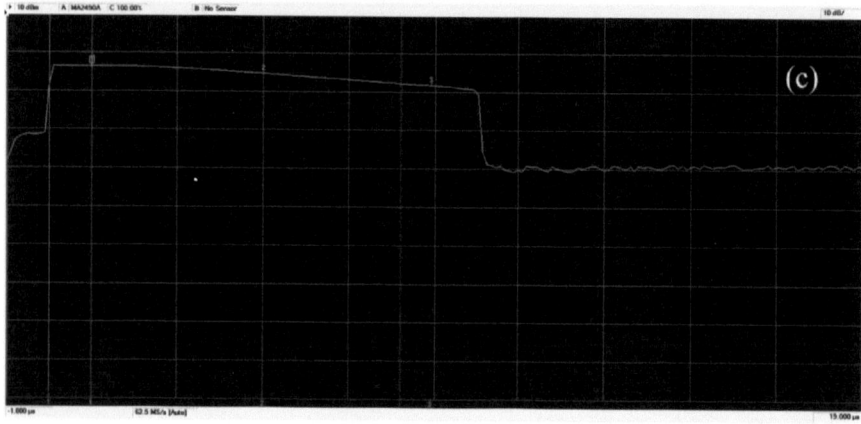

FIGURE 7.7   (Continued)

TABLE 7.2

Transmitter's Output Power Calculation (Measured Using Power Meter)

Chirp Parameters (Bandwidth/Pulse Width)	f (GHz)	P_m (dBm)	L_T (dB)	P_{tx} (dBm)	P_{tx} (Watt)
100 MHz/10 μs	5.26	-4.38	61.80	57.42	552.08
	5.30	-5.18	61.80	56.62	459.20
	5.34	-5.96	61.80	55.84	383.71
200 MHz/10 μs	5.22	-3.73	61.80	58.07	641.21
	5.30	-5.09	61.80	56.71	468.81
	5.38	-6.61	61.80	55.19	330.37
400 MHz/10 μs	5.14	-3.48	61.80	58.32	679.20
	5.30	-5.37	61.80	56.43	439.54
	5.46	-8.39	61.80	53.41	219.28

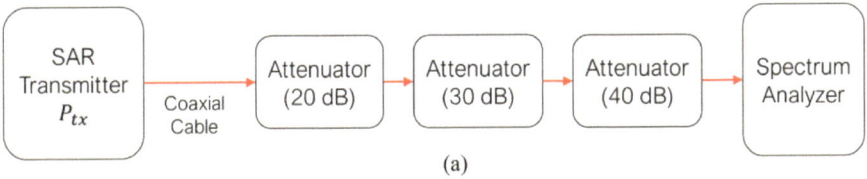

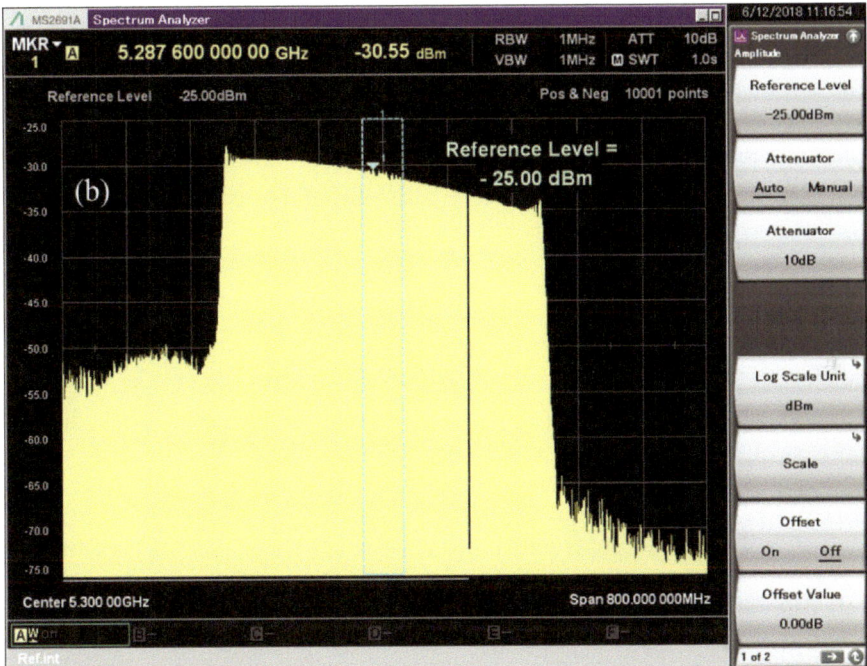

FIGURE 7.8 Bandwidth measurement using spectrum analyzer (chirp with 400MHz bandwidth and 10μs pulse width); (a) measurement setup, (b) chirp spectrum.

be attenuated to an appropriate power level before feeding it into the equipment. Figure 7.8(a) shows the measurement setup for bandwidth measurement using a spectrum analyzer. The measured spectrum in Figure 7.8(b) shows that the transmitting chirp pulse has a bandwidth of approximately 400MHz.

7.2.3   RF RECEIVER

The primary function of the RF receiver is to detect weak echoes. The receiver amplifies the echo to an appropriate power level and downconverts it before digitizing the signal [3]–[5]. The following tests and measurements are necessary to confirm the performance of a receiver.

7.2.3.1   1dB Compression Point

One of the primary functions of a receiver is to amplify the received weak signal before the subsequent hardware processes it. The amplifiers that amplify the weak signal in the receiver work in two regions, namely the linear region and the compression region. The linear region of operation is where the gain is constant and independent of the input power level. On the other hand, the compression region is where the gain compression occurs. It happens when the input power of an amplifier is increased to a level that reduces the gain of the amplifier and causes a nonlinear increase in output power. The border separating these two regions is called the 1dB compression point, or P1dB.

It is crucial to ensure the signal's power level applied to the amplifier is below P1dB to avoid saturation because a saturated amplifier will not have the expected gain and produces more spurious effects. P1dB can be determined from the gain compression plot, which could be measured by applying a swept power signal (P_{in}) to the receiver starting from the lowest power level, record the output power level (P_{out}), and repeat the measurement by increasing the power level at the step size of 1dB until the P1dB point is found. Figure 7.9(a) illustrates the measurement setup to determine the P1dB of the Hinotori-C SAR receiver, and Figure 7.9(b) plots the gain compression of the receiver.

7.2.3.2   Receiver Dynamic Range and Bandwidth

The dynamic range of a receiver is essentially the range of signal levels over which it can operate. The low end of the range is governed by its sensitivity, while the high end is the receiver's gain compression point. The same setup used for measuring P1dB can also measure the receiver's dynamic range, but the output power has to be converted to gain. Figure 7.10 plots the dynamic range of the Hinotori-C SAR receiver. In the plot, the upper limit of the receiver's dynamic range is the 1dB compression point.

In contrast to the receiver's dynamic range, the receiver bandwidth is the range of frequencies it opens for reception. The receiver bandwidth is usually bounded to its front-end filter's upper and lower cut-off frequencies to limit the noise bandwidth and suppress the out-of-band interference signal detected by the receiving antenna. The receiver bandwidth can be measured using a similar setup illustrated in Figure 7.9(a), but sweeping for the frequency instead of the power level in this case. Figure 7.11 plots the receiver's gain versus frequency. The plot shows that the receiver's reception is in the frequency range of 5,100MHz to 5,600MHz.

7.2.3.3   Noise Floor Measurement

The receiver's noise floor is the sum of all the noise sources and unwanted signals produced within the receiver. It can be plotted by terminating the receiver's input with a 50Ω terminator and obtaining the spectrum of its output. Figure 7.12(a) illustrates the measurement setup to plot the noise floor of the Hinotori-C SAR receiver, while Figure 7.12(a) shows the measured noise spectrum. From the plot, the receiver's noise floor was approximately at the power level of -90dBm, with some noticeable weak spurious effects at frequencies of 150MHz and 250MHz.

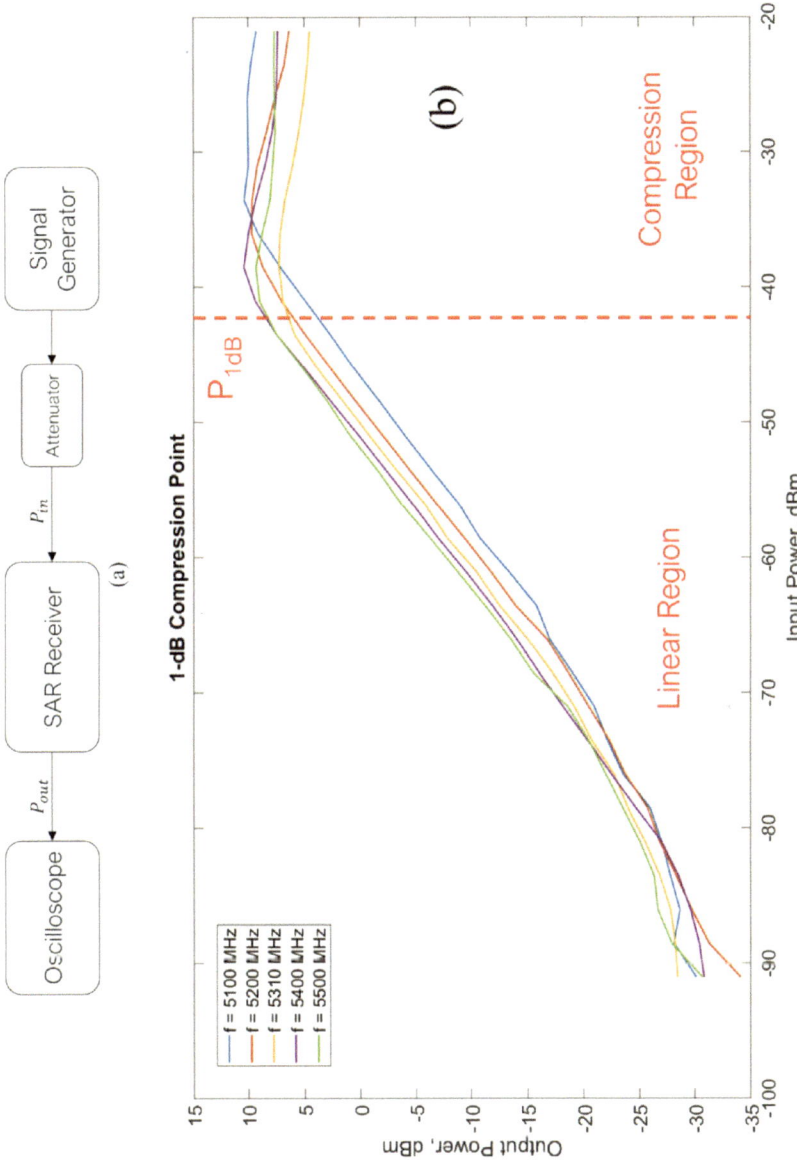

FIGURE 7.9 1dB compression point of Hinotori-C SAR receiver; (a) measurement setup. (b) P_{out} versus P_{in}.

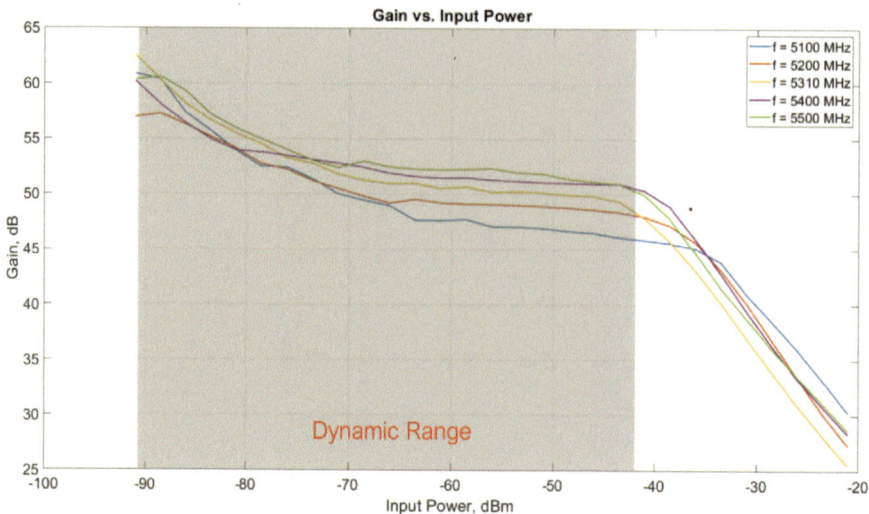

FIGURE 7.10 Dynamic range of Hinotori-C SAR receiver.

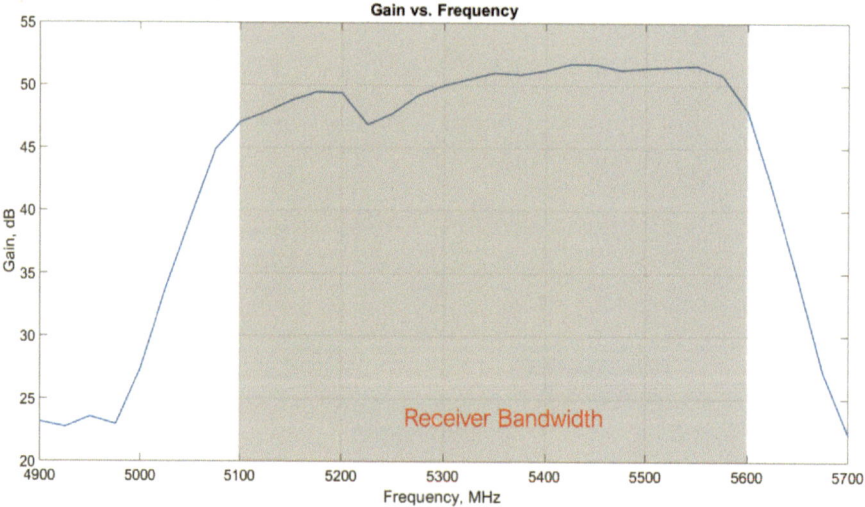

FIGURE 7.11 The Hinotori-C SAR receiver bandwidth.

7.2.4 CHIRP GENERATOR AND TIMING AND CONTROL UNIT

The signal produced by the chirp generator is a baseband signal. Similarly, the timing signals are low-frequency control signals [6]–[7]. The signal's specifications – such as the bandwidth, pulse width, PRF, and peak amplitude – have to be validated before connecting them to other subsystems. These signals can be plotted in the time domain using a high-speed oscilloscope. Figure 7.13 and Figure 7.14 show the plotted chirps

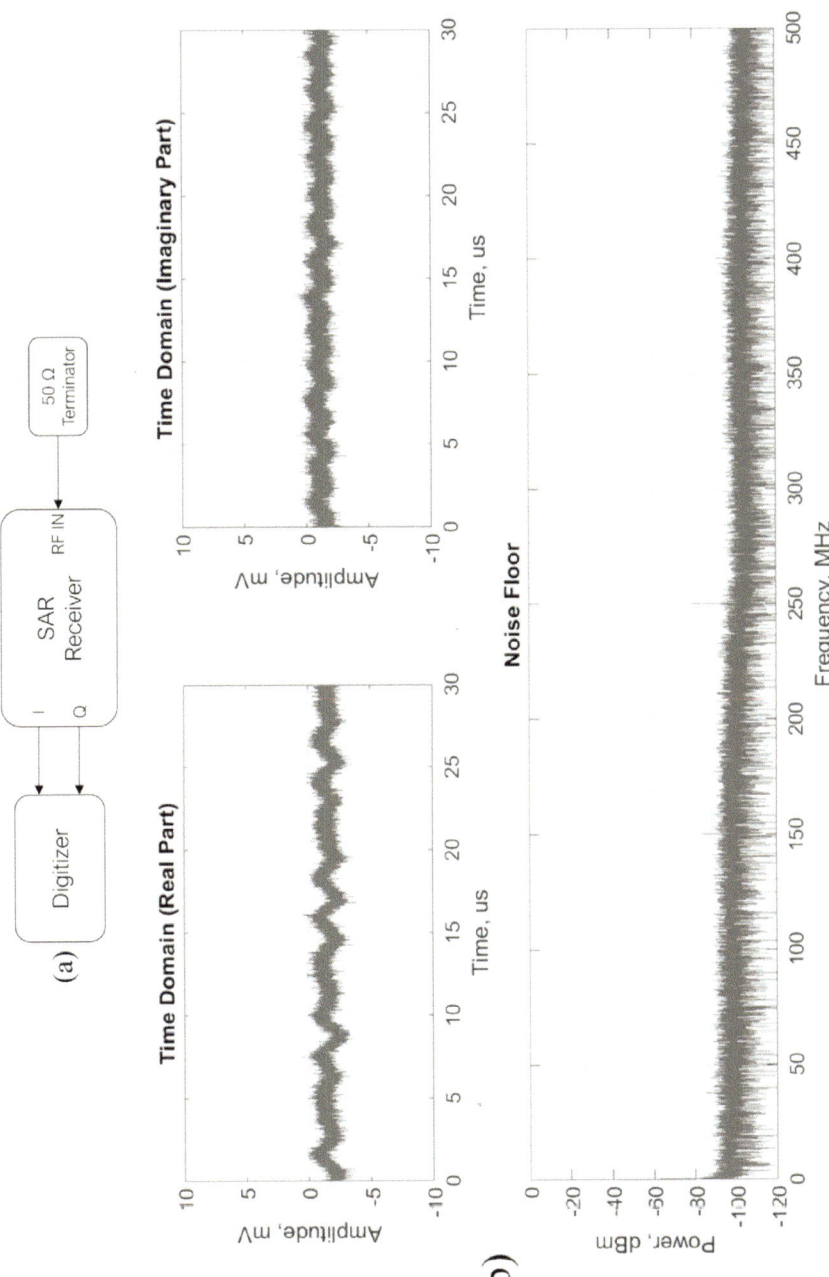

FIGURE 7.12 Receiver noise floor measurement; (a) measurement setup, (b) receiver noise floor.

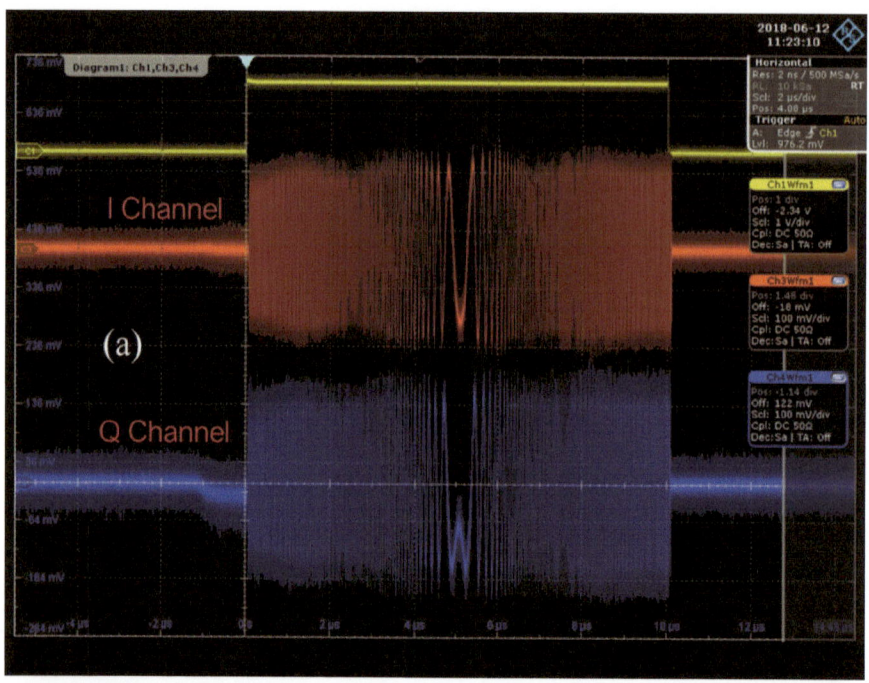

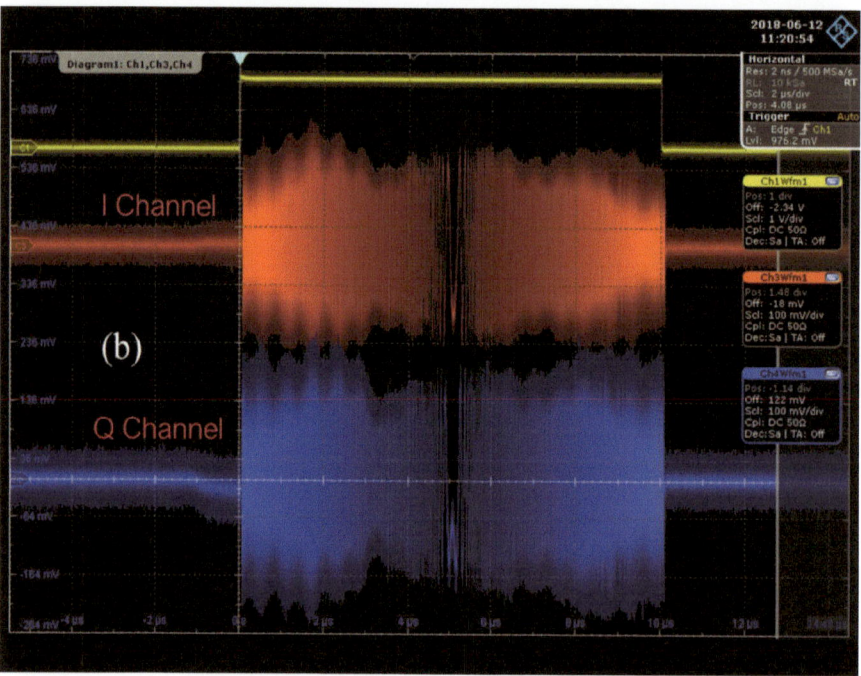

FIGURE 7.13   Measured chirp signals; (a) 100MHz bandwidth, 10μs pulse width, (b) 400MHz bandwidth, 10μs pulse width.

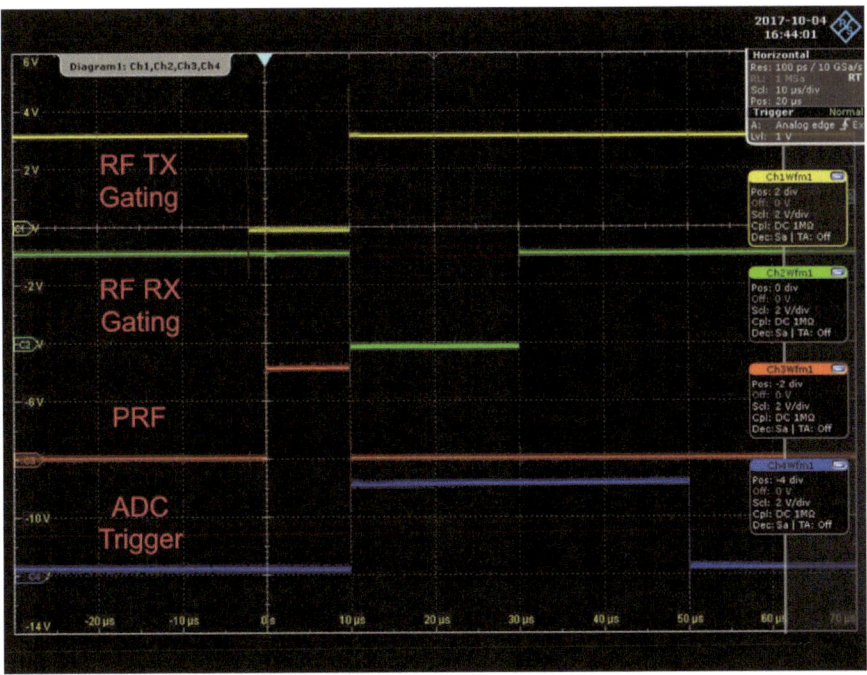

FIGURE 7.14 The measured control signals; RF transmitter gating signal (yellow), RF receiver gating signal (green), PRF pulse (orange), ADC trigger signal (blue).

and control signals using Rohde & Schwarz RTO 1044 high-speed oscilloscope. The baseband spectrum, which can be used to check the chirp bandwidth, cannot be plotted directly by the oscilloscope. However, it can be plotted easily using numerical analysis software from the recorded waveform data.

7.2.5 DATA ACQUISITION SYSTEM

The data acquisition system functions to record the baseband echoes from the receiver and store the data in storage for offline processing. The minimum sampling rate of the digitizer must be at least twice the highest frequency of the echoes and use the control signal from the TCU to trigger the acquisition [8]–[9]. The data acquisition system can be tested quickly by sampling the baseband chirp from the chirp generator. Figure 7.15 shows a sample plot of the recorded data.

7.3 SYSTEM INTEGRATION AND FULL SYSTEM TEST

After the functionality of each subsystem was verified, all subsystems were integrated into a complete SAR system. Figure 7.16 shows the complete Hinotori-C SAR system and the electrical connections among them. An integrated user control software was developed and installed in the SAR workstation to control and monitor the SAR system. Figure 7.17 shows a screenshot of the user control software.

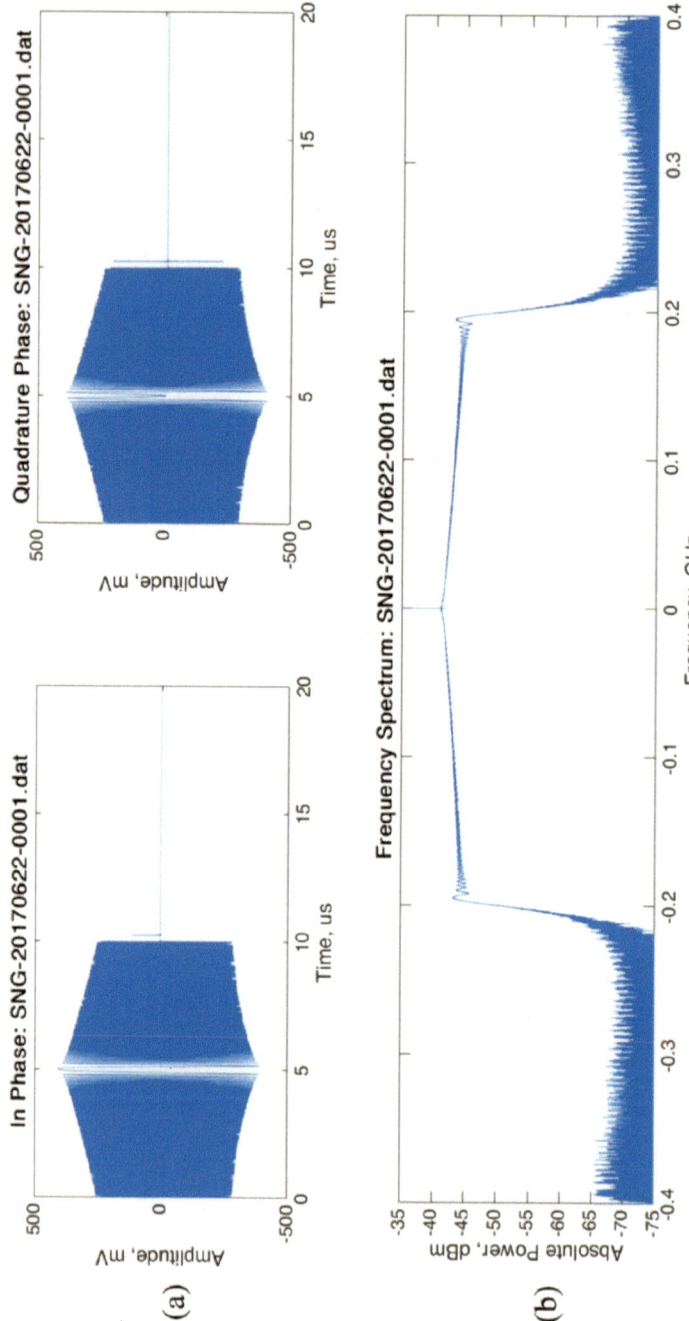

FIGURE 7.15 The recorded data: (a) time domain plot, (b) spectrum, (c) auto-correlation plot.

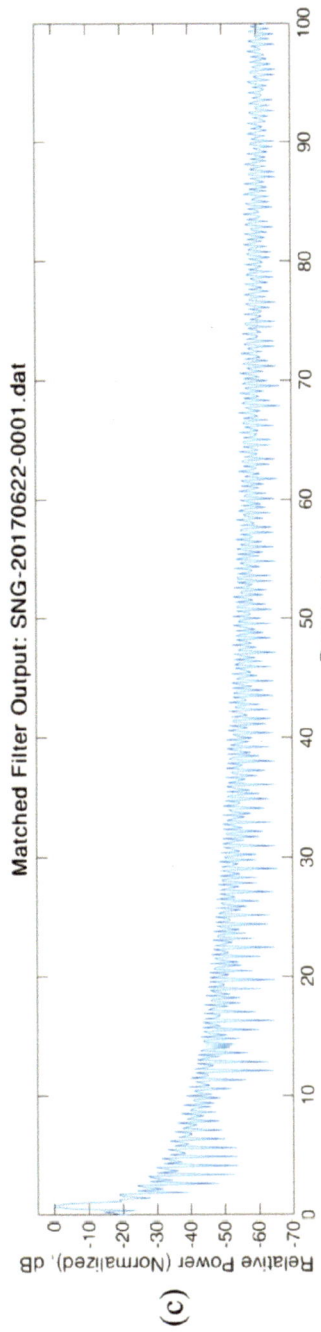

FIGURE 7.15   (Continued)

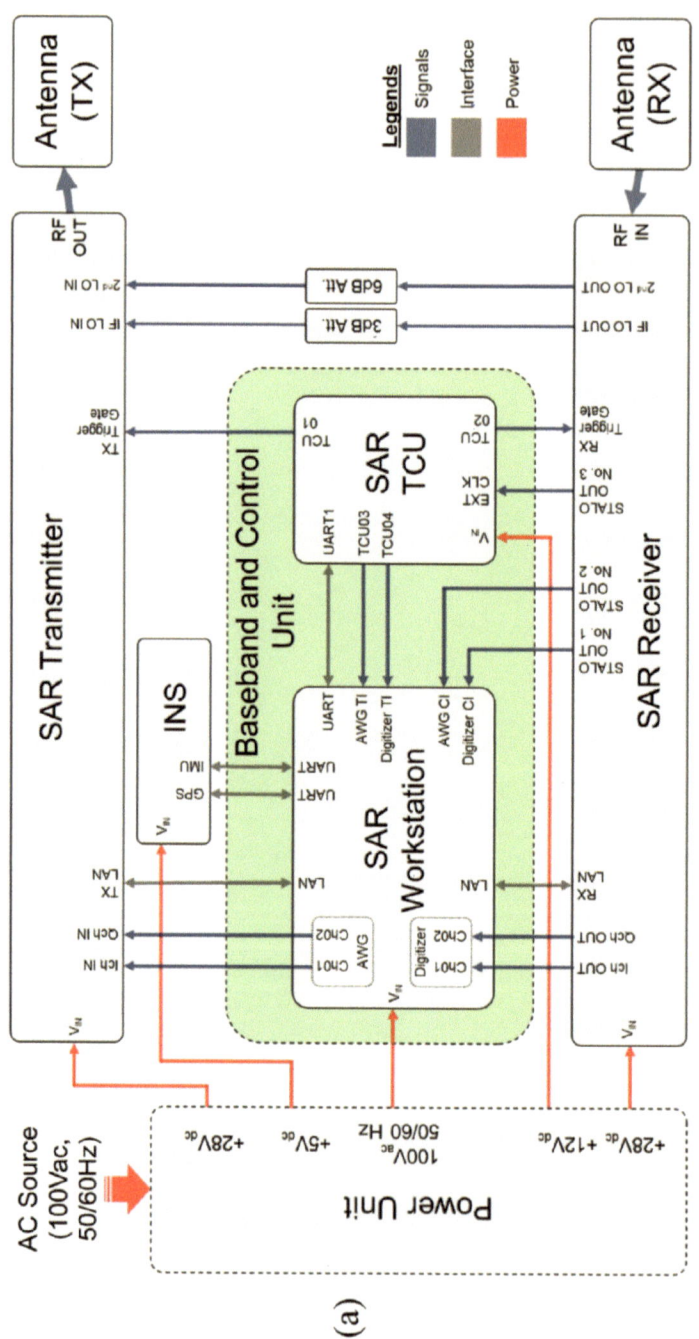

FIGURE 7.16 The integrated Hinotori-C SAR; (a) electrical connection, (b) antenna subsystem, (c) SAR electronics, (d) inertia navigation system.

(b)

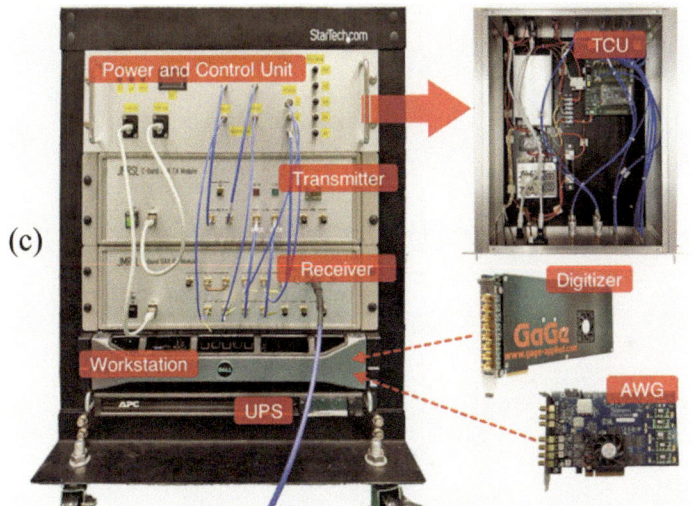

(c)

(d)

FIGURE 7.16 (Continued)

FIGURE 7.17 The Hinotori-C SAR user control software.

After the system integration, some full system tests must be carried out to confirm the functionality of the complete SAR system. Two main tests must be performed – the full system loopback test and the ground test.

7.3.1 FULL SYSTEM LOOPBACK TEST

A loopback test is an evaluation in which a signal is sent from the system and returned (looped back) to it to determine whether the system is working correctly. In a loopback test, the transmitter's output is connected to the receiver's input so that the sent-out chirp will become a hardware-emulated replica and be received by the receiver.

The primary goal of the loopback test for an SAR system is to verify the system's ability to work seamlessly during the mission, which is: (i) to produce the desired baseband chirp, (ii) to receive, amplify, and demodulate the received chirp replica, and (iii) to digitize and record the stream of chirp replicas into the onboard storage without a miss. Besides that, the loopback test also helps to verify the system's triggering mechanism and the applicability of the timing applied to each subsystem.

Figure 7.18 shows the setup of the loopback test performed on the Hinotori-C SAR. Since the transmitter generates approximately 400W (~ +56dBm) peak power chirp pulses and the compression gain of the receiver lies around -42dBm, looping the high-power pulse directly into the receiver will cause the receiver to saturate, or in the worst case, permanently damage the receiver. In the test, the added 110dB attenuators (a cascaded 40dB, 30dB, 20dB, and 20dB attenuator) between the transmitter and the receiver weaken the transmitter's output signal to an appropriate level before it goes into the receiver.

A looped back chirp replica is a time-shifted and amplitude-weighted chirp. Figure 7.19 shows a snapshot of some recorded looped-back replicas using an oscilloscope with various bandwidths. The total time shift in the replica due to the delay in the attenuators is shown in Figure 7.19(d). On the other hand, Figure 7.20 plots the processed replicas for various bandwidths and pulse widths using matched filtering processing. The purpose is to verify that the recorded signal can be processed for range compression. The plot shows that the matched filter detected a target (from the replica) at 13m emulated from the delay from the attenuators. The actual system's

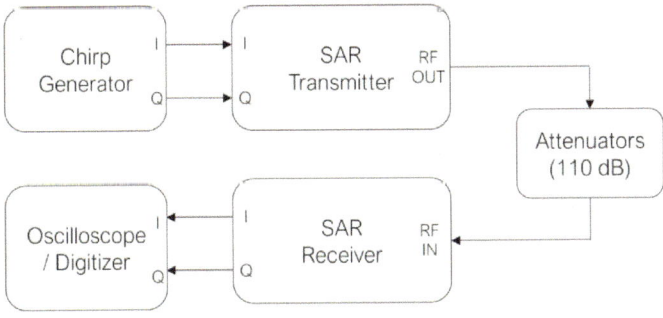

FIGURE 7.18   The loopback test measurement setup.

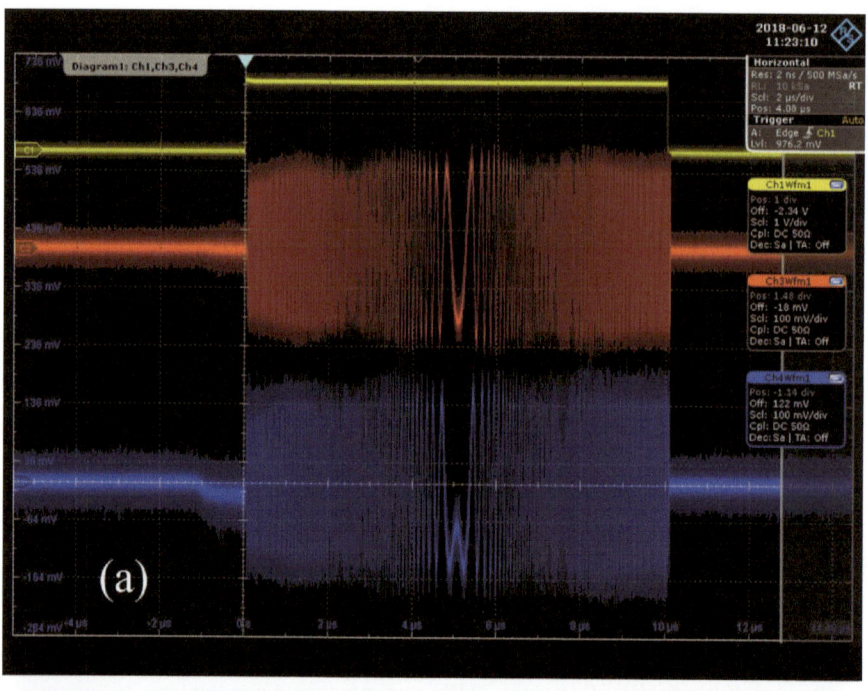

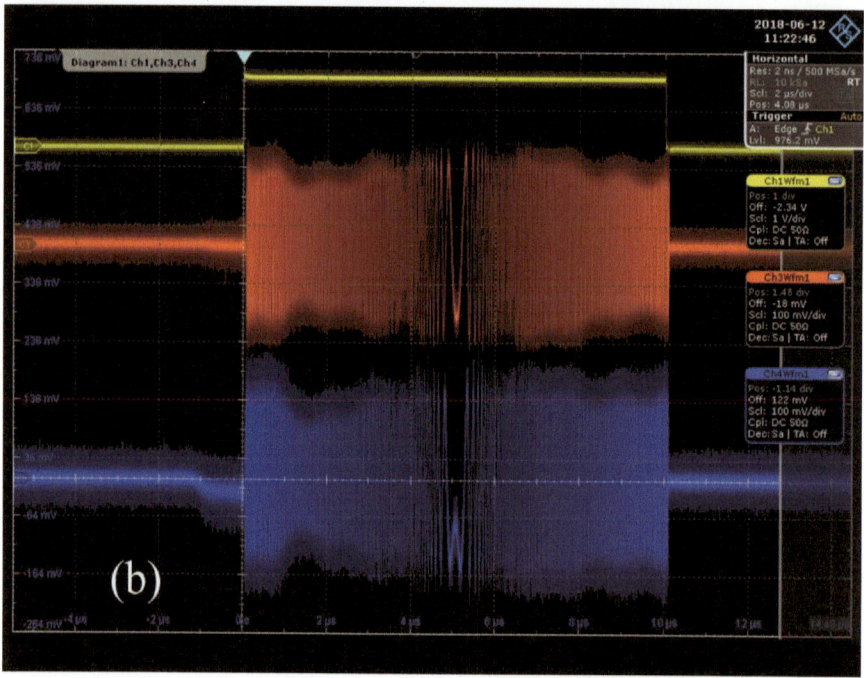

FIGURE 7.19 The downconverted chirp from loopback test (yellow: PRF timing, orange: I channel, blue: Q channel); (a) 100MHz bandwidth 10μs pulse width, (b) 200MHz bandwidth 10μs pulse width, (c) 400MHz bandwidth 10μs pulse width, (d) a zoom in view.

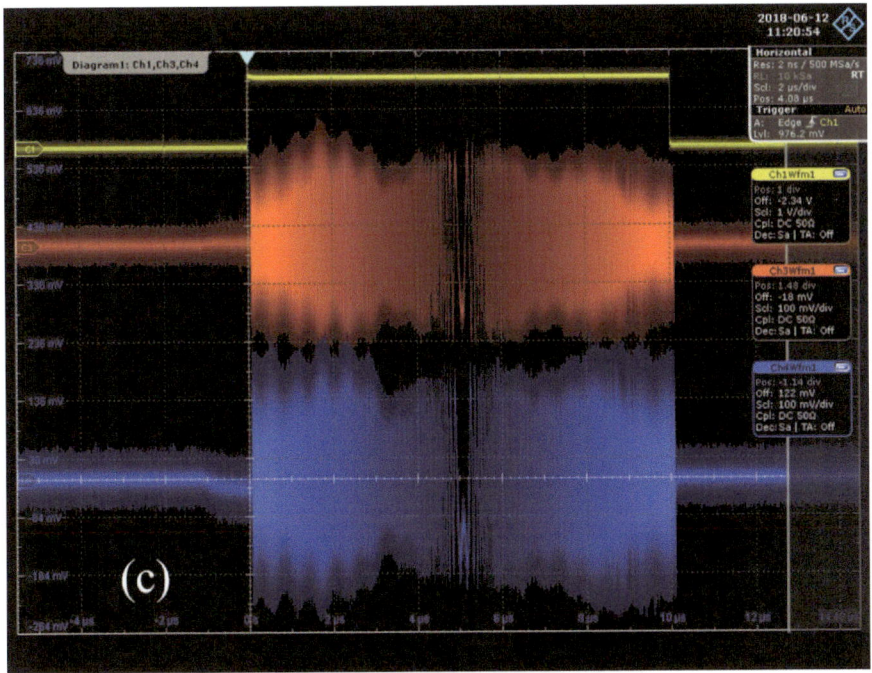

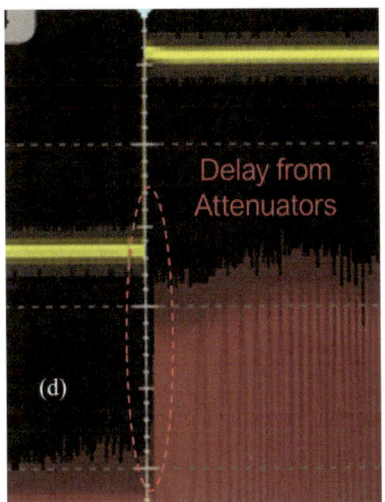

FIGURE 7.19   (Continued)

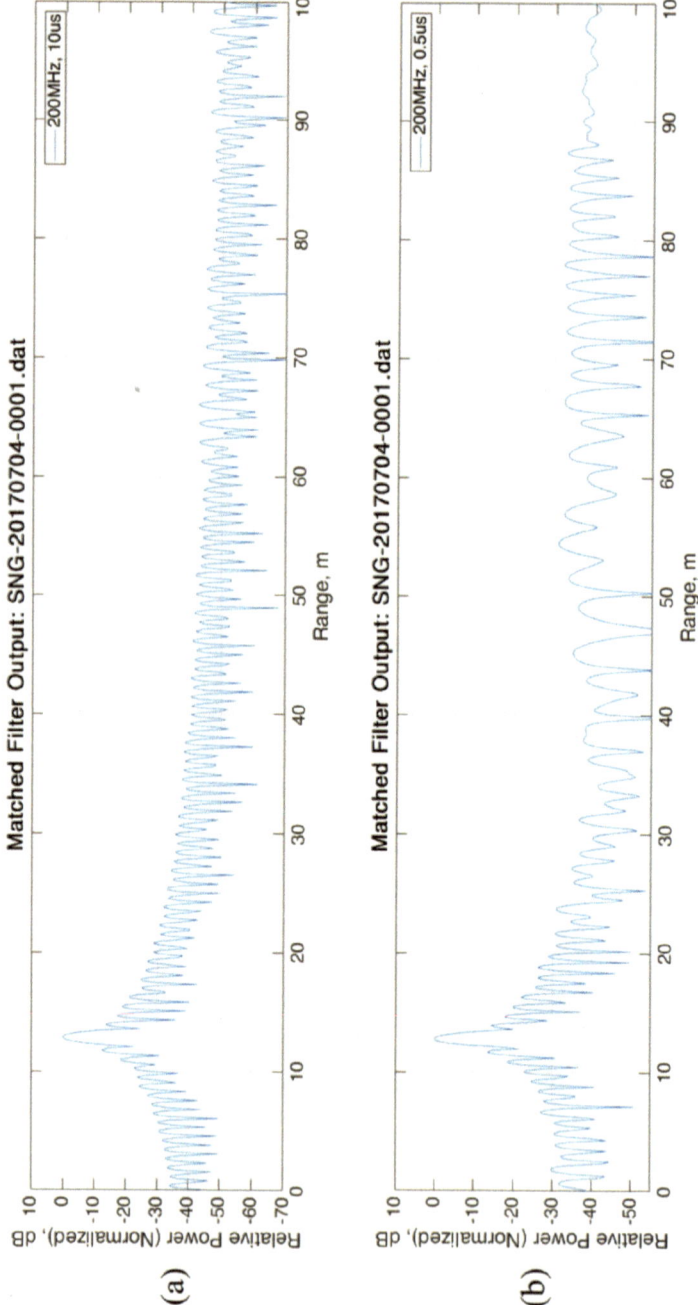

FIGURE 7.20 The matched filtered chirp; (a) 200MHz bandwidth 10μs pulse width, (b) 200MHz bandwidth 0.5μs pulse width, (c) 400 MHz bandwidth 10μs pulse width, (d) 400MHz bandwidth 0.5μs pulse width.

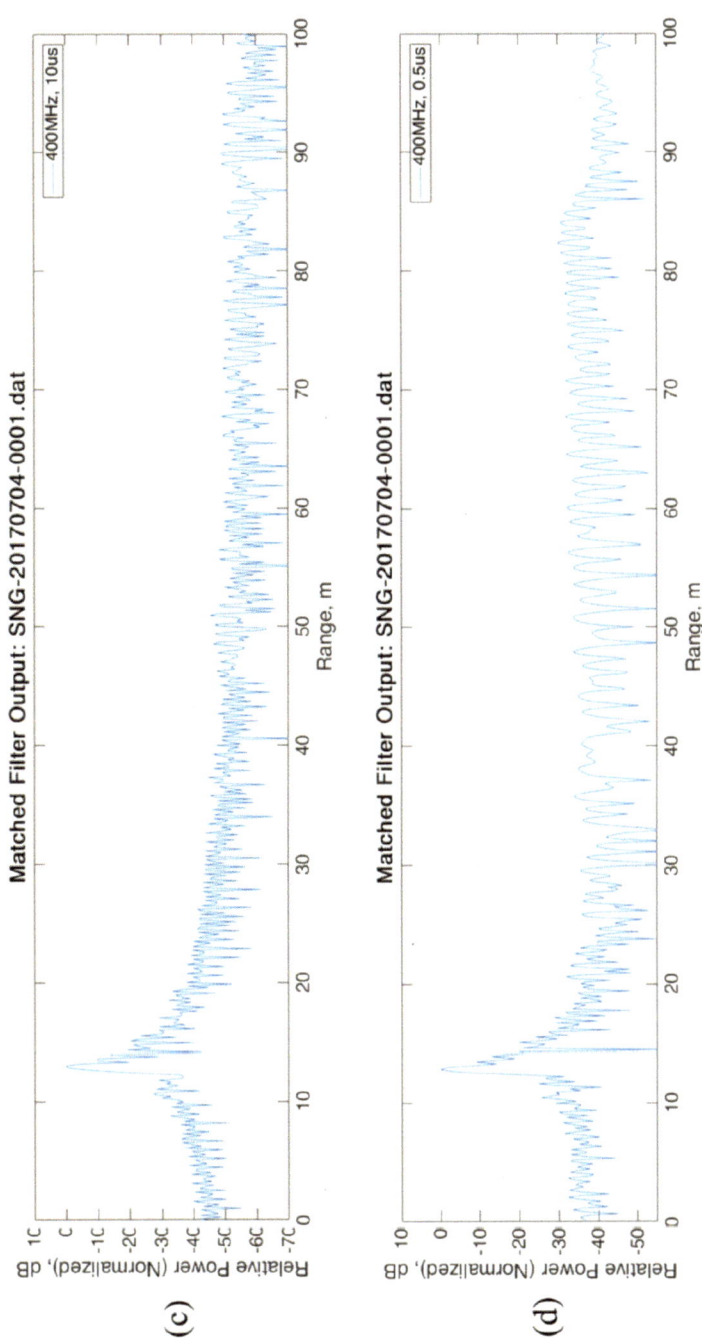

FIGURE 7.20 (Continued)

range resolution can be measured from the Impulse Response Width (IRW) of the impulse in the plots. The plots show that the IRW will be narrower at higher bandwidth resulting in finer range resolution. It also shows that the range resolution of the system is independent of the pulse width of the chirp.

7.3.2 GROUND TEST

A system loopback test can only validate the SAR system's functionality on range domain processing for a point target. The test alone will not prove that the system works well for distributed target detection in range and azimuth processing (azimuth compression). With these uncertainties, it would be too risky to mount the system on an aircraft for a flight mission.

A number of outdoor ground tests were conducted for the Hinotori-C SAR to address these uncertainties. Figure 7.21(a) depicts a panoramic view of the test site, while Figure 7.21(b) illustrates the test setup. In the test, both the SAR antenna was installed on a specially designed track robot and elevated 2m above the ground. The SAR electronics were put on a trolley because the total weight of the SAR system was

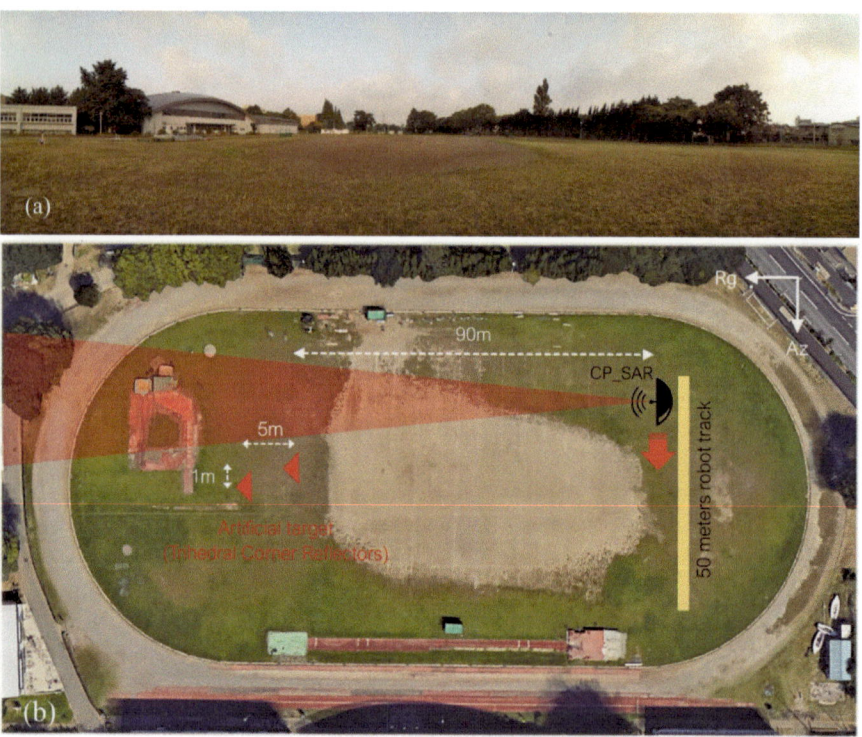

FIGURE 7.21 Ground test for the Hinotori-C SAR; (a) panoramic view of the test site, (b) test setup.

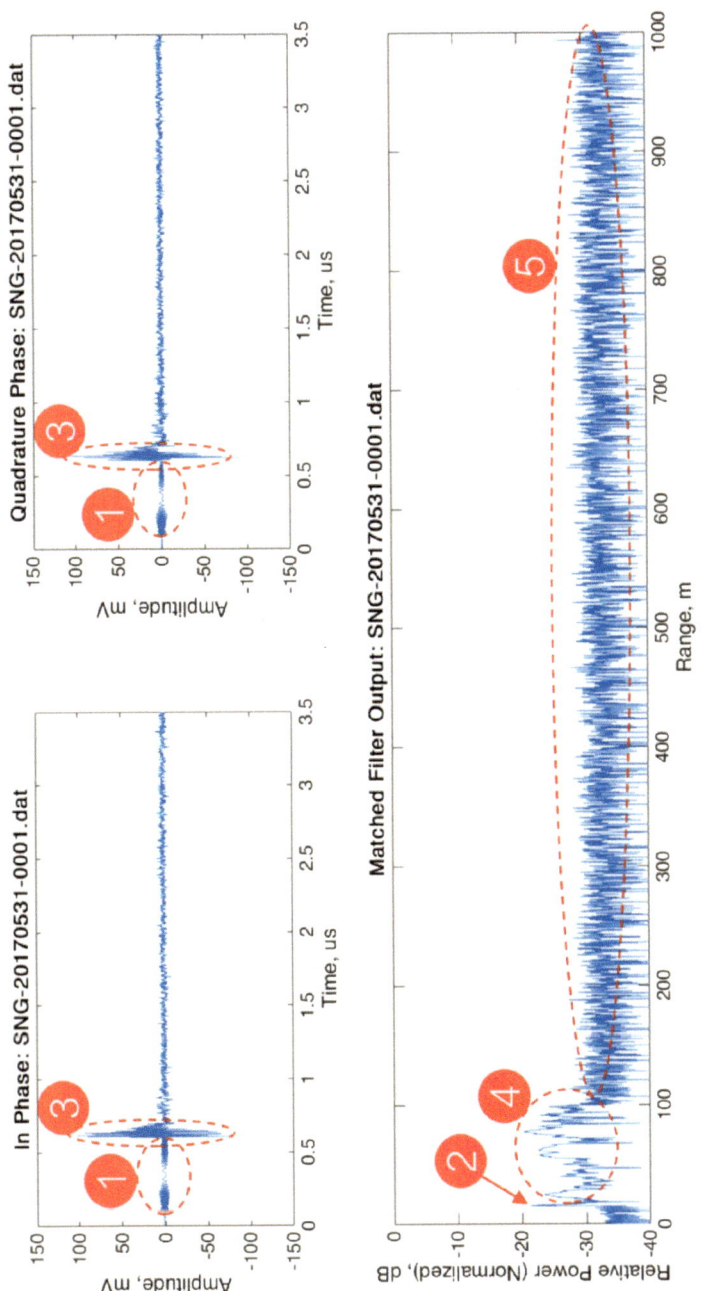

FIGURE 7.22 Noise floor measurement of complete SAR system; (a) recorded time domain signal (top), (b) range compressed signal (bottom)

about 90kg (including the server rack housing all the modules), which far exceeds the track robot's payload limit.

During the test, the entire system was placed on one side of the field (right side, as shown in Figure 7.21(b)) while the antenna was adjusted pointing to the other side. Two corner reflectors were installed 90m from the SAR systems, perpendicular to the robot's track. The corner reflectors were separated by a distance of 5m in the range direction and 1m in the azimuth direction. Since the synthetic targets were too close to the SAR system, the transmitter was configured to emit 5W, 0.5µs, and 200MHz chirp pulses.

7.3.2.1 SAR System (with Antenna) Noise Floor Measurement

The noise floor measured in Section 7.2.3 was not for the complete SAR system because the receiving antenna was not mounted to the receiver, and the emitting transmitter system (transmitting antenna and the RF transmitter) was not in its place. The noise floor for the complete system setup can only be measured during the ground test in which both the transmitting and receiving antenna must be pointed to a clear sky, emulating a scene without any targets.

Figure 7.22 shows the recorded data and its range compressed signal. The weak amplitude chirp replica at 0.1–0.6µs (labeled 1) was the result of the antenna's coupling, which also can be seen in the range compressed plot (labeled 2). The short burst of around 0.6µs (labeled ③) was caused by the internal reflection when the transmitter was being switched off while the receiving was re-enabled. The same internal signal's reflection in the system appeared at the range compressed signal plot in the same figure (labeled ④) at a microwave distance lower than 100m. Beyond 100m (labeled ⑤), the noise floor was flat and low in power, indicating that the noises picked up by the receiver were merely Gaussian noise.

7.3.2.2 Static SAR Distributed Target Test

When an SAR system operates during a mission, the system continuously emits the chirp at every pulse repetition interval (PRI). At the same PRI, the receiver collects and records a bin of echoes (range bin) returned from the distributed targets in the illuminating footprint of the transmitting antenna. A static SAR test scanning a scene with the distributed target can verify the system's functionality in this aspect.

With the antenna adjusted parallel with the ground, a range bin of echoes was recorded and analyzed during the ground test with the same test setup, except that the SAR system was moved to the center of the robot's track. The high radar cross section (RCS) distributed targets (labeled ①–⑦) along the illuminating beam, and their measured distance from the antenna is detailed in Figure 7.23(a). The processed range bin (distance uncalibrated) in Figure 7.23(b) shows that the SAR system "saw" these distributed targets, and the detected distance matched with the ground truth measured data. The strong chirp-alike signal at 1.1–1.6µs in the time domain plot was the coherent sum replica from the two corner reflectors.

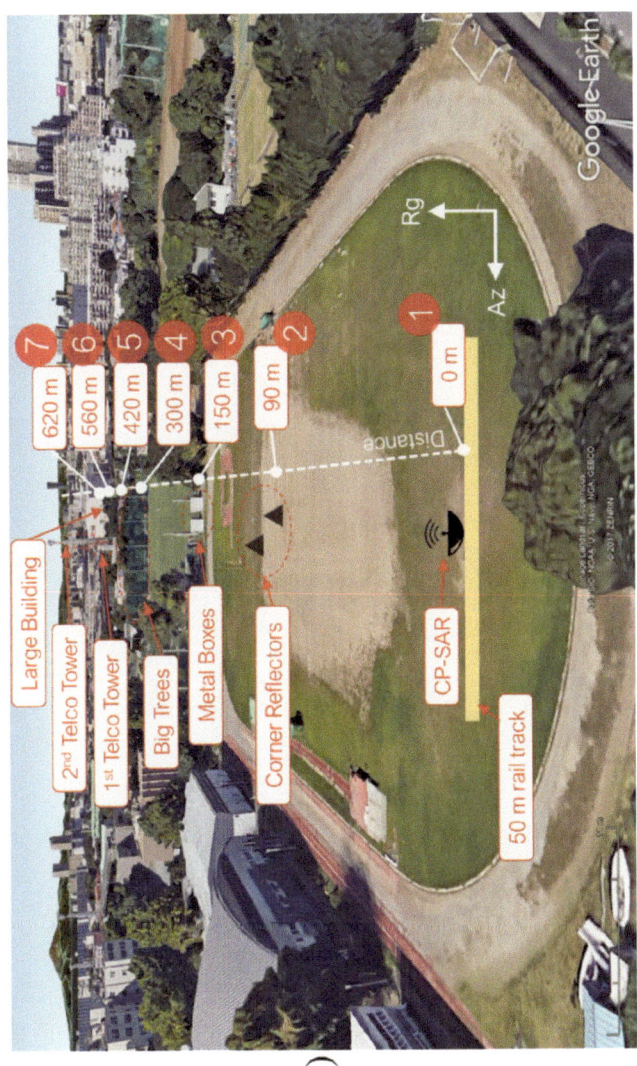

FIGURE 7.23 The Hinotori-C SAR distributed target test: (a) distributed targets within the illumination of antenna, (b) the recorded echoes and the processed range bin.

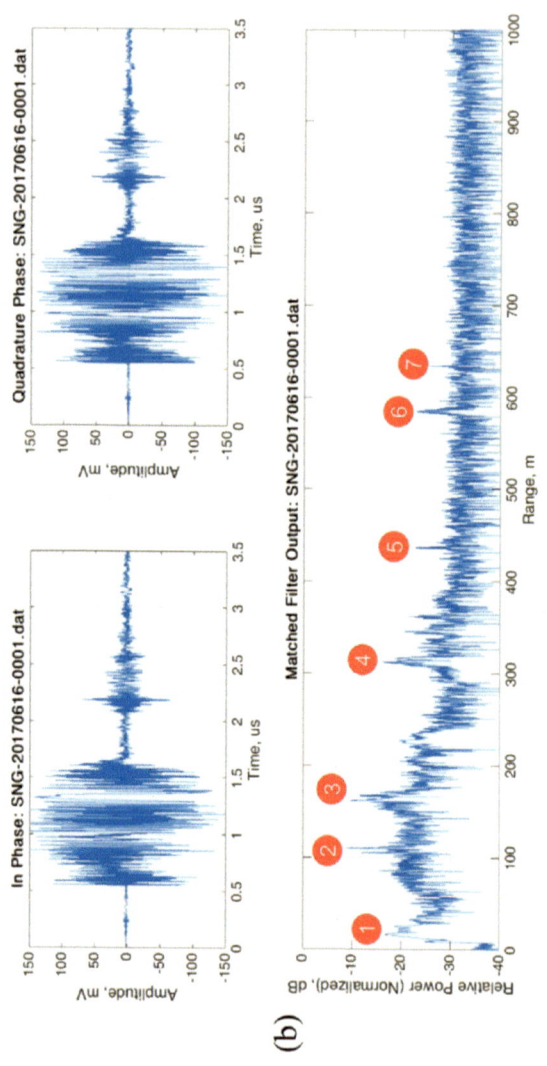

FIGURE 7.23   (Continued)

7.3.2.3 SAR Point Target Ground Imaging Experiment

A processed range bin signal provides only one-dimensional information in the range direction. An SAR image has two-dimensional information with an addition in the azimuth direction. Contrary to a real aperture radar (RAR) system, SAR uses the forward motion of the radar antenna over a target region to synthesize a huge size antenna for getting a finer resolution in the azimuth direction. Hence, the system has to be able to "synthesize" the azimuth chirp where the coherency of the echoes collected at every PRI interval is critical. A ground imaging experiment with (known) point targets could effectively validate it.

In the Hinotori-C SAR ground imaging experiment that used the same setup illustrated in Figure 7.21(b), the robot and the SAR system were programmed to scan the targets (the two trihedral corner reflectors) starting at the right side of the track for a total azimuth distance of 50m. During the scanning, the robot carried the antenna, moved and stopped with a displacement of 50cm until it reached the left end of the track. At every stop, the system recorded the collected signals and accumulated a total number of 1001 bins of echoes.

Figure 7.24(a) shows the image of the range compressed signal. The image clearly shows the curvature range profile of the two-corner reflector at the range distance of 100m and 110m. The azimuth compressed image is shown in Figure 7.24(b), in which

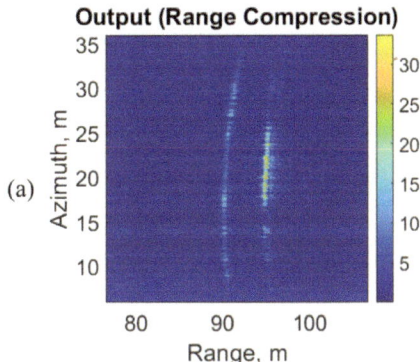

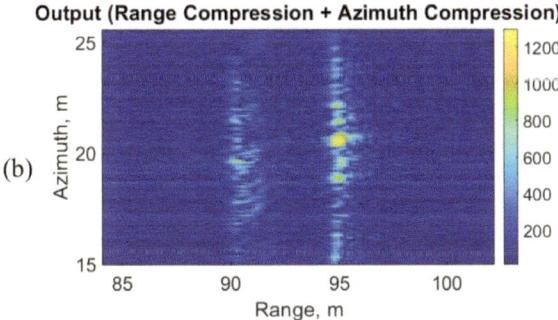

FIGURE 7.24 The Hinotori-C SAR gound imaging experiment (antenna polarization: LR); (a) range compressed image, (b) point target image, (c) range compressed signal at cross-range center, (d) azimuth compressed signal at range center.

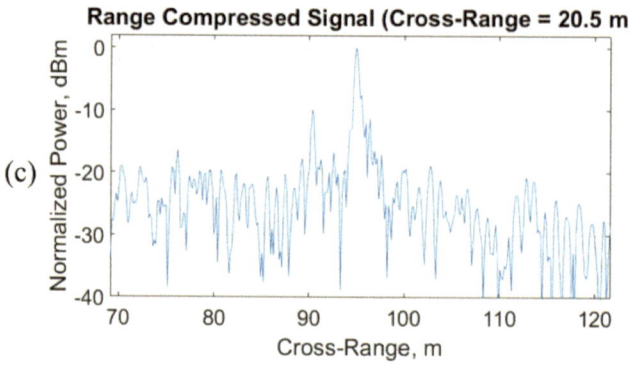

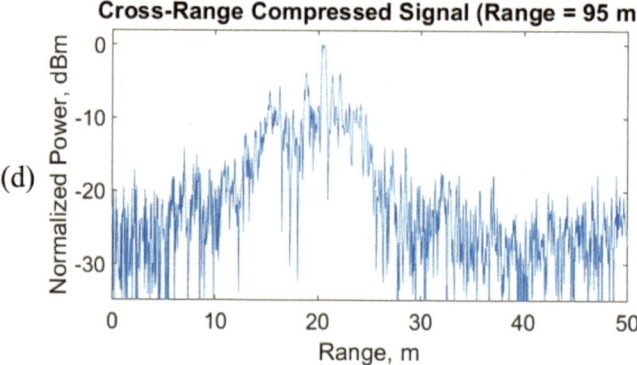

FIGURE 7.24   (Continued)

the signal has been compressed in the azimuth direction. Meanwhile, Figures 7.24(c) and 7.24(d) plot the slices of range compressed signal at the cross-range center and azimuth compressed signal at the range center, respectively. A detailed discussion on the signal processing techniques used for SAR images formation is in Chapter 8.

7.4   THE HINOTORI-C2 FLIGHT TEST

The Hinotori-C SAR had its maiden flight test in March 2018 at Makassar, South Celebes island, Indonesia. The aircraft that carried the Hinotori-C SAR in the flight test was CN-235 maritime patrol aircraft (MPA), a medium-range twin-engine transport aircraft built by CASA/IPTN shown in Figure 7.25(a). The main objective of the flight test was to verify the functionality of the circularly polarized SAR system and to acquire full polarimetric circularly polarized SAR images that cover several types of natural and artificial targets, such as corner reflectors for calibration, man-made buildings, water body and ship, an agricultural area, forest, and so forth.

7.4.1   SAR SYSTEM INSTALLATION

The Hinotori-C SAR antennas were mounted on a supporting bracket, and the entire structure was installed in the nosecone of the aircraft, as depicted in Figure 7.25(b).

On the other hand, the SAR electronics were stacked up and firmly placed on the floor in the cabin behind the pilot's cockpit area near the plane's central entrance door. Figure 7.25(c) shows the picture of the installed SAR electronics in the aircraft. The antennas were connected to the SAR electronics using 7m of coaxial cables with approximately 5.5dB of transmission loss each. The installation setup not only reduced the transmitting power by 5.5dB but also significantly degraded the receiver's

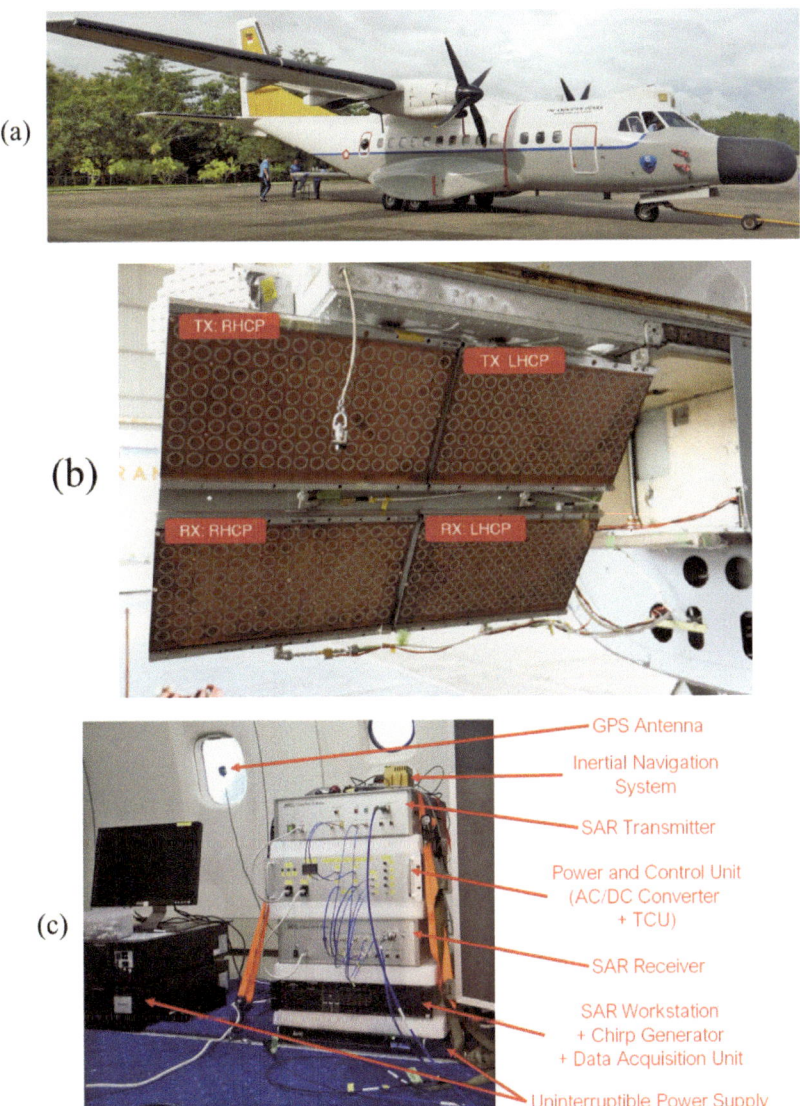

FIGURE 7.25 Hardware installation of the Hinotori-C SAR in Hinotori-C2 mission; (a) the CN235MPA aircraft for Hinotori-C2 mission, (b) SAR antennas, (c) SAR electronics.

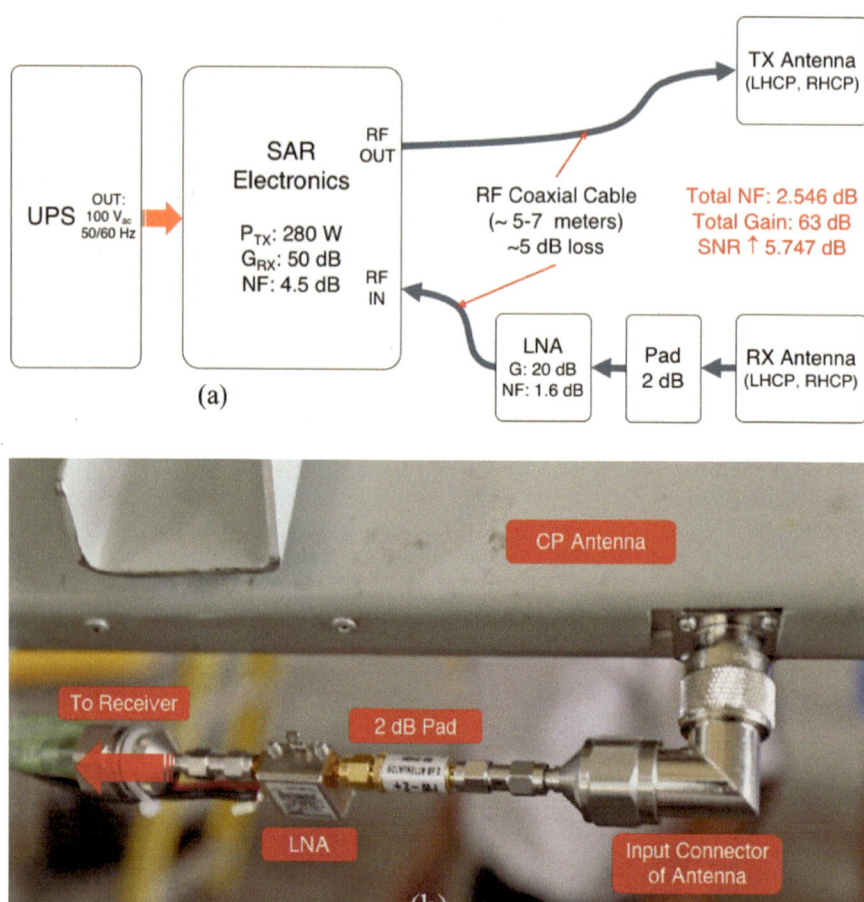

FIGURE 7.26 Front-end setup for NF improvement.

noise figure and, thus, lowered the noise equivalent sigma zero (NESZ) of the SAR system by 8.8dB. The use of a waveguide for better interconnection is not applicable due to the limited space in the aircraft for waveguide installation.

To further improve the quality of the final SAR imagery, a slight modification to the installation setup in the receiving chain was proposed and implemented. As illustrated in Figure 7.26, it was proposed to add a low noise amplifier (LNA) at each receiving antenna's input port so that the receiver's NF can be lowered. The intent of placing a 2dB pad between the LNA and the input connector of the antenna was to minimize the impedance mismatch between the LNA and the antenna.

The estimated NF and several determinant parameters that will affect the image quality of the final SAR imagery for each system setup were calculated and tabulated in Table 7.3. The setups were: (i) the initial setup without the long interconnection coaxial cable, (ii) the system setup with long coaxial cables (~ 5dB cable loss)

TABLE 7.3

Determinant Parameters Comparison for Various Setups

System Setup	RX Gain (dB)	NF (dB)	ΔNF (dB)	NESZ (dB)	TX Power (Watt)	Path (RX:TX)	Measured		Target Delay (ns) **
							SNR (dB)	ΔSNR (dB) *	
Setup ① (Initial)	50	4.50	—	-47.57	280	LL	53.35	—	69.5
						RR	52.97	—	69.5
						LR	52.86	—	69.5
						RL	53.66	—	69.5
Setup ② (Initial + Cable)	43	8.29	-3.79	-38.78	79	LL	43.68	-9.67	115.5
						RR	43.92	-9.05	114.0
						LR	43.89	-8.97	114.5
						RL	43.70	-9.96	115.0
Setup ③ (Initial + Cable + LNA)	63	2.55	1.95	-44.52	79	LL	56.30	2.95	116.5
						RR	56.09	3.12	114.5
						LR	56.13	3.27	115.5
						RL	56.21	2.55	115.0

connecting the RF system and the antennas, and (iii) the proposed system setup as illustrated in Figure 7.26(a). From the estimation, for setup (ii), in addition to a 5.5dB weakening in the transmitting power, the receiver's NF will deteriorate by 3.79dB. As a result, the NESZ will be substantially degraded for as much as 9dB, causing the targets with low radar cross section (RCS) (< 30dBsm) to be possibly buried by the noise floor of the final SAR imagery. Meanwhile, for setup (iii), although operating at weakened transmitting power, the NF of the receiver is substantially improved, with NESZ at 3dB lower than the initial system setup.

7.4.1.1 Pre-Flight Tests

After the installation, a loopback test with a 100dB attenuation between the TX-RX chain was performed for each system setup to quantify the signal-to-noise ratio (SNR) of the emulated target (from the loopback chirp replica) in addition to examining the internal reflection in the RF system. Figure 7.27 shows the emulated target's normalized power plots (to the peak magnitude) after the range compression. The plots show that the SNR was substantially improved by adopting the system setup (iii). The SNR degradation caused by the internal reflection in the RF system is noticeable in the range of 0m to 800m.

The SNR readings for each system setup and the delay incurred by the emulated target were extracted and tabulated in Table 7.3. In the table, ΔSNR quantifies the difference of the SNR for setup (ii) and setup (iii) to the initial setup, whereby a positive value infers an improvement in the NF of the receiver while a negative value indicates a degradation. The mean value across four RX: TX polarization for ΔSNR_{12} is -9.41dB (degradation) while for ΔSNR_{13} it has 2.97dB (improvement). The notable variation in the target's delay is due to the different lengths of cable used in the setups.

After the loopback test, the aircraft was towed to a taxiway for a ground test. With the system being set up on the taxiway, a few tests were performed, mainly to inspect the hardware system's background noise and quantify the system's time delay and amplitude error through external point target calibration. Figure 7.28(a) shows the moment when the aircraft was being towed to the taxiway, the arrangement of the SAR antennas for external point target calibration in Figure 7.28(b), and the radomes loss test in Figure 7.28(c).

The antennas were positioned with their beam pointing almost parallel to the ground, as shown in Figure 7.28(b). The antennas were tilted up at approximately 5° to minimize the reflection from the ground surface while at the same time ensuring that the reflectors were still within the illumination area of the main beam. At 640m and 655m away from the SAR antennas, two different sized and shaped calibration targets (a large trihedral corner reflector and a medium-sized trihedral corner reflector) were put on the taxiway, elevated by 1m off the ground. The reflected EM waves reflected by the calibration targets and the objects behind the targets were collected by the SAR system and recorded. In the Hinotori-C2 mission, two rounds of external point target calibration were successfully conducted. The first round was with a large trihedral corner reflector (CR_L: 600mm of edge length) and a medium-sized trihedral (CR_M: 55 mm of edge length), while in the second round, the smaller trihedral corner reflector was replaced by a cylinder-shaped reflector (CY_S: height of 900mm and radius of 136mm).

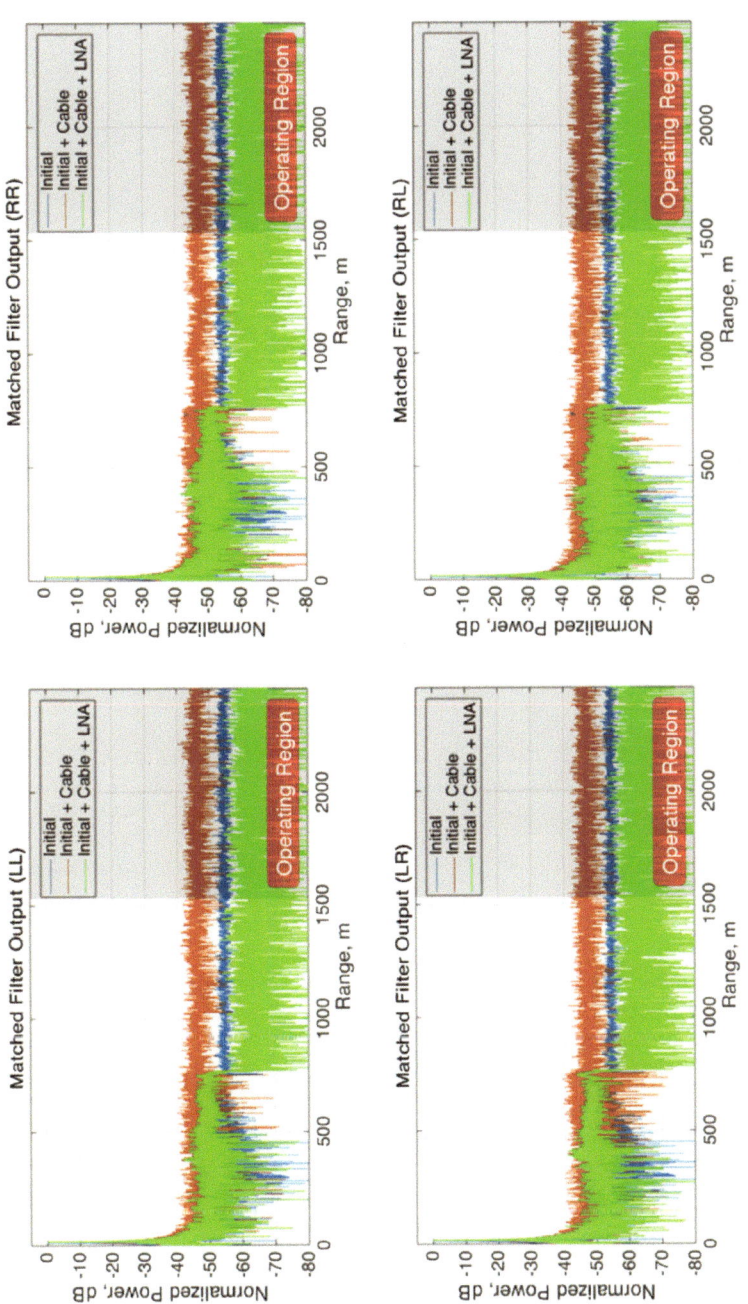

FIGURE 7.27 Loopback test of Hinotori-C (with and without additional front-end LNA).

FIGURE 7.28 Ground test; (a) moving aircraft to the runway, (b) position of antenna for ground test, (c) the radome loss test.

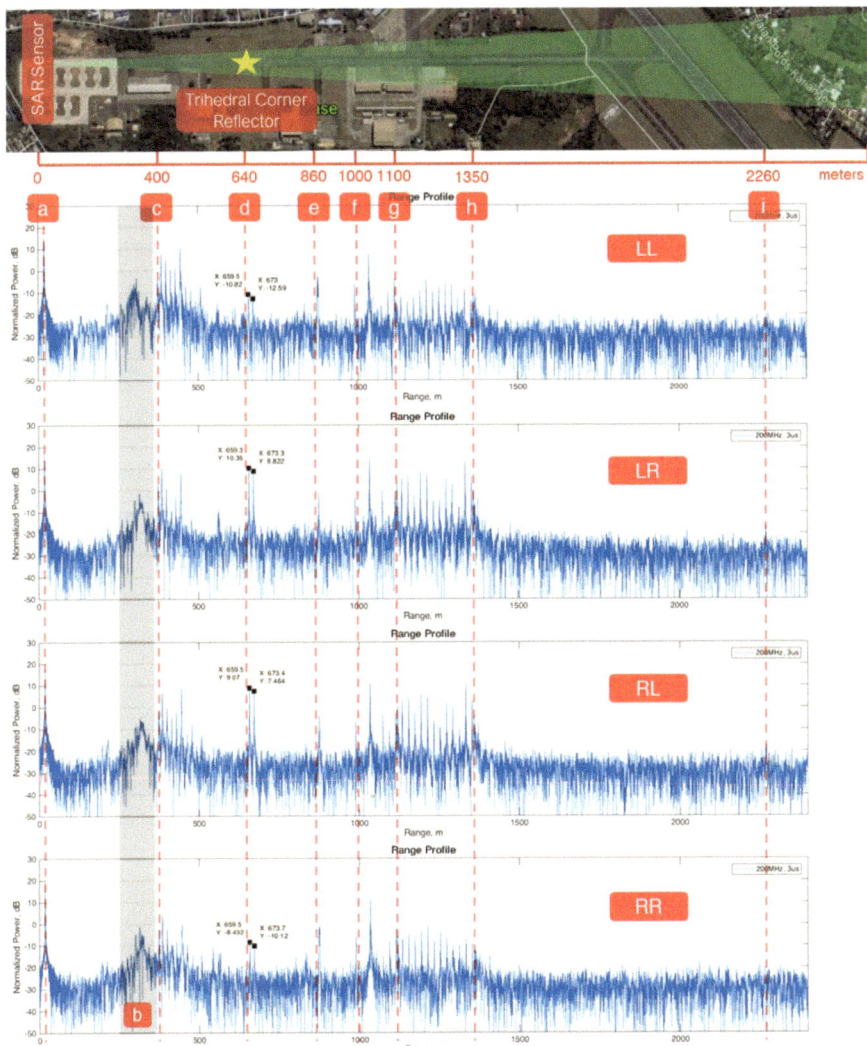

FIGURE 7.29 Ground range profile test and external point target calibration; (a) coupled signal from TX-RX antenna, (b) system internal reflection, (c) runway edge light, (d) trihedral corner reflectors (CR_L and CR_M), (e) buildings, (f-g) aircraft hangar, (h) buildings, and (i) residential area.

Figure 7.29 shows one range profile plot (with CR_L and CR_M) for all combinations of different polarization in the transmitting and receiving antennas. Similar responses induced by the antenna-to-antenna coupling effect and the internal reflection in the RF electronics can be observed in the plot, denoted as line (a) and region (b). The reflected signals from the reflectors were detected by the SAR system, labeled as line (d) in the plot showing a weaker amplitude response for co-polarization (LL and RR) and stronger in cross-polarization (LR and RL). This is because the incident wave

TABLE 7.4

Hinotori-C2 Ground Point Target Calibration

Polarization	Reflector Type	Range (m)			RCS (dBsm)		
		R_I	R_M	Δ_R	P_I	P_M	Δ_P
LL	CR_L	637.8	659.3	21.5	22.29	-10.82	-33.12
	CR_M	651.8	673.0	21.2	20.79	-12.59	-33.39
	CY_S	651.8	673.4	21.6	7.35	-12.52	-19.87
RR	CR_L	637.8	659.5	21.7	22.29	-8.43	-30.73
	CR_M	651.8	673.7	21.9	20.79	-10.12	-30.92
	CY_S	651.8	673.0	21.2	7.35	-14.80	-22.15
LR	CR_L	637.8	659.3	21.5	22.29	10.35	-11.95
	CR_M	651.8	673.3	21.5	20.79	8.82	-11.98
	CY_S	651.8	673.0	21.2	7.35	-3.12	-10.47
RL	CR_L	637.8	659.5	21.7	22.29	9.07	-13.23
	CR_M	651.8	673.4	21.6	20.79	7.46	-13.34
	CY_S	651.8	673.1	21.3	7.35	-1.48	-8.83

was reflected three times by the trihedral corner reflector, and the polarization of the returned waves mirrored each time the wave was reflected (LHCP to RHCP, RHCP to LHCP). The targets that were detected by the SAR system in the scan are as described in the caption in the figure. Meanwhile, the range and amplitude error contributed by the hardware system for different calibration targets (CR_L, CR_M, and CY_S) were calculated and tabulated in Table 7.4.

7.4.1.2   Maiden Flight of Hinotori-C

In the mission, three rounds of flight tests were completed on 14 March 2018 afternoon, and 15 March 2018 morning and afternoon, with different flight altitudes ranging from 1,000m to 1,500m above sea level. Two SAR system setups for SAR geometry of different bandwidth, pulse width, and altitude were used during the Hinotori-C2 flight tests whereby Setup ① is a normal resolution imaging mode with a bandwidth of 200MHz and pulse width of 5μs, while setup ② is a high-resolution imaging mode at higher altitude with operating bandwidth of 400MHz and pulse width of 10μs. The PRF was kept constant at 1,000Hz, being at least twice the Doppler bandwidth to avoid aliasing problems in azimuth sampling and not too high to overburden the data acquisition system. The INU recorded the aircraft's altitude and the information was used in the post-processing of the SAR data.

The Hinotori-C adopted airborne repeat-pass polarimetry to acquire full polarimetric circularly polarized SAR data. In the Hinotori-C2 flight test, three scenes were planned for full polarimetric SAR data acquisition using the hardware setup ①, covering diverse types of objects such as forests, water bodies, artificial structures, and the calibration targets. For each scene, the aircraft passed the same area of interest four times, and different polarization of SAR data was acquired in each pass. The swath of the scene and the planned and actual flight path are shown in Figure 7.30.

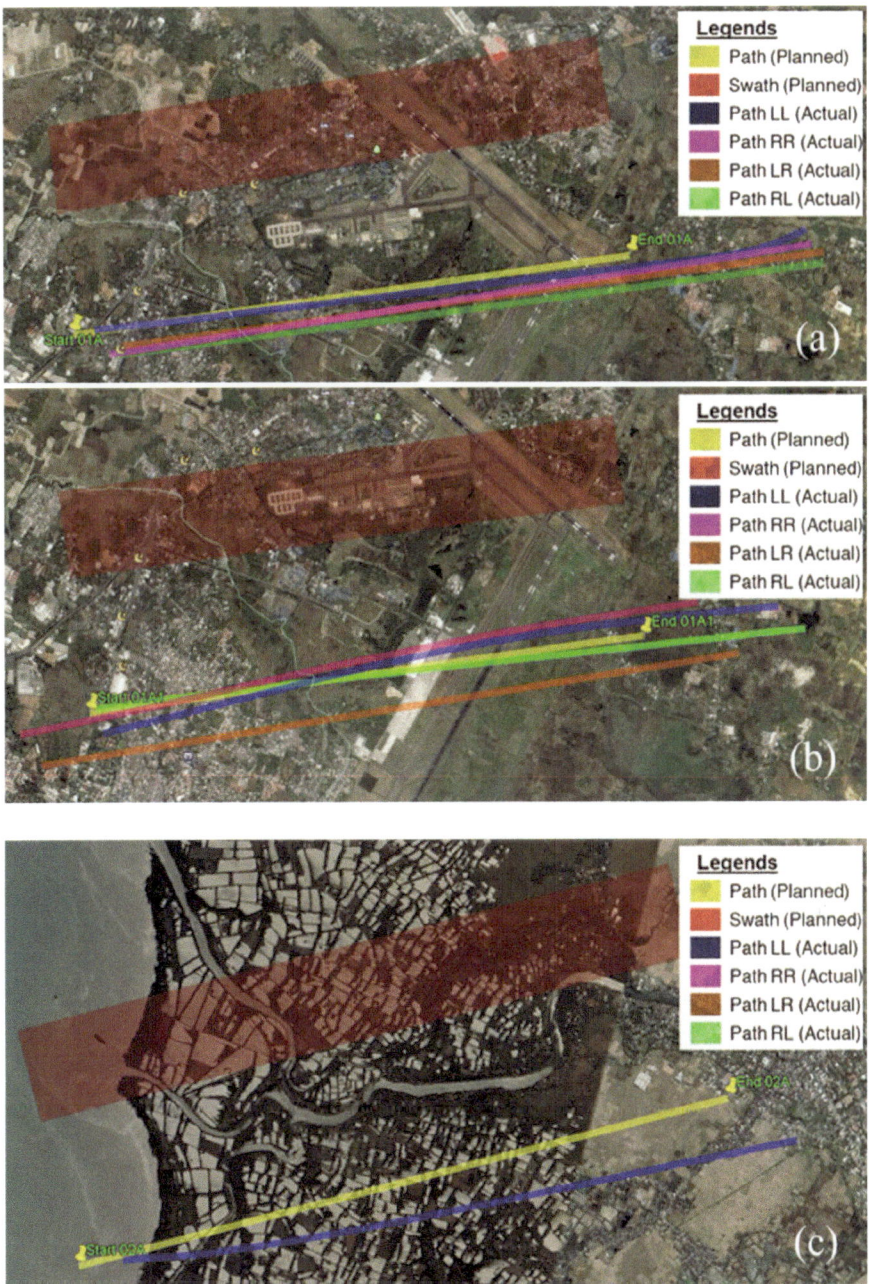

FIGURE 7.30 Flight path; (a) scene S01A, (b) scene S01A1, and (c) scene S02A.

The scene shown in Figure 7.30(b) was for external point target calibration from which the scene's swath covered the reflectors' location.

Figure 7.31 shows the geometrically uncorrected full polarimetric circularly polarized SAR images for each pass (with different polarization) in every scene. The comparison of a small area in the circularly polarized SAR image to the optical image from Google Earth™ is illustrated in Figure 7.32. Apart from the full polarimetric SAR data acquisition mission, two special missions were accomplished in the same flight test to acquire higher resolution SAR images and sparse SAR data (using ships in the sea). The special missions used hardware Setup ②, and the aircraft was cruised at a higher altitude (1,500m above sea level). The acquired high-resolution SAR images are shown in Figure 7.33. Meanwhile, Figure 7.34 shows the sparse SAR data, with Figure 7.34(a) showing the ships cruising nearby the Biringkassi Port, South Sulawesi, located in the northwest of Makassar, Indonesia, and Figure 7.34(b) shows a large cargo ship.

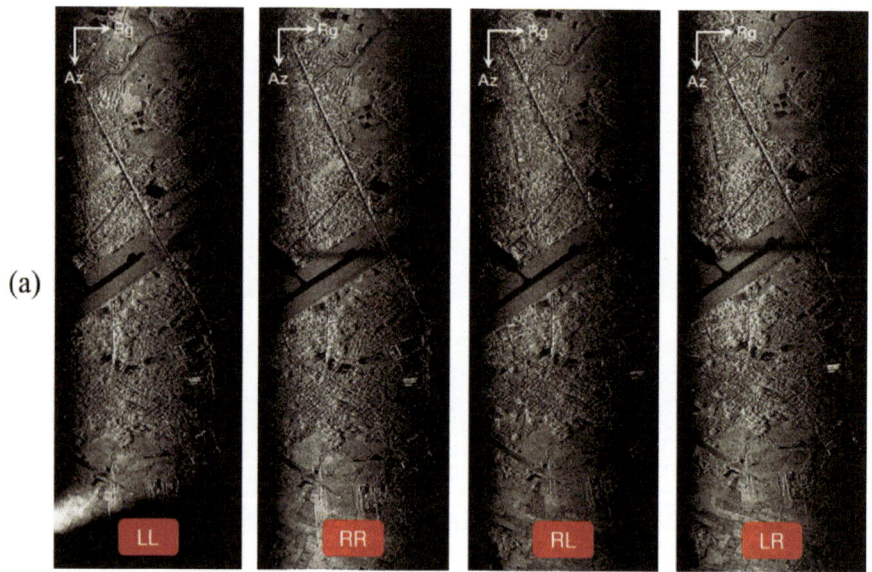

FIGURE 7.31   Full polarimetric circularly polarized SAR images (geometrically uncorrected); (a) scene S01A, (b) scene S01A1, and (c) scene S02A.

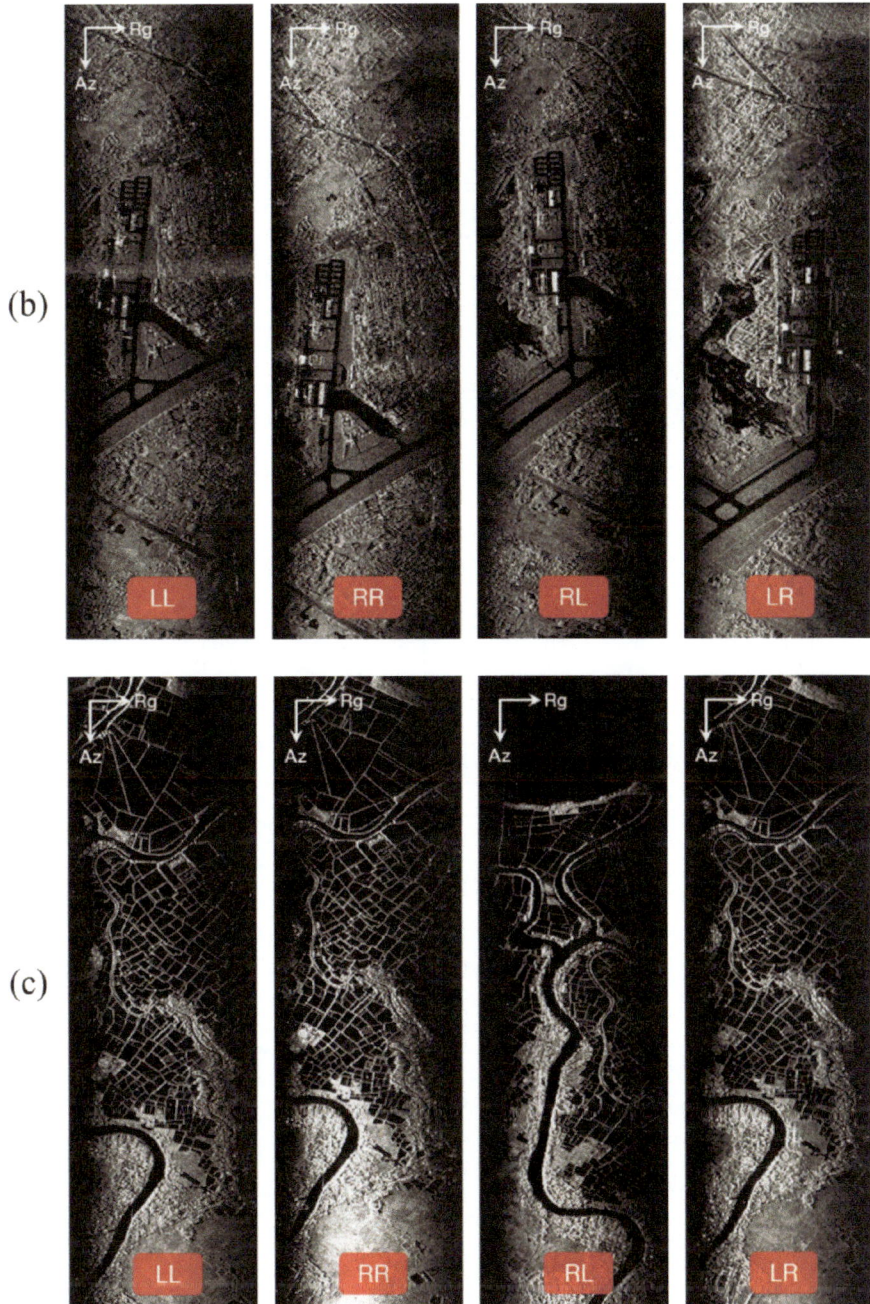

FIGURE 7.31   (Continued)

FIGURE 7.32 Comparison with Google Earth image; (a) scene S01A (LR), (b) enlarged view (light brown rectangle), (c) Google Earth digital globe image dated on 27 September 2017 (©2017 Google LLC, used with permission).

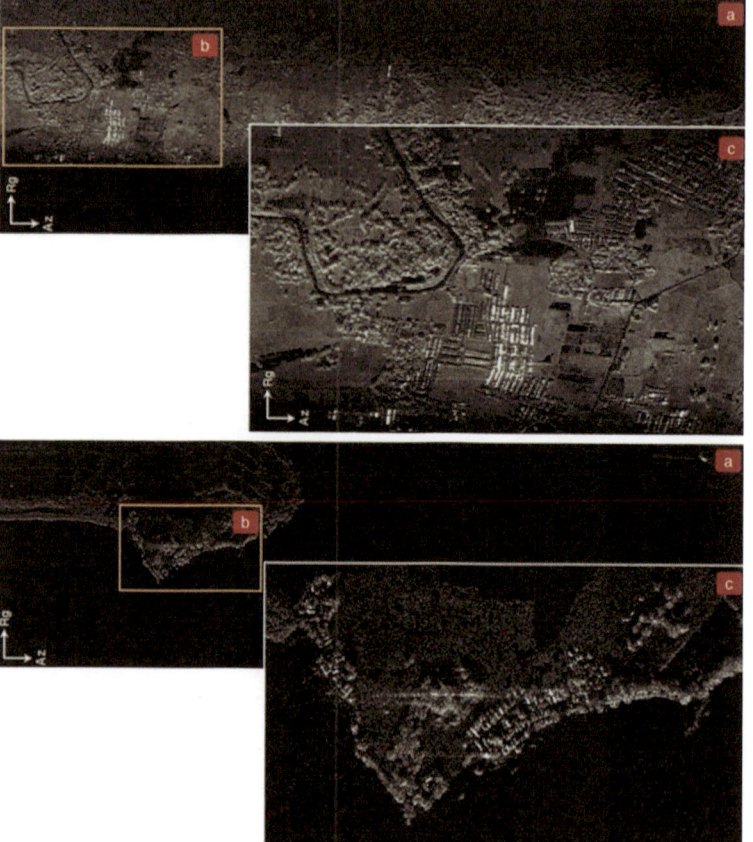

FIGURE 7.33 High-resolution SAR images from the Hinotori-C.

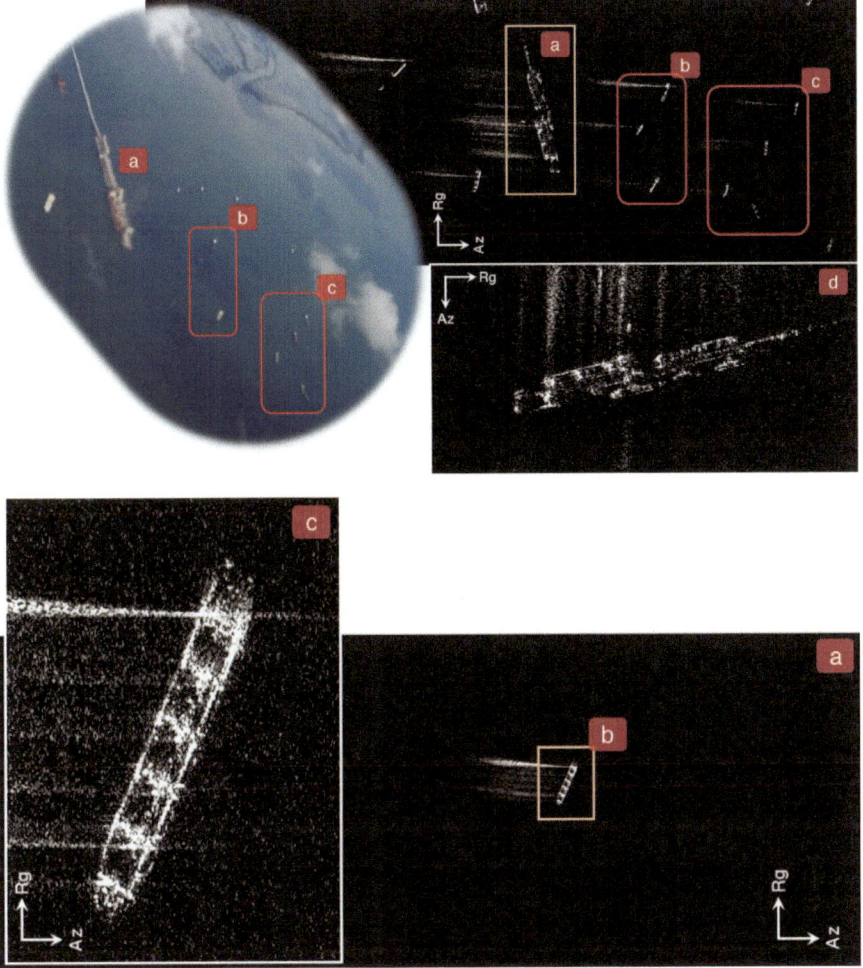

FIGURE 7.34   Ships detection; (a) ships nearby Biringkassi port, (b) a large cargo ship.

REFERENCES

[1] JT Sri Sumantyo, MY Chua, CE Santosa, GF Panggabean, T Watanabe, B Setiadi, FD Sri Sumantyo, K Tsushima, K Sasmita, A Mardiyanto, E Supartono, ET Rahardjo, G Wibisono, MA Marfai, RH Jatmiko, S Sudaryatno, TH Purwanto, BS Widartono, M Kamal, D Perissin, S Gao, and K Ito, "Airborne Circularly Polarized Synthetic Aperture Radar," IEEE Selected Topics in Applied Earth Observations and Remote Sensing (JSTARS), Vol.14, pp.1676–1692, January 2021, DOI:10.1109/JSTARS.2020.3045032.

[2] MY Chua, JT Sri Sumantyo, CE Santosa, GF Panggabean, FD Sri Sumantyo, T Watanabe, YQ Ji, Peberlin, P Sitompul, M Nasucha, F Kurniawan, A Babag B Purbantoro, A Awaludin,, K Sasmita, ET Rahardjo, W Gunawan, RH Jatmiko, S Sudaryatno, TH Purwanto, BS Widartono, and M Kamal, "The Maiden Flight of

Hinotori-C: The First C Band Full Polarimetric Circularly Polarized Synthetic Aperture Radar in the World", IEEE Aerospace and Electronic Systems Magazine, Vol. 34, Issue 2, pp. 24–35, June 2019.

[3] JT Sri Sumantyo, MY Chua, CE Santosa, GF Panggabean, T Watanabe, B Setiadi, K Tsushima, FD Sri Sumantyo, K Sasmita, A Mardiyanto, E Supartono, ET Rahardjo, G Wibisono, RH Jatmiko, TH Purwanto, BS Widartono, M Kamal, RH Triharjanto, S Gao, and K Ito, "Hinotori-C2 mission: CN235MPA Aircraft Onboard Circularly Polarized Synthetic Aperture Radar (CP-SAR)", 2019 IEEE International Geoscience and Remote Sensing Symposium (IGARSS2019), pp. 8538–8541, July 2019.

[4] JT Sri Sumantvo, MY Chua, CE Santosa, A Takahashi, and K Ito, "Aircraft and High-Altitude Platform System Onboard Circularly Polarized Synthetic Aperture Radar (CP-SAR)", 2021 IEEE International Geoscience and Remote Sensing Symposium (IGARSS2021), pp. 8515–8518, July 2021.

[5] JT Sri Sumantvo, MY Chua, CE Santosa, GF Pariguabean, K Taushima, T Watanabe, K Sasmita, A Mardiyanto, FD Sri Sumantyo, Eko Tjipto Rahardjo, Gunawan Wibisono, E Supartono, S Gao, SP Parulian, M Nasucha, F Kurniawan, A Awaludin, B Purbantoro, YQ Ji, and N Imura, "Hinotori-C: A Full Polarimetric C Band Airborne Circularly Polarized Synthetic Aperture Radar for Disaster Monitoring", 2018 Progress in Electromagnetics Research Symposium (PIERS-Toyama), pp. 1466–1473, August 2018.

[6] MY Chua, JT Sri Sumantyo, and YQ Ji, "An 8-Channels FPGA-Based Reconfigurable Chirp Generator for Multi-Band Full Polarimetric Airborne/Spaceborne CP-SAR", 2018 Progress in Electromagnetics Research Symposium (PIERS-Toyama), pp. 876–881, Aug 2018.

[7] MY Chua, VC Koo, HS Lim, JT Sri Sumantyo, "FPGA-based Reconfigurable Chirp Generator for L-band UAV CP-SAR", International Symposium on Remote Sensing 2017, 2017.

[8] CH Lim, CS Lim, MY Chua, YK Chan, TS Lim, VC Koo, "A New Data Acquisition and Processing System for UAVSAR", IEICE Electronics Express, Vol. 8, Issue 20, pp. 1716–1722, 2011.

[9] MY Chua, JT Sri Sumantyo, CE Santosa, GF Panggabean, YQ Ji, SP Parulian, and M Nasucha, "An PC-based Airborne SAR Baseband System", 2018 Progress in Electromagnetics Research Symposium (PIERS-Toyama), pp. 882–888, Aug 2018.

8 SAR Image Formation

Ming Yam Chua

8.1 INTRODUCTION

Most people visualize a satellite image as an optical image, that is, a photograph taken by a very powerful camera, as shown in Figure 8.1(a). However, this way of thinking is not entirely true. Depending on the type of remote sensor used to produce the image, a satellite image could be a radar image, as shown in Figure 8.1(b) – an image formed from spatially overlapped radar phase histories, more commonly known as an SAR image.

An optical instrument produces an optical image using an image sensor. The instrument records the focused light rays bounced onto it through the lens system in the camera. The major drawback of an optical instrument is that the spatial resolution of an optical image it produces is highly affected by its distance to the target and the instrument's instantaneous field of view (IFOV). The further the distance is, the poorer the spatial resolution of the optical image obtained.

Contrary to the optical instrument, synthetic aperture radar (SAR) uses a different technique to produce an image. SAR actively illuminates its illumination source to the ground and fully uses the echoes (or returns from the ground) to form a two-dimensional radar image [1]. In SAR, the instrument itself does not directly output a radar image. The recorded echoes (the individual return signal) by the instrument are raw, in that they are unfocused and overlap each other. Plotting the data at this stage results in a coarse image that does not give much meaningful information to the user. To resolve this, SAR employs additional postdata processing that compresses the echoes and makes the individual signals from the target resolvable in both the range and the azimuth direction [2].

There are two categories of SAR postprocessing algorithms. These algorithms differ in the data processing domain, hence providing: (i) the frequency-domain algorithms, and (ii) the time domain algorithms. The Range-Doppler algorithm (RDA), Chirp Scaling algorithm (CSA), and Omega-K algorithm (ωKA) are the three most commonly used frequency domain algorithms that are proven to work well for stripmap SAR [3]–[5]. On the other hand, the back-projection algorithm is a time domain algorithm that works with lesser imaging modes and geometry limitations but with the tradeoff of high computational requirements. Nevertheless, these algorithms must use a common signal processing algorithm called matched filtering.

DOI: 10.1201/9781003282693-8

FIGURE 8.1 (a) Optical image (©2017 Google LLC, used with permission) and (b) SAR image.

8.2 MATCHED FILTER IN PULSE COMPRESSION

Modern radar systems detect the presence of a target using the pulse compression technique. In pulse compression radar, the received echo is correlated with the reference transmitted signal by passing it through a special filter at the receiver, known as the "matched filter." Typically, the transfer function for a matched filter is the

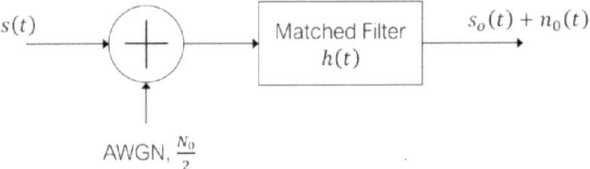

FIGURE 8.2 Functional block diagram of a matched filter.

complex conjugate function of the transmitter signal in the frequency domain and is implemented through a digital signal processor (DSP) [6]–[8].

In matched filtering, the detection probability is always related to the system's SNR rather than the physical shape of the received waveform. Thus, it is more of a concern to maximize the system's SNR rather than to maintain its waveform shape. An adequately designed matched filter shall exhibit an impulse response where the maximum SNR is attainable at the filter's output when the signal and white noise are passed through the filter, as illustrated in Figure 8.2. The matched filter's input is a composite of the signal, $s(t)$, and two-sided power spectral density additive white gaussian noise (AWGN). The filter is a function that gives the best output SNR at a pre-set delay t_0 where

$$\left(\frac{S}{N}\right)_{out} = \frac{\left|s_o\left(t_0\right)^2\right|}{n_0^2\left(t\right)}$$ (Equation 8.1)

The matched filter's output when a signal, $s(t)$ is received at the input can be determined with the formula derived from its transfer function, $h(t)$. To obtain the time domain filter output, a convolution integral is employed. The matched filter's output, $g(t)$ can be formulated as

$$g(t) = h(t)*s(t) = \int_{-\infty}^{\infty} h(t-u)s(u)\,du$$ (Equation 8.2)

When a signal with a narrow pulse width of $-\frac{\tau}{2}$ to $\frac{\tau}{2}$ is received at the receiver, the filter's output is

$$g(t) = \int_{-\frac{\tau}{2}}^{\frac{\tau}{2}} h(t-u)s(u)\,du$$ (Equation 8.3)

By changing the time variable, the above equation becomes

$$g(t) = h(t)*s(t) = \int_{-\frac{\tau}{2}}^{\frac{\tau}{2}} h(t)s(\tau-t)\,dt$$ (Equation 8.4)

Writing $g(t)$ in the frequency domain gives $G(j\omega)$, where

$$G(j\omega) = H(j\omega)S(j\omega) \qquad \text{(Equation 8.5)}$$

The time domain output, $g(t)$, can be obtained by applying Inverse Fourier Transform $(\mathcal{F}^{-1})$ to $G(j\omega)$, and can be expressed as

$$g(t) = \mathcal{F}^{-1}\{G(j\omega)\} \qquad \text{(Equation 8.6)}$$

When the returned signal is a weakened replica of the transmitted signal with an additional time delay of t_0, the filter is matched to $s(t)$ and that yields an impulse response of

$$h(t) = Ks(t_0 - t) \qquad \text{(Equation 8.7)}$$

where

$K =$ the amplitude attenuation factor
$t_0 =$ the associated delay

The matched filter transfer function, $H(j\omega)$, can be obtained by taking the Fourier Transform $(\mathcal{F})$ of the impulse response $h(t)$ and can be written as,

$$H(j\omega) = \mathcal{F}\{h(t)\} = \int_{-\infty}^{\infty} h(t)e^{-j\omega t}\, dt \qquad \text{(Equation 8.8)}$$

Hence

$$H(j\omega) = K \int_{-\infty}^{\infty} s(t_0 - t)e^{-j\omega t}\, dt \qquad \text{(Equation 8.9)}$$

Changing the time variable to $\tau = t_0 - t$, we get

$$H(j\omega) = K \int_{-\infty}^{\infty} s(\tau)e^{j\omega(\tau - t_0)}\, dt \qquad \text{(Equation 8.10)}$$

The $\mathcal{F}$ of $s(t)$ is written as

$$S(j\omega) = \int_{-\infty}^{\infty} s(\tau)e^{-j\omega t}\, dt \qquad \text{(Equation 8.11)}$$

The complex conjugate of $S(j\omega)$, $S^*(j\omega)$ can be written as

$$S^*(j\omega) = S(-j\omega)$$ (Equation 8.12)

From the equations above, $H(j\omega)$ is

$$H(j\omega) = -Ke^{-j\omega t_0}S^*(j\omega)$$ (Equation 8.13)

Mathematically, the nominal transfer function of the matched filter, $H(j\omega)$, is the complex conjugate of the transmitted signal spectrum. The impulse response of the matched filter, $h(t)$, is a delayed, time-reversed, and scaled version of its transmitting signal.

The output of the system, $G(j\omega)$, can be formulated as

$$G(j\omega) = H(j\omega)S(j\omega) = -Ke^{-j\omega t_0}S^*(j\omega)S(j\omega)$$ (Equation 8.14)

Applying $\mathcal{F}^{-1}$ to $G(j\omega)$, the time domain output, $g(t)$, can be obtained. Note that if $h(t) = s(t)$, $g(t)$ is then the autocorrelation function of $s(t)$.

There are two reasons for employing such processing to SAR raw data. The first is that the radar system's resolution, δR, can be improved with a factor of time bandwidth product (TBP) compared to a conventional pulse radar. The TBP can be quantified as

$$TBP = B \times \tau$$ (Equation 8.15)

where

B = the bandwidth of the chirp
τ = the pulse width of the chirp

The second reason is that a pulse compression radar has a Signal-to-Noise Ratio (SNR) improvement with a factor of TBP and, thus, requires less transmitting power. In both instances, a pulse compression radar can use a wider-width pulse with higher pulse energy to achieve the same resolution as the conventional pulse radar system. Figure 8.3 compares the chirp (100MHz bandwidth, 10µs pulse width) before and after pulse compression. The figure clearly shows that the pulse width has been "compressed" from 10µs to approximately 10ns with an improvement factor of 1,000.

8.3   SAR POINT TARGET ANALYSIS

The performance of a SAR system can be evaluated by characterizing the impulse response obtained through measuring the system response to an isolated, single target scatterer, such as a trihedral corner reflector. The system's response usually exhibits a sinc-like impulse response for a linear FM modulated SAR signal. In

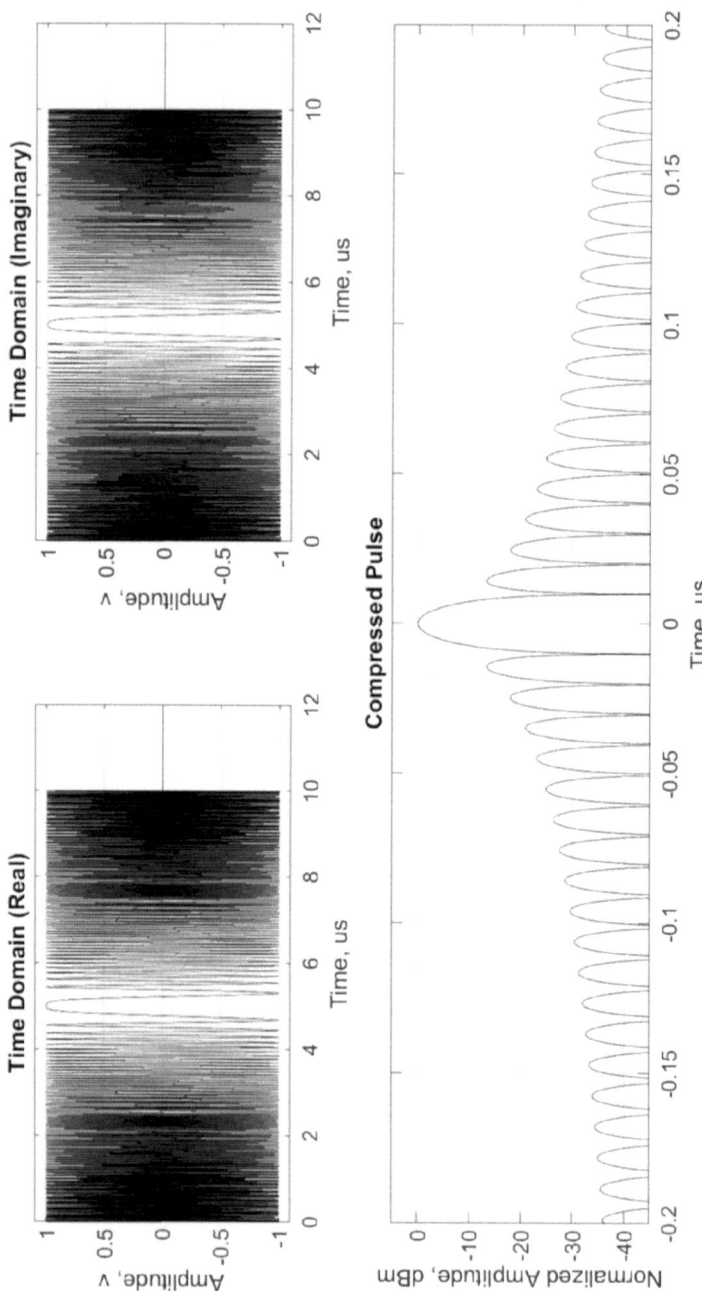

FIGURE 8.3   Pulse compression; (top) chirp in time domain, (bottom) compressed pulse.

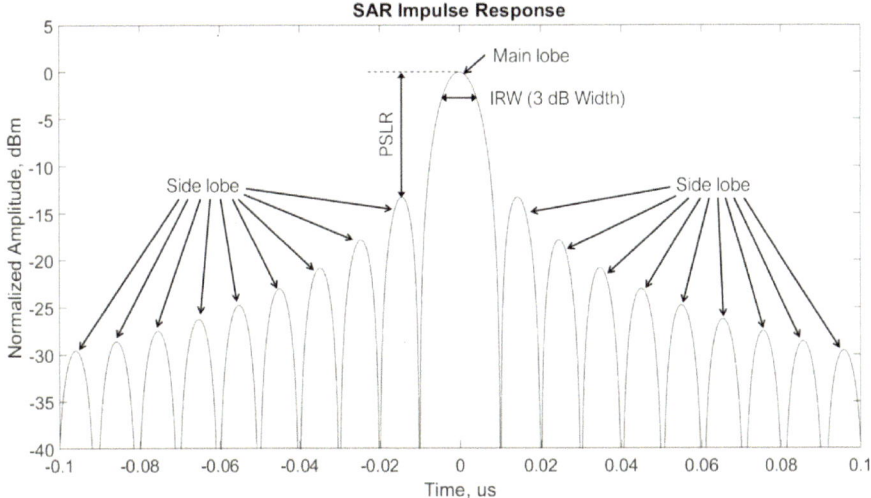

FIGURE 8.4 LFM SAR system response.

characterizing the system's response, one usually quantifies the quality parameters that can be measured from the point target, such as the pulse compression ratio (PCR), the impulse response width (IRW), and the peak or integrated sidelobe ratio (PSLR or ISLR). In short, IRW defines the SAR system's best achievable range and azimuth resolution, while PSLR and ISLR directly impact the reconstructed SAR image contrast. Figure 8.4 shows a typical impulse response for a linear frequency modulate (LFM) SAR system and its main components.

The definitions of the key parameters associated with the SAR signal quality are:

PCR: The pulse compression ratio (PCR) is the improvement factor in the range resolution compared to an un-modulated pulse. Some literature refers to it as the time bandwidth product (TBP). A high PCR waveform indicates the finer resolution a pulse compression radar can obtain.

$$PCR = TBP = B \times \tau = \frac{\left(c_0 \cdot \dfrac{\tau}{2}\right)}{\left(\dfrac{c_0}{2B}\right)} \qquad \text{(Equation 8.16)}$$

IRW: The impulse response width (IRW) is the main lobe width of the impulse response, measured at 3dB below the peak value. IRW is also the actual system's range resolution.

PSLR: The peak sidelobe ratio (PSLR) is the ratio between the amplitude of the largest sidelobe to the amplitude of the main lobe. Usually, it is expressed in decibels (dB). A compressed LFM pulse exhibits a sinc function with a PSLR of -13dB. PSLR has to be as small as possible so that a strong target will not mask the return from small targets. An acceptable level of PSLR is approximately -20dB.

ISLR: The integrated sidelobe ratio (ISLR) can be obtained by integrating the power of the impulse response over suitable regions. For example, let P_{main} quantify the main lobe power and P_{total} quantify the total power, the ISLR is then

$$ISLR = 10\log_{10}\left(\frac{P_{total} - P_{main}}{P_{main}}\right) \qquad \text{(Equation 8.17)}$$

ISLR value may vary depending on the sidelobe range limits. A typical value for ISLR is -17dB, with the main lobe being within the null-to-null limits.

8.4   SAR POINT TARGET MODEL

SAR uses its platform's forward motion to synthesize an extremely large synthetic antenna to improve the image resolution in the azimuth domain. Figure 8.5 depicts the top view of the geometry for a stripmap SAR system with a single point target at the center of the scanning area. While the SAR system is moving in the slow time domain,u, at each pulse repetition interval (PRI), a pulsed chirp will be transmitted at the vector direction in parallel to the fast time domain, t. A portion of the echo returns to the system, and the system will record a large array of the echoes at

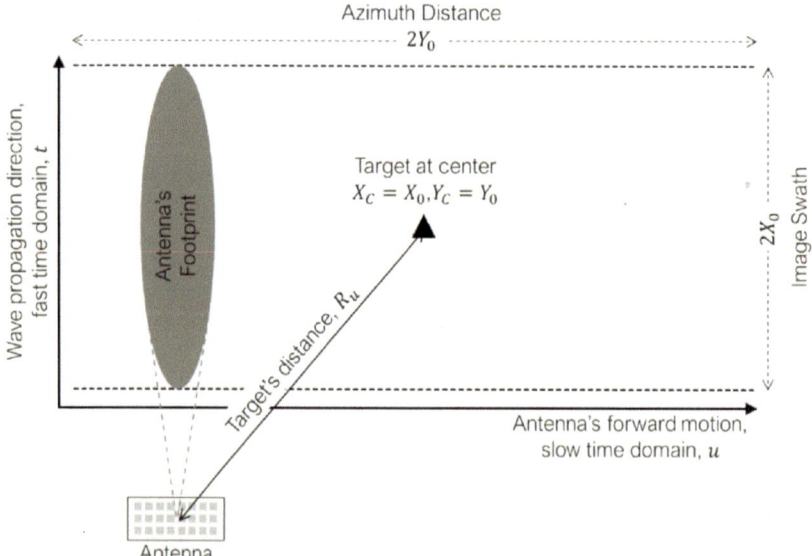

FIGURE 8.5   Geometry of a stripmap SAR (top view).

different azimuth positions (different distances between the antenna and the target) for postsignal processing.

The echo is a replica of the weakened amplitude delayed version of its transmitted chirp pulse [9]–[10]. Its associated delayed time function can be expressed by

$$t_d = t - \tau \qquad \text{(Equation 8.18)}$$

where $\tau = \dfrac{2R}{c}$ is the round-trip time delay of the target, R is the range of the target, and c is the speed of light.

Substituting Equation 8.18 into Equation 5.5, the echo from the single target can be formulated as

$$x_{echo\_c}(t_d) = A_r \cdot \Pi\left(\frac{t-\tau}{T}\right) \cdot e^{j\pi\left(\alpha t_d^2 - B t_d\right)} \cdot e^{j 2\pi f_c t_d} \qquad \text{(Equation 8.19)}$$

where A_r is the reduced amplitude of the echo.

Let ϕ_{echo_c} be the expanded phase terms of Equation 8.19

$$\phi_{echo\_c} = \alpha t^2 - Bt + 2f_c t - 2\alpha t \tau + \alpha \tau^2 + B\tau - 2f_c \tau \qquad \text{(Equation 8.20)}$$

Mixing Equation 8.19 with a carrier, the baseband echo becomes

$$\begin{aligned} x_{echo_b}(t_d) &= A_r \cdot \Pi\left(\frac{t-\tau}{T}\right) \cdot e^{j\pi\left(\phi_{echo_c}\right)} \cdot e^{-j2\pi f_c t} \\ &= A_r \cdot \Pi\left(\frac{t-\tau}{T}\right) \cdot e^{j\pi\left(\phi_{echo\_b}\right)} \end{aligned} \qquad \text{(Equation 8.21)}$$

where $\phi_{echo\_b}(t) = \alpha t^2 - Bt - 2\alpha t \tau + \alpha \tau^2 + B\tau - 2f_c \tau$.

The instantaneous phase of the baseband echo is composed of

$$\phi_{echo\_b}(t) = \left(\underbrace{\alpha t^2 - Bt}_{[a]} \quad \underbrace{-2\alpha t \tau}_{[b]} \quad \underbrace{+\alpha \tau^2 + B\tau - 2f_c \tau}_{[c]} \right) \qquad \text{(Equation 8.20)}$$

where $[a]$ is the baseband replica, $[b]$ is the frequency shift component and $[c]$ is the phase shift component. Compared to the originally transmitted signal, two additional components are introduced in the instantaneous phase of the baseband echo signal as a result of the additional round-trip time delay in the transmitting SAR signal.

If there are n numbers of point targets within the illumination, the echo signal will then become the coherent summation of each echo pulse. Mathematically, it can be written as

$$x_{echo}(t_d) = \sum_{k=0}^{n} A_k \cdot \Pi\left(\frac{t-\tau_k}{T}\right) \cdot e^{j\pi\left(\alpha t^2 - Bt - 2\alpha t \tau_k + \alpha \tau_k^2 + B\tau_k - 2f_c \tau_k\right)} \qquad \text{(Equation 8.21)}$$

Considering a two-dimensional case where the slow-time domain, u, is included in Equation 8.21, the echo signal can be expressed in (u, t_d) domain as

$$x_{echo}(u, t_d) = \sum_{k=0}^{n} A_{u,k} \cdot \Pi\left(\frac{t - \tau_{u,k}}{T}\right) \cdot e^{j\pi\left(\alpha t^2 - Bt - 2\alpha t \tau_{u,k} + \alpha \tau_{u,k}^2 + B\tau_{u,k} - 2f_c \tau_{u,k}\right)} \qquad \text{(Equation 8.22)}$$

In summary, the baseband echo SAR signal is synthesizable by introducing three components into the transmitting SAR signal, namely:

i.   A frequency shifts component
ii. A phase shift component
iii. A time shift component (delayed replica of the originally transmitted signal)

These additional components are related to the target's distance, R, and origin from the antenna.

8.5 RANGE-DOPPLER ALGORITHM (RDA)

The Range-Doppler algorithm (RDA) was developed in 1976 for processing SEASAT SAR data, with the first digitally processed spaceborne SAR image produced in 1978 processed using this algorithm. Although it has been available for almost 50 years, it is still a widely used algorithm to process SAR data due to its simplicity of one-dimensional block processing and frequency domain operations in both range and azimuth.

As depicted in Figure 8.6, the basic RDA algorithm has three processing blocks: range compression, range cell migration correction (RCMC) and azimuth

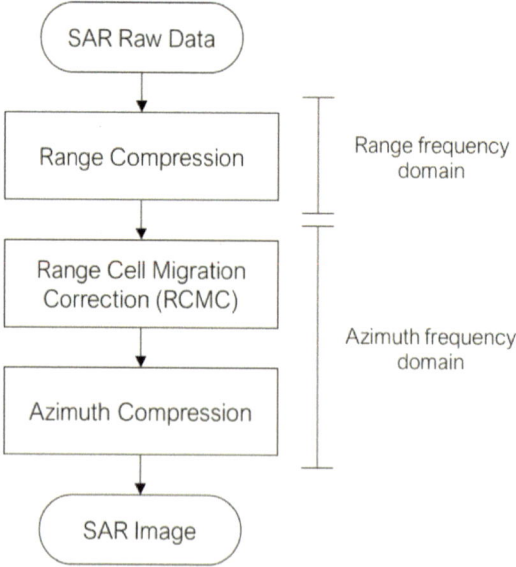

FIGURE 8.6   Processes in Range-Doppler Algorithm (RDA).

compression. This algorithm is suitable for processing SAR data with small squint angles and short aperture lengths, such as airborne/UAV-borne SAR data. However, RDA can still process squinted data requiring an additional process called secondary range compression (SRC) before the RCMC. All processes in RDA process data in either range (range compression) or azimuth (RCMC and azimuth compression) frequency domain for processing efficiency. Each processing block can be carried out independently, and its intermediate output can be plotted/viewed for interpretation and analysis.

8.5.1 Range Compression

The range compression for pulse chirp deals with the matched filtering of the received echo with the waveform of the transmitted chirp at the range frequency domain. Figure 8.7 illustrates the subprocesses in the range compression.

For a given an $m \times n$ array of SAR raw data with m range bins, and each bin containing n points of data, the range compression performs the following three subprocesses for each range bin.

i. Convert the m-th range bin to the frequency domain using fast Fourier transform (FFT).
ii. Multiply it with the range-matched filter. The range-matched filter is the complex conjugate of the transmit chirp in the frequency domain. The transmit

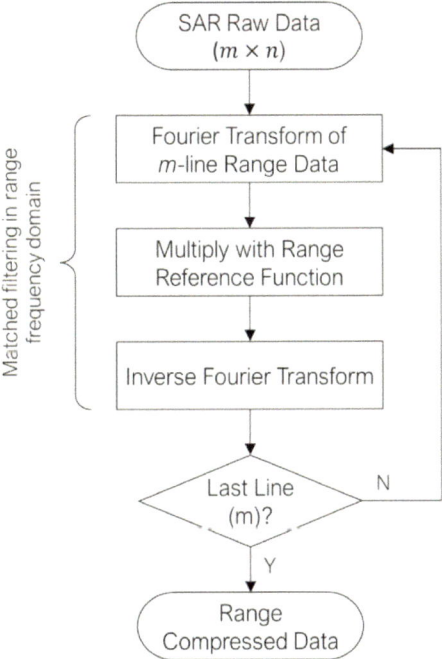

FIGURE 8.7 Detailed processes in range compression.

chirp data can be generated mathematically or recorded using the system by looping the attenuated transmit signal back to the receiver.

iii. Transform the resultant data back to the time domain by performing inverse Fourier transform (IFFT).

8.5.2 RANGE CELL MIGRATION CORRECTION

During SAR data collection, the signal energy from a point target follows a trajectory in the two-dimensional SAR data. This trajectory needs to be corrected to a constant range cell with a process called range cell migration correction (RCMC), and it has to be done before the azimuth compression but after the range compression. Figure 8.8 illustrates the subprocesses in RCMC.

For the same $m \times n$ array of range compressed data, the RCMC performs the following five subprocesses.

i.   Convert all n bin of data in the azimuth domain to the frequency domain using fast Fourier transform (FFT).

ii. Zero pad each range bin. The amount of data depends on the maximum amount of range cells to migrate.

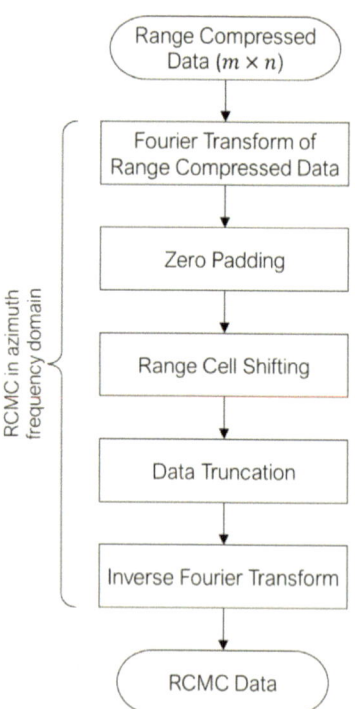

FIGURE 8.8   Detailed processes in RCMC.

iii. Shift each range bin. The amount of range cell to shift for each line depends on

$$\Delta R\left(f_\eta\right) = \frac{\lambda^2 R_0 f_\eta^2}{8 V_r^2}$$
(Equation 8.23)

where

λ = carrier wavelength
R_0 = slant range of closest approach
f_η = azimuth frequency
V_r = effective radar velocity

iv. Truncate the amount of data added in step (ii).
v. Transform the resultant data back to the time domain by performing inverse Fourier transform (IFFT).

8.5.3 AZIMUTH COMPRESSION

Similar to range compression, the azimuth compression process involves matched filtering of the range compressed, or range cell migration corrected data with the azimuth matched filter. Figure 8.9 illustrates the sub-processes in the azimuth compression.

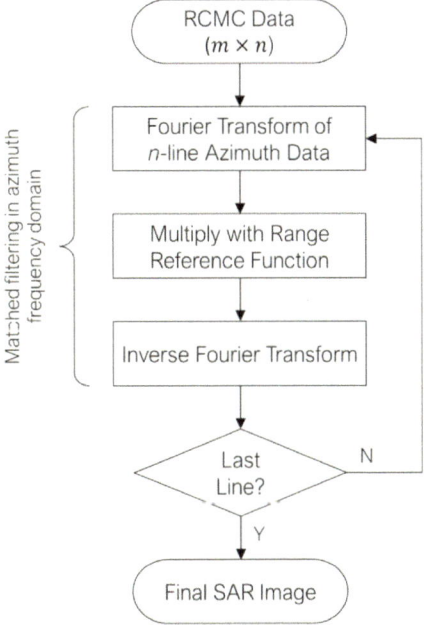

FIGURE 8.9 Detailed processes in azimuth compression.

For the same $m \times n$ array of range cell migration corrected data, the azimuth compression performs the following three sub-processes for each bin of data in the azimuth domain.

i.   Convert the n-th bin of data in the azimuth domain to the frequency domain using fast Fourier transform (FFT).
ii. Multiply it with the azimuth-matched filter. The azimuth-matched filter is

$$H_{az}\left(f_{\eta}\right) = e^{-j\pi\frac{f_{\eta}}{K_a}} \qquad \text{(Equation 8.24)}$$

where

$$K_a \approx \frac{2V_r^2}{\lambda R_0}$$

f_{η} = azimuth frequency
V_r = effective radar velocity
R_0 = slant range of closest approach

iii. Transform the resultant data back to the time domain by performing inverse Fourier transform (IFFT).

8.6   THE HINOTORI-C SAR PROCESSOR

This book provides the complete source code of the Hinotori-C SAR processor written in MATLAB®. The user interfaces of the SAR processor link all processing scripts used to process the SAR raw data acquired by the Hinotori-C SAR and to generate the point target simulated raw data. Figure 8.10 shows the snapshot of the modules in the SAR processor. The tool's operating manual will be supplied with the source code. The source code is open source, which is possible for modification and redistribution, but acknowledging the authors and citing this book would be much appreciated if it is used in any work [11]–[16].

8.6.1   GENERATING POINT TARGET SIMULATION RAW DATA

To explain and visualize the processing of SAR raw data using the RDA algorithm, first, let us generate simulated raw data consisting of five-point targets with different radar cross sections at different positions using the SAR simulator tool. Figure 8.11(a) visualizes the target's information (position and RCS) and its slant range profile in Figure 8.11(b). Meanwhile, Figures 8.11(c) and (d) plot the magnitude and phase of the raw data in a 2-D image, and the time domain echoes at the center of the range and azimuth distance in Figure 8.11(e). The images and the time domain plots show that the echoes from the target are blended and unsolvable. The weaker targets (e.g., targets with -15dB and -30dB of RCS) are not visible.

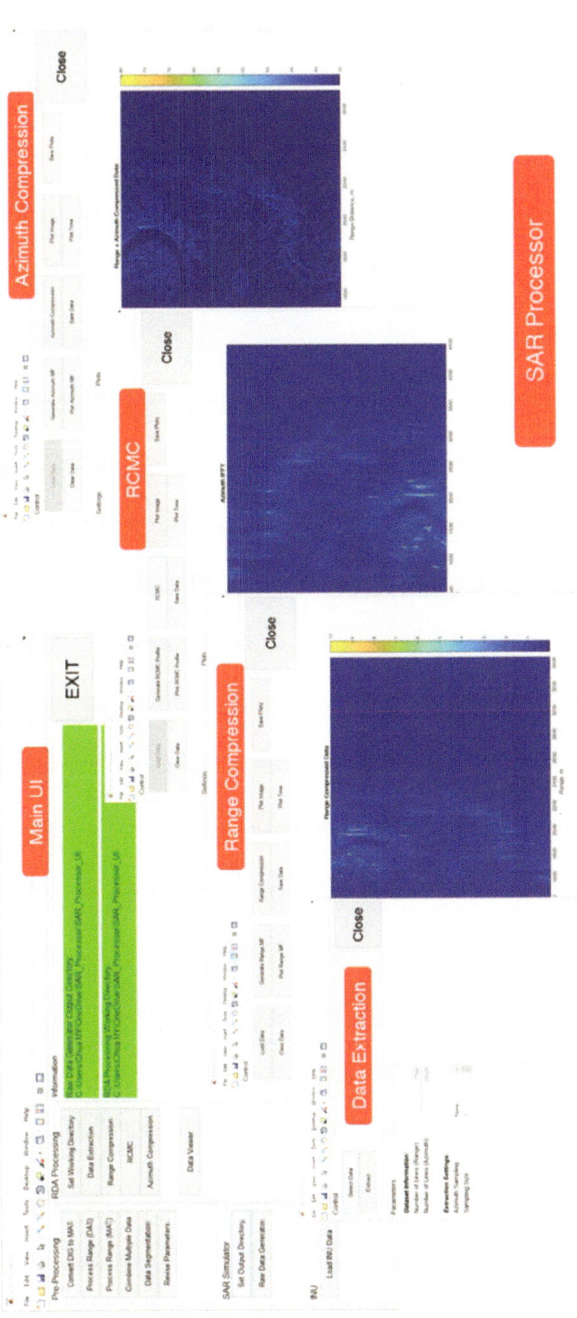

FIGURE 8.10 The Hinotori-C SAR processor.

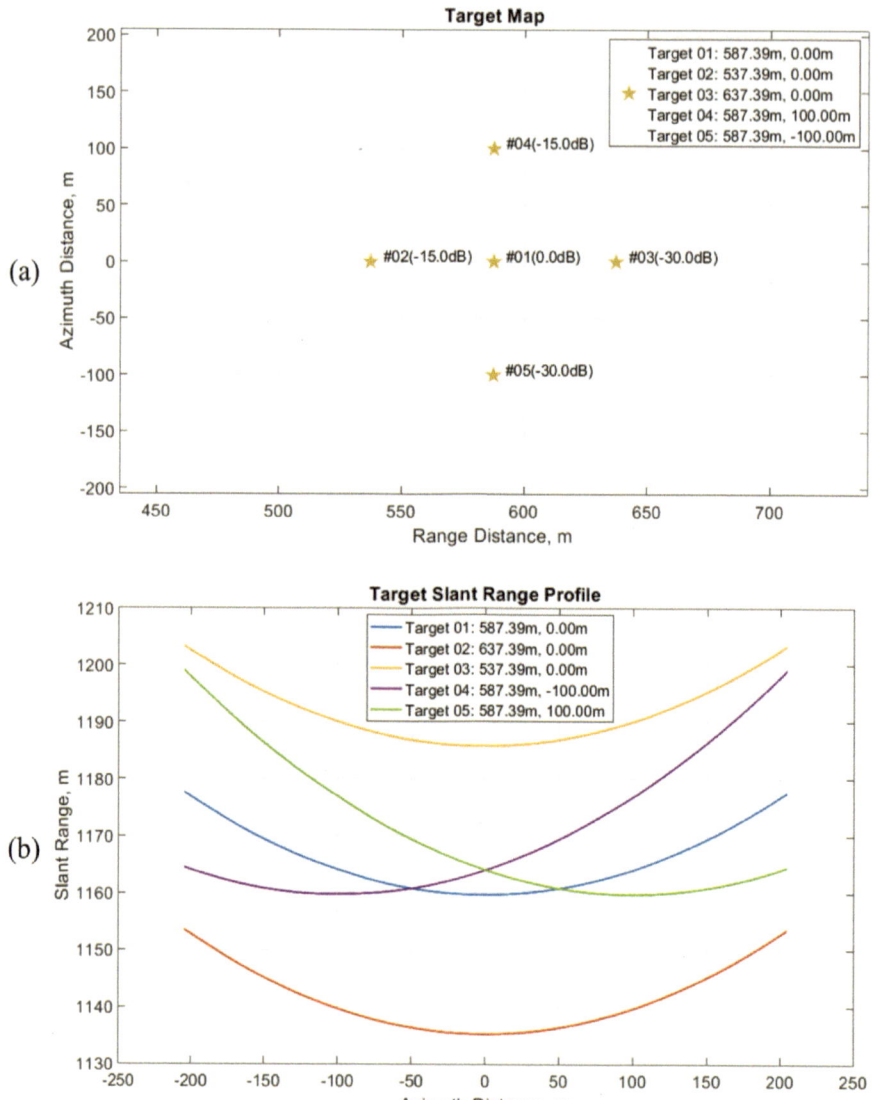

FIGURE 8.11 Point targets simulation; (a) target's information, (b) target's slant range profile, (c) image of raw data (magnitude), (d) image of raw data (phase), (e) range and azimuth signals at the center of the image.

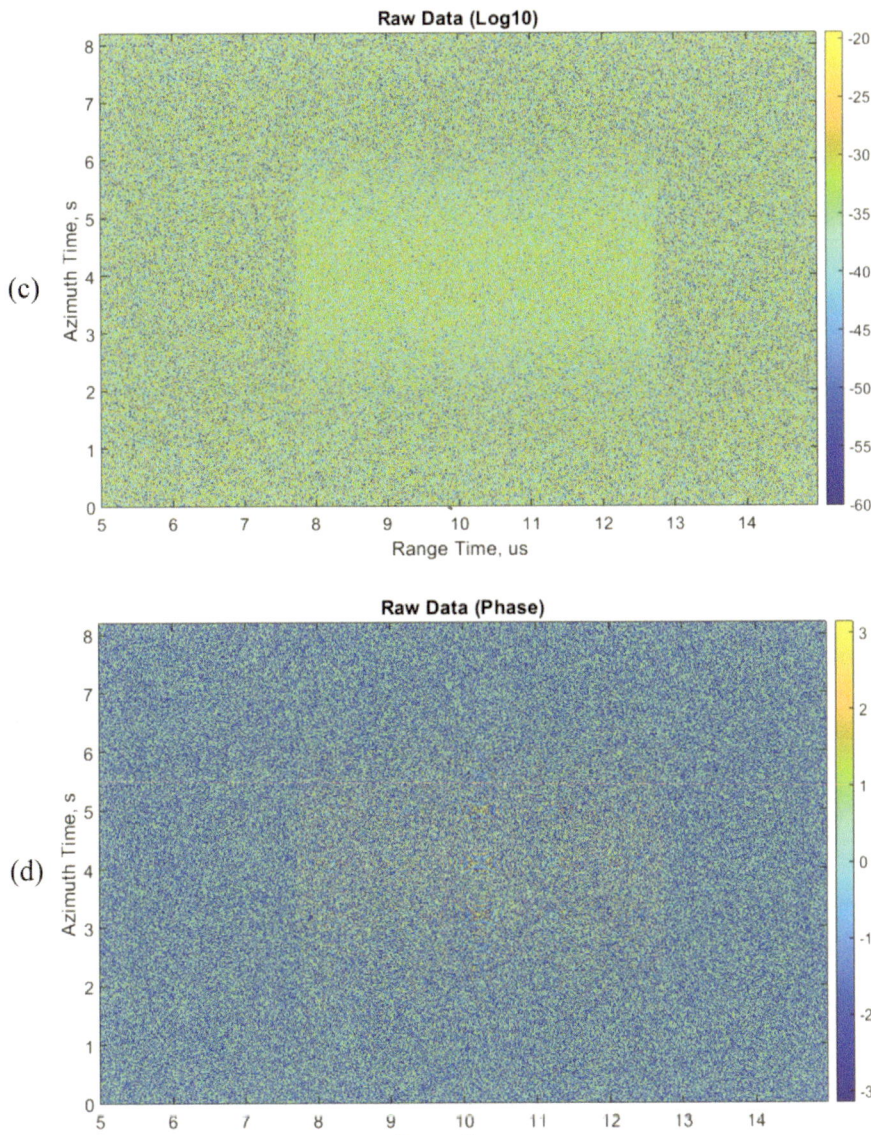

(c)

(d)

FIGURE 8.11 (Continued)

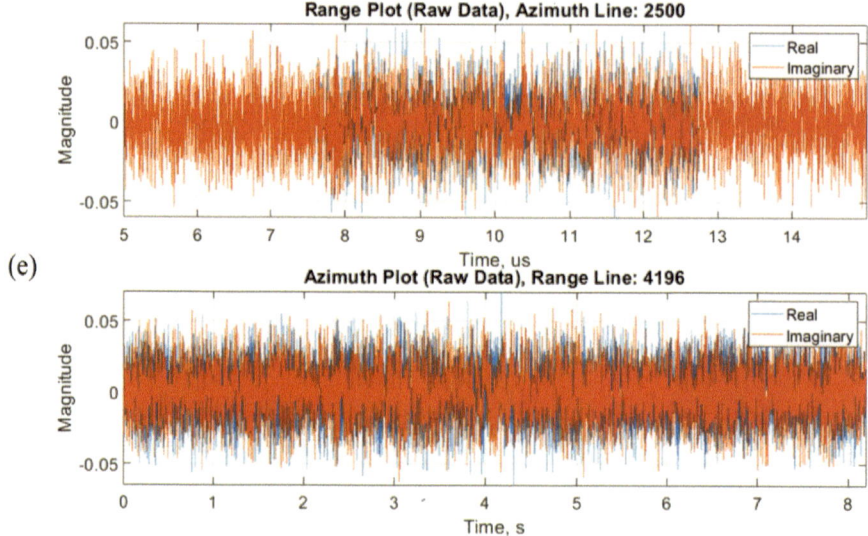

FIGURE 8.11   (Continued)

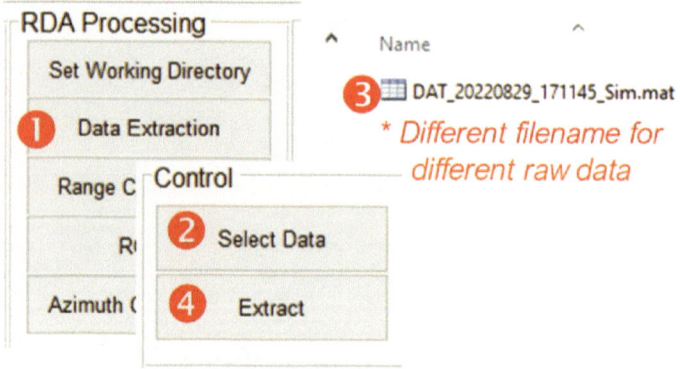

FIGURE 8.12   Data extraction in Hinotori-C SAR processor.

8.6.2   PROCESSING POINT TARGET SIMULATION RAW DATA USING RDA

This section describes the processing of the simulated point target SAR data using the RDA processor in the Hinotori-C SAR processor. A copy of the raw data is provided in this book.

8.6.2.1   Data Extraction

After setting the working directory of RDA processing to the folder containing the raw data file, the simulated raw data with the point targets were extracted. The steps to extract the data are shown in Figure 8.12. A specific file will be created in the working

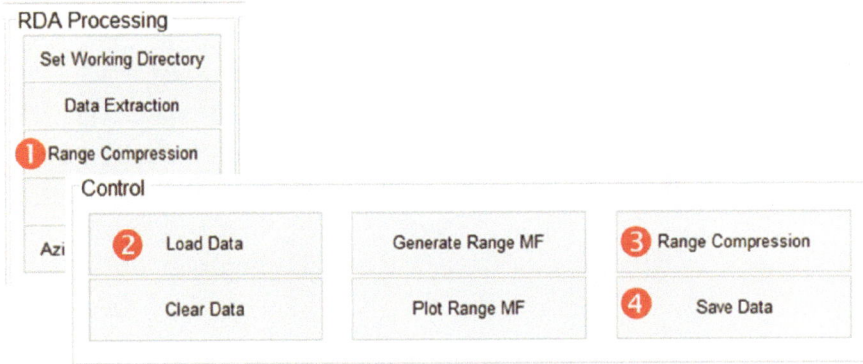

FIGURE 8.13   Range compression in Hinotori-C SAR processor.

directory. Once the data is loaded, the image or the time plot can be viewed in the data viewer module in the SAR processor. The image or the time domain plots would be similar to the plots in Figure 8.11.

8.6.2.2   Range Compression

After the raw data was extracted, the process in the range compression could be carried out in the SAR processor in the range compression module. It is simple and easy to do with the minimum steps shown in Figure 8.13. It is a must to save the data after each processing for use in the next processing step.

The range compression focuses the spreading energy from the echo into an impulse response. By doing so, the range resolution and the echo SNR are significantly improved. As shown in Figure 8.14(a), the trajectory for Target #2 and Target #5 with a lower RCS of -15dB is now visible, but not before the compression as in Figures 8.11(c) and (d). The wide pulse width from the echo has been compressed into a sharper impulse response, as shown in Figure 8.14(b).

8.6.2.3   Range Cell Migration Correction

The next process after range compression is RCMC. This process is essential in RDA as it corrects the target's trajectory into a straight line before the azimuth compression. By doing so, the energy could be focused better in the azimuth domain, thus achieving better SNR improvement to get the optimum impulse response. The minimum steps to perform RCMC in the Hinotori-C SAR processor are described in Figure 8.15. The result after RCMC in Figure 8.16(b) shows that the trajectory of the targets has been straightened.

8.6.2.4   Azimuth Compression

The final process in SAR image formation using RDA is azimuth compression. Azimuth compression applies another matched filtering but at azimuth domain with azimuth matched filter. The minimum steps to perform azimuth compression in the

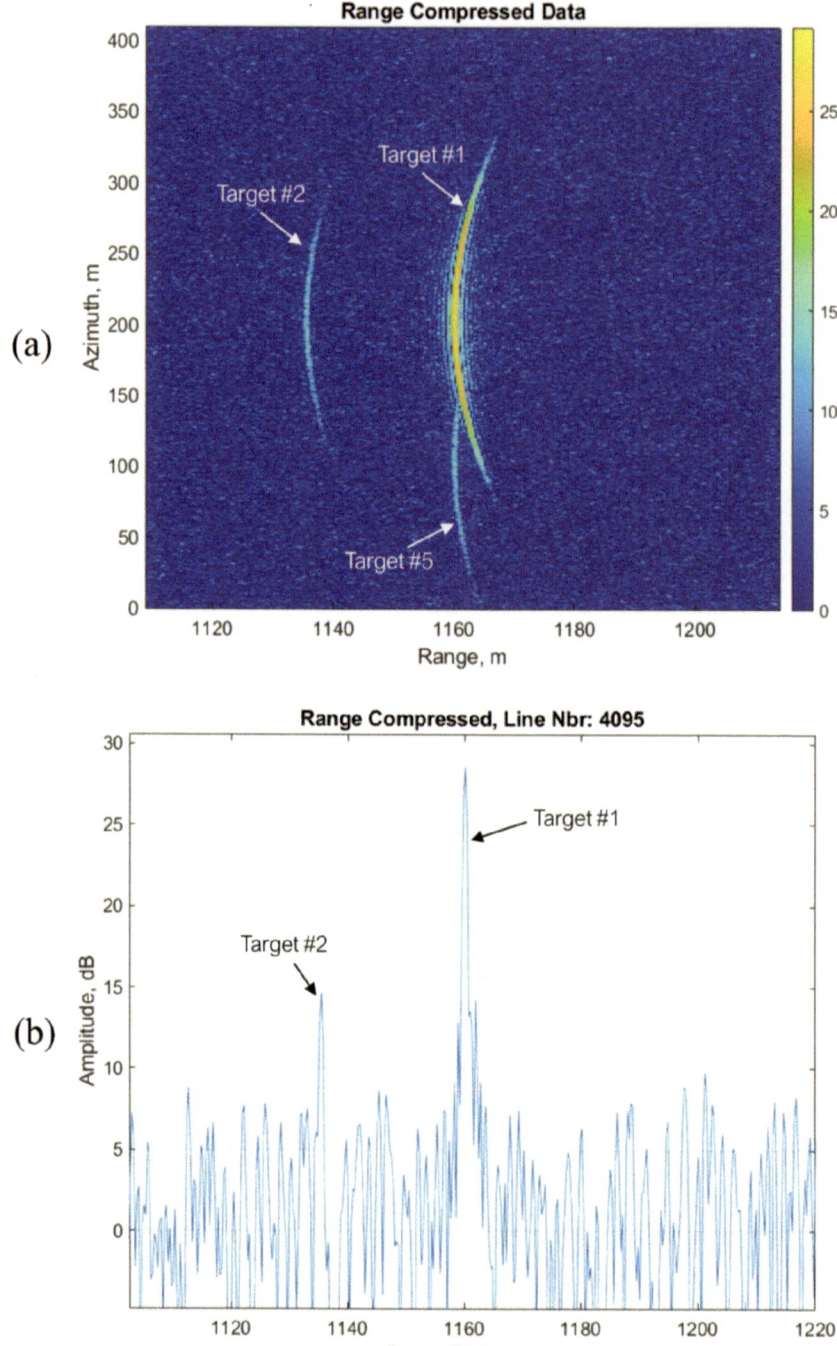

FIGURE 8.14 Range compression of simulated SAR data; (a) image after range compression, (b) signal plots at the center of azimuth distance.

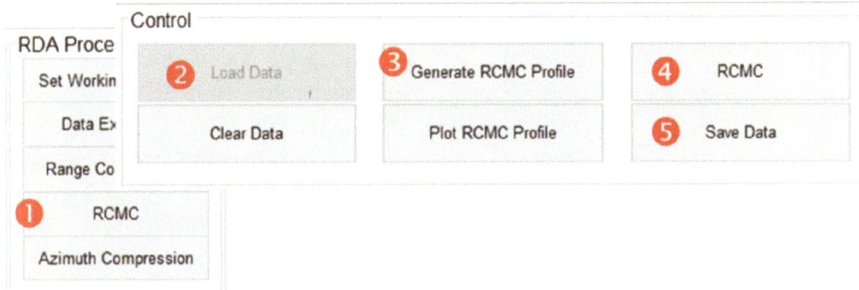

FIGURE 8.15   RCMC in Hinotori-C SAR processor.

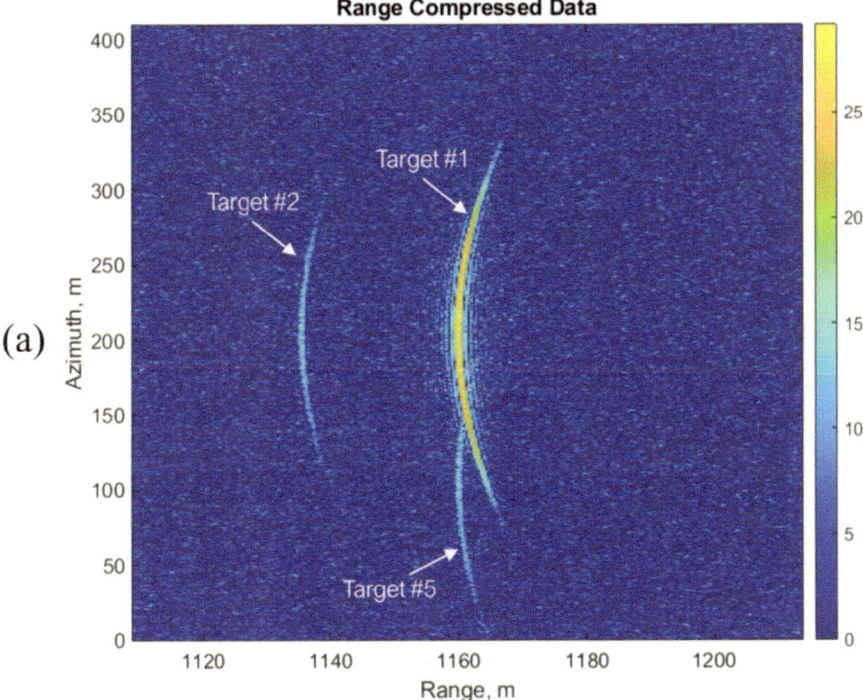

FIGURE 8.16   RCMC of simulated SAR data; (a) before RCMC, (b) after RCMC.

Hinotori-C SAR processor are described in Figure 8.17. The azimuth compression module in the Hinotori-C SAR processor can process the SAR data with and without range cell migration correction. Figure 8.18 compares the SAR image formed with and without the RCMC process, showing that the point target's energy improved by about 6dB with RCMC processing. Meanwhile, Figure 8.19 compares the signals at the image center on both distances before and after the azimuth compression. Targets #3 and #4 are visible after the azimuth compression due to the SNR gain.

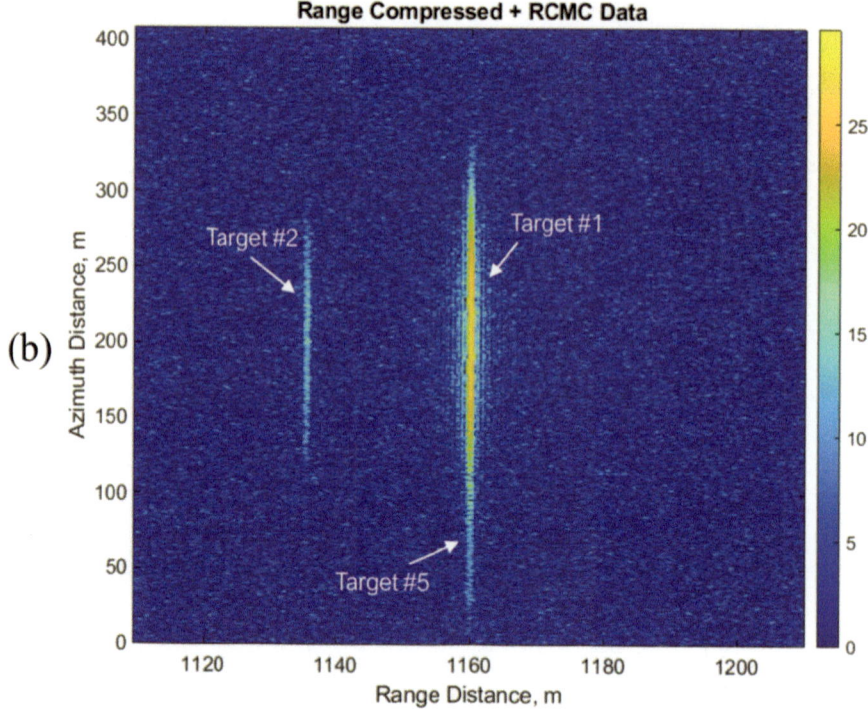

FIGURE 8.16 (Continued)

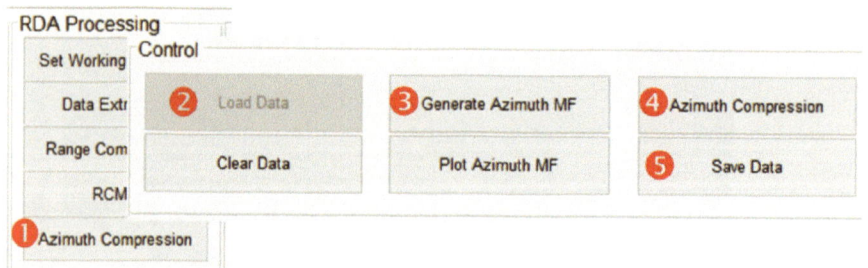

FIGURE 8.17 Azimuth compression in Hinotori-C SAR processor.

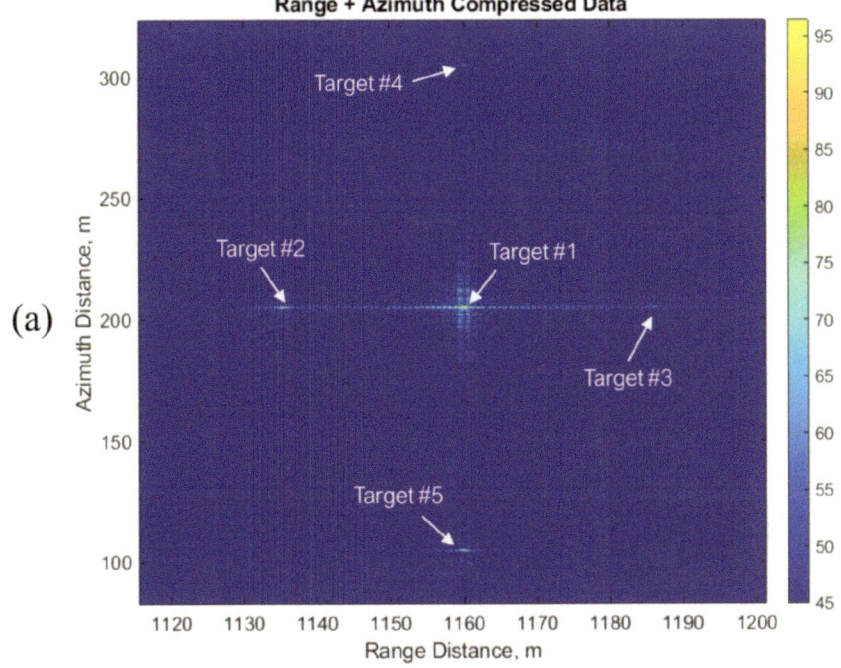

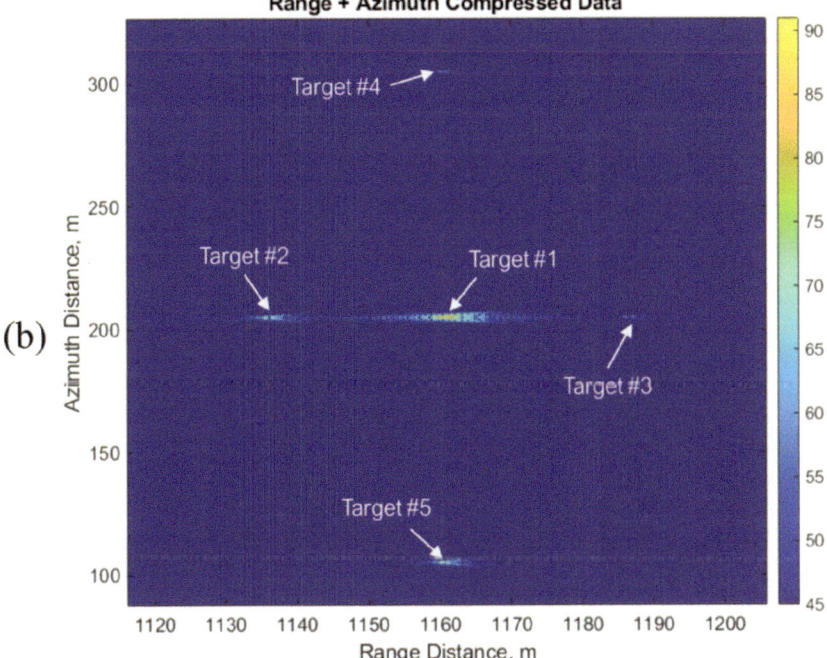

FIGURE 8.18 Azimuth compression of simulated SAR data; (a) with RCMC, (b) without RCMC.

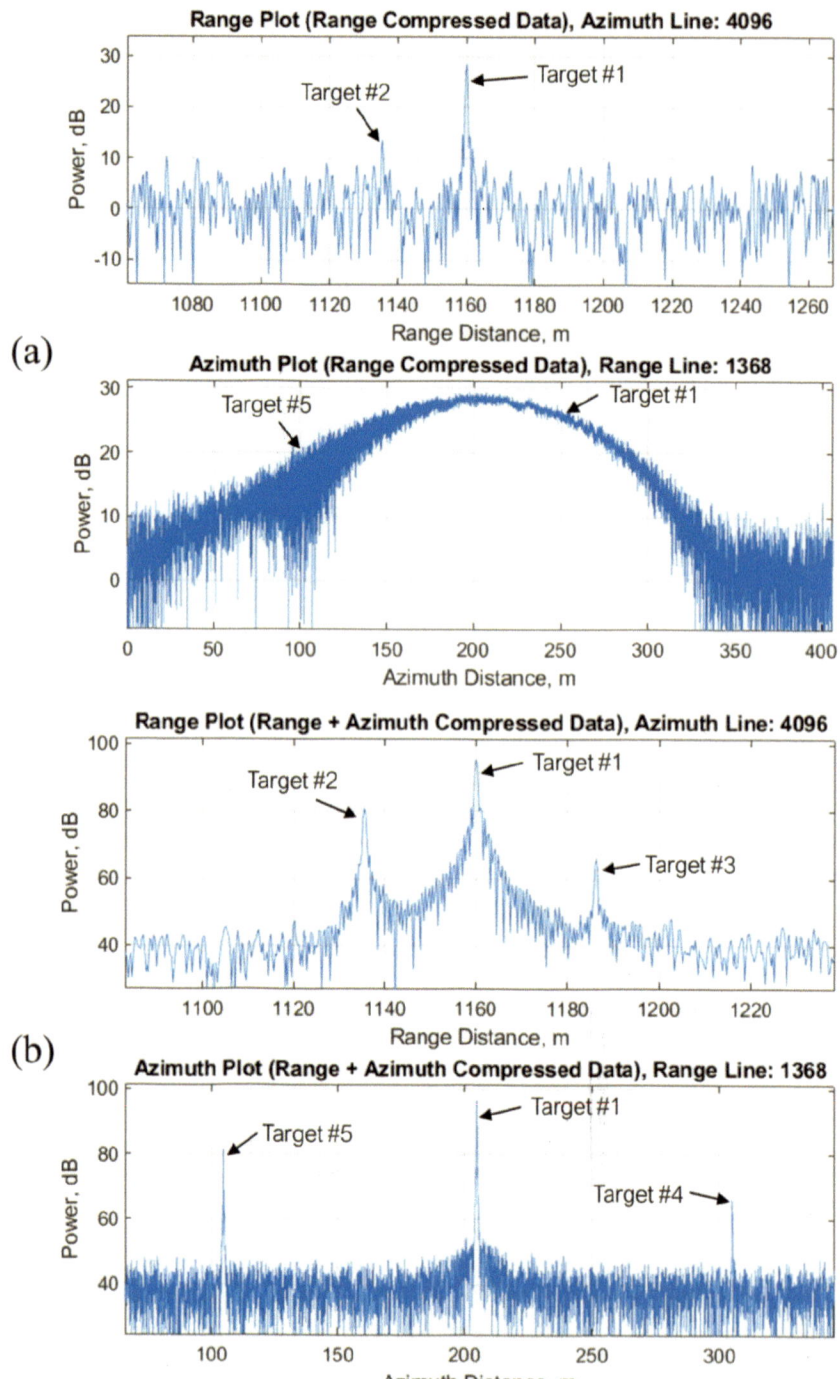

FIGURE 8.19 Signal plots at the center of the image; (a) before azimuth compression, (b) after azimuth compression.

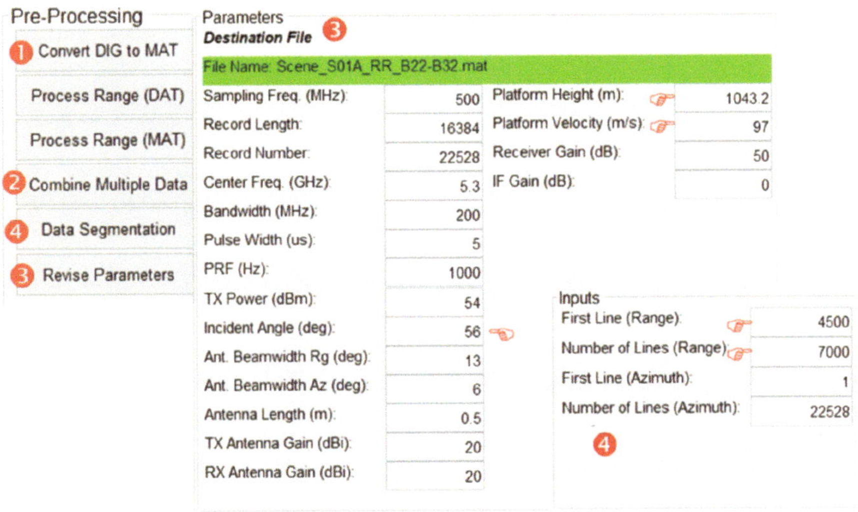

Pre-Processing	Parameters

Destination File ❸

File Name: Scene_S01A_RR_B22-B32.mat

Pre-Processing	Parameters		
❶ Convert DIG to MAT			
Process Range (DAT)	Sampling Freq. (MHz):	500	Platform Height (m): ☞ 1043.2
Process Range (MAT)	Record Length:	16384	Platform Velocity (m/s): ☞ 97
	Record Number:	22528	Receiver Gain (dB): 50
❷ Combine Multiple Data	Center Freq. (GHz):	5.3	IF Gain (dB): 0
❹ Data Segmentation	Bandwidth (MHz):	200	
	Pulse Width (us):	5	
❸ Revise Parameters	PRF (Hz):	1000	
	TX Power (dBm):	54	Inputs
	Incident Angle (deg):	56	First Line (Range): ☞ 4500
	Ant. Beamwidth Rg (deg):	13	Number of Lines (Range): ☞ 7000
	Ant. Beamwidth Az (deg):	6	First Line (Azimuth): 1
	Antenna Length (m):	0.5	Number of Lines (Azimuth): 22528
	TX Antenna Gain (dBi):	20	❹
	RX Antenna Gain (dBi):	20	

FIGURE 8.20   Data preparation for processing Hinotori-C2 SAR data.

8.6.3 PROCESSING HINOTORI-C2 MISSION DATA

This section describes the processing of the SAR data acquired during the Hinotori-C2 flight mission using the RDA processor in the Hinotori-C SAR processor. The SAR raw data is provided in this book.

8.6.3.1 Data Preparation

The SAR echoes recorded by the Hinotori-C SAR are saved in blocks of 2048 binary data files by the digitizer. These files need to be converted to MAT-file following the SAR processor data structure so the processing can be carried out in the Hinotori-C SAR processor. Figure 8.20 illustrates the steps need for data preparation. Detailed instructions on what to do in each step are available in the operating booklet of the SAR processor. The processes involve converting the data blocks into the appropriate file format, combining several data blocks, revising the processing parameters associated with the file, and segmenting the data to discard the unuseful data points in the raw data to speed up the processing. Once the data is prepared, the RDA processor can use it for image formation.

8.6.3.2 RDA Processing

The steps for processing the SAR data are similar to the steps in processing the simulated data in Section 8.5 Figure 8.21 shows the intermediate outputs from each RDA sub-processes.

8.6.3.3 Multilooking

Multilooking is a postimage processing technique that generates images with lower speckle and increased image quality. A multilooked image will have less noise and

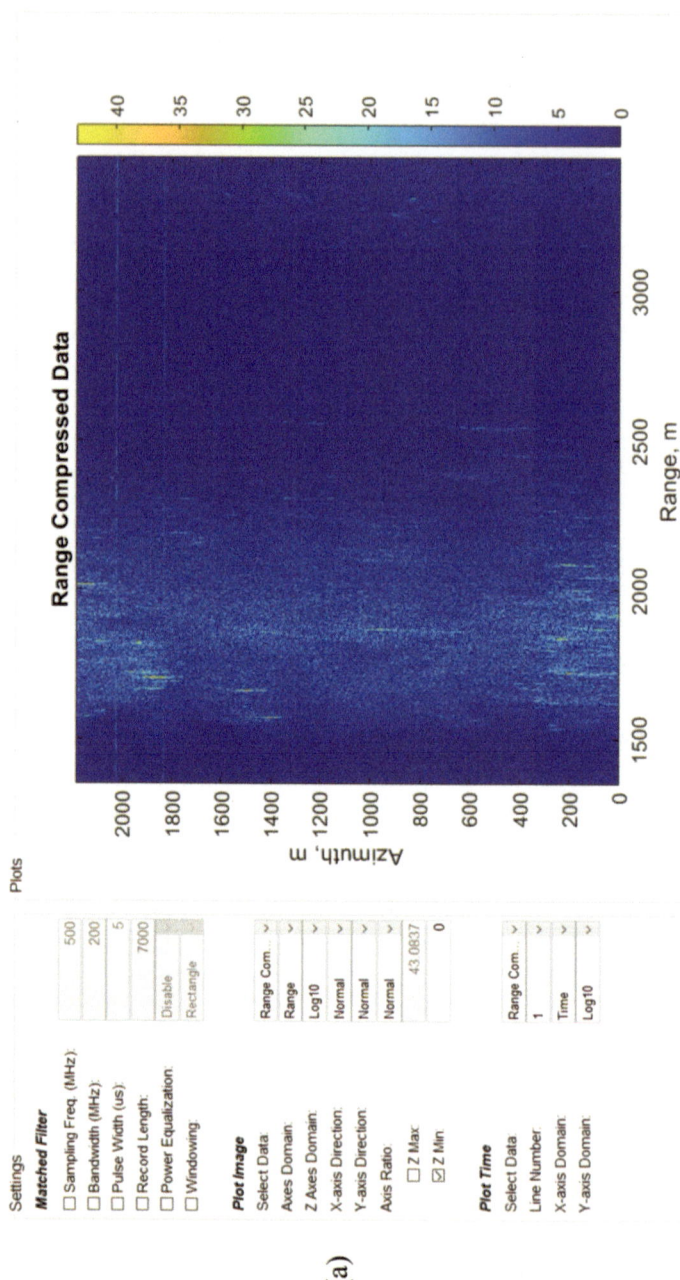

FIGURE 8.21 Processing a block of Hinotori-C SAR data: (a) range compression, (b) range cell migration correction, (c) azimuth compression.

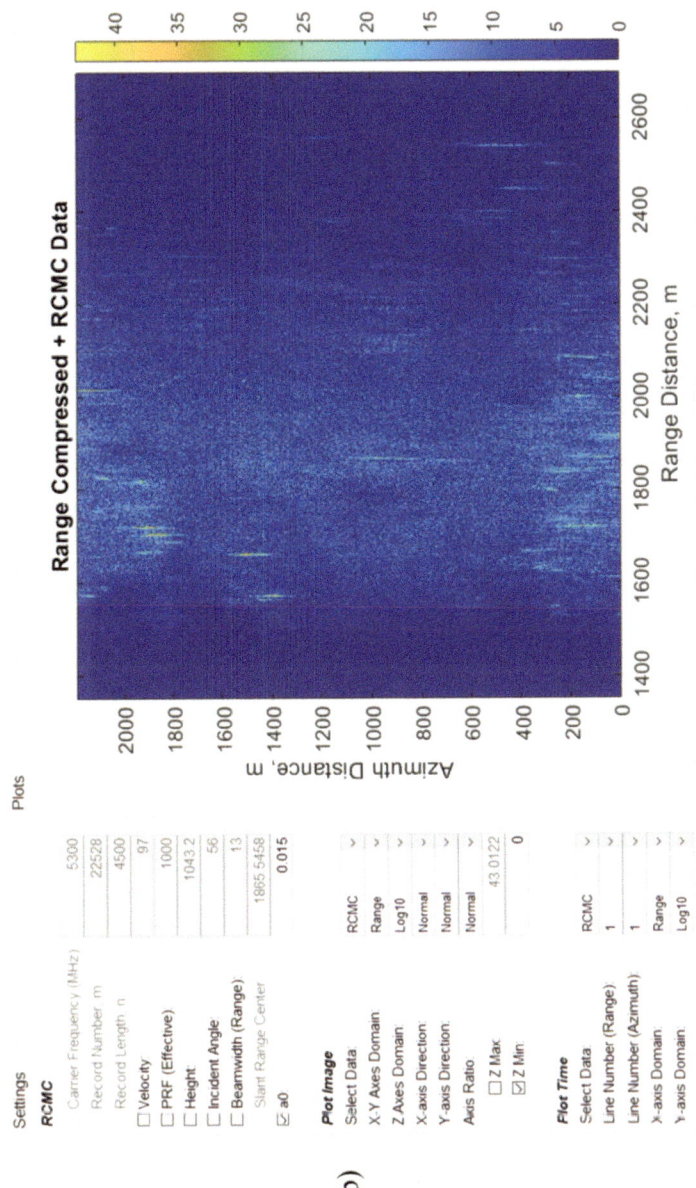

FIGURE 8.21 (Continued)

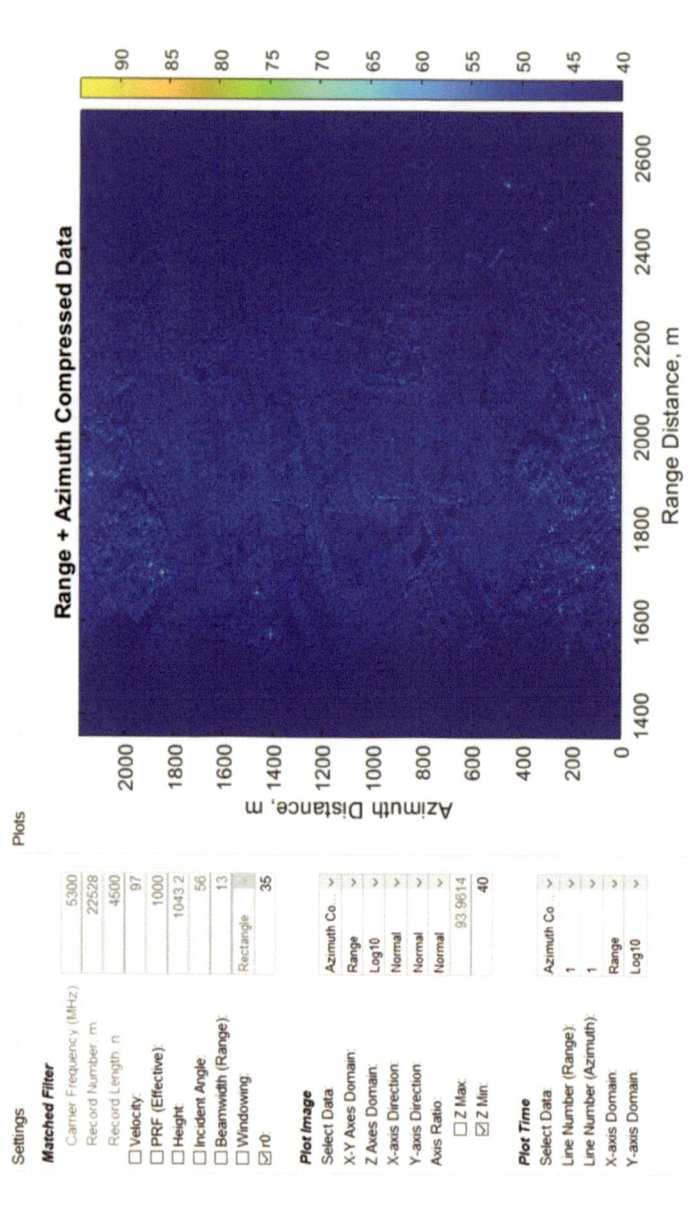

FIGURE 8.21 (Continued)

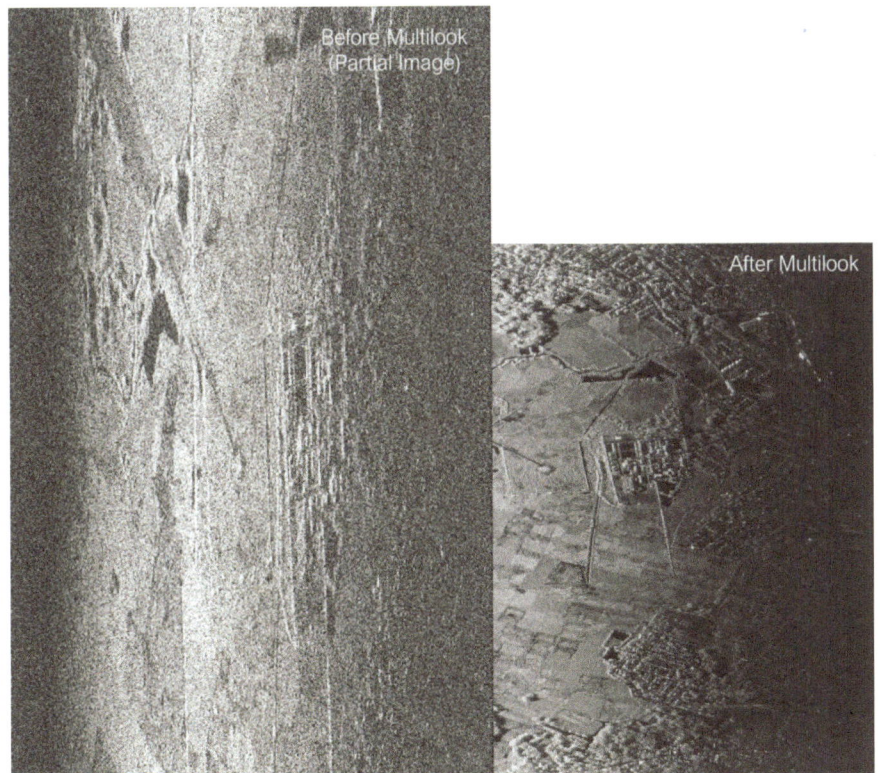

FIGURE 8.22 SAR image before and after multilooking.

approximate square pixel spacing after being converted from slant range to ground range. The SAR processor generates multi-look images by averaging over range and/ or azimuth resolution cells. Figure 8.22 compares the SAR image produced by the Hinotori-C SAR processor before and after multilooking.

REFERENCES

[1] MI Skolnik, *Radar Handbook*, New York, McGraw-Hill, 1970.
[2] WR August., *High Resolution Radar*, Boston, Artech House, 1996.
[3] IG Cumming and F.H. Wong, *Digital Processing of Synthetic Aperture Radar Data: Algorithms and Implementation*, Boston, Artech House, 2005.
[4] M Soumekh, *Synthetic Aperture Radar Signal Processing with MATLAB Algorithms*, New York, Wiley, 1999.
[5] BF Mahafza, *Introduction to Radar Analysis*, New York, CRC Press, 1998.
[6] YK Chan, *Transmitter and Receiver Design of an Airborne Synthetic Aperture Radar*, Master Dissertation, Multimedia University, Cyberjaya, Malaysia, 2002.
[7] MY Chua, *Reconfigurable Waveform Synthesis Techniques for Synthetic Aperture Radar*, Ph.D. Dissertation, Multimedia University, Malaysia, 2016.

[8] YK Chan, MY Chua, VC Koo, "Sidelobes Reduction Using Simple Two and Tri-stages Nonlinear Frequency Modulation (NLFM)", Progress in Electromagnetics Research, Vol. 98, pp. 33–52, 2009.

[9] MY Chua, VC Koo, HS Lim, YK Chan, "An FPGA Based Real-Time Multi-Target Synthetic Aperture Radar Echoes Synthesizer", Journal of Engineering Technology and Applied Physics, Vol. 2, Issue 2, pp. 10–16, Dec 2020.

[10] MY Chua, VC Koo, HS Lim, JT Sri Sumantyo, "Phase-coded Stepped Frequency Linear Frequency Modulated Waveform Synthesis Technique for Low Altitude Ultra-Wideband Synthetic Aperture Radar", IEEE Access, Vol. 5, pp. 11391–11403, May 2017.

[11] JT Sri Sumantyo, MY Chua, CE Santosa, GF Panggabean, T Watanabe, B Setiadi, FD Sri Sumantyo, K Tsushima, K Sasmita, A Mardiyanto, E Supartono, ET Rahardjo, G Wibisono, MA Marfai, RH Jatmiko, S Sudaryatno, TH Purwanto, BS Widartono, M Kamal, D Perissin, S Gao, and K Ito, "Airborne Circularly Polarized Synthetic Aperture Radar," IEEE Selected Topics in Applied Earth Observations and Remote Sensing (JSTARS), Vol. 14, pp.1676–1692, January 2021, DOI:10.1109/JSTARS.2020.3045032.

[12] MY Chua,, JT Sri Sumantyo, CE Santosa, GF Panggabean, FD Sri Sumantyo, T Watanabe, YQ Ji, PP Sitompul, M Nasucha, F Kurniawan, B Purbantoro, Awaludin, K Sasmita, ET Rahardjo, G Wibisono, RH Jatmiko, S Sudaryatno, T Purwanto, BS Widartono, and M Kamal, "The Maiden Flight of Hinotori-C: The First C Band Full Polarimetric Circularly Polarized Synthetic Aperture Radar in the World", IEEE Aerospace and Electronic Systems Magazine, Vol. 34, Issue 2, pp. 24–35, June 2019.

[13] JT Sri Sumantyo, MY Chua, CE Santosa, GF Panggabean, T Watanabe, B Setiadi, K Tsushima, FD Sri Sumantyo, K Sasmita, A Mardiyanto, E Supartono, ET Rahardjo, G Wibisono, RH Jatmiko, TH Purwanto, BS Widartono, M Kamal, RH Triharjanto, S Gao, and K Ito, "Hinotori-C2 mission: CN235MPA Aircraft Onboard Circularly Polarized Synthetic Aperture Radar (CP-SAR)", 2019 IEEE International Geoscience and Remote Sensing Symposium (IGARSS2019), pp. 8538–8541, July 2019.

[14] JT Sri Sumantvo, MY Chua, CE Santosa, A Takahashi, and K Ito, "Aircraft and High-Altitude Platform System Onboard Circularly Polarized Synthetic Aperture Radar (CP-SAR)", 2021 IEEE International Geoscience and Remote Sensing Symposium (IGARSS2021), pp. 8515–8518, July 2021.

[15] JT Sri Sumantvo, MY Chua, CE Santosa, GF Panggabean, K Tsushima, T Watanabe, K Sasmita, A Mardiyanto, FD Sri Sumantyo, Eko Tjipto Rahardjo, Gunawan Wibisono, E Supartono, S Gao, SP Parulian, M Nasucha, F Kurniawan, A Awaludin, B Purbantoro, YQ Ji, and N Imura, "Hinotori-C: A Full Polarimetric C Band Airborne Circularly Polarized Synthetic Aperture Radar for Disaster Monitoring", 2018 Progress in Electromagnetics Research Symposium (PIERS-Toyama), pp. 1466–1473, August 2018.

[16] MY Chua, JT Sri Sumantyo, CE Santosa, GF Panggabean, YQ Ji, SP Parulian, and M Nasucha, "An PC-based Airborne SAR Baseband System", 2018 Progress in Electromagnetics Research Symposium (PIERS-Toyama), pp. 882–888, Aug 2018.

9 Applications

Yuta Izumi and Josaphat Tetuko Sri Sumantyo

9.1 INTRODUCTION

Fully polarimetric circularly polarized SAR has been developed. Polarimetric data gives a chance to retrieve target information and scattering mechanism, as demonstrated in Chapter 2. To better understand the scattering phenomenon on circularly polarized SAR, this chapter presents several experiments regarding polarimetric analysis for circularly polarized SAR. We first introduce the indoor circularly polarized SAR scattering experiment under a controlled environment with several canonical targets and an aircraft model as a complex target. Furthermore, as applications of the circularly polarized SAR, we present indoor rice phenology monitoring and compare the performances among several polarimetric modes introduced in Chapter 2. Finally, we show the polarimetric parameters estimation results by the airborne circularly polarized SAR through the Hinotori-C2 mission.

9.2 INDOOR CIRCULARLY POLARIZED SAR SCATTERING EXPERIMENT

This section aims to demonstrate the circularly polarized SAR properties on polarimetric data and target decomposition performance through the actual ground-based indoor measurement. We collect the circularly polarized SAR data in a controlled laboratory environment to experiment with high precision. We first demonstrate the system description of the employed circularly polarized SAR system, imaging, and polarimetric calibration method. We then present the circular polarization backscattering and target decomposition experiment on some canonical targets and aircraft models, highlighting the difference from linearly polarized SAR.

9.2.1 Imaging Geometry and System Description

In this section, we conduct a radar experiment within an anechoic chamber. We adopt the inverse SAR (ISAR) data collection approach where the target is moved for the synthetic aperture processing because of the size restriction of an anechoic chamber. Furthermore, we introduce two types of data collection geometry – linear ISAR geometry, where the target moves linearly along the rail, and circular ISAR geometry,

DOI: 10.1201/9781003282693-9

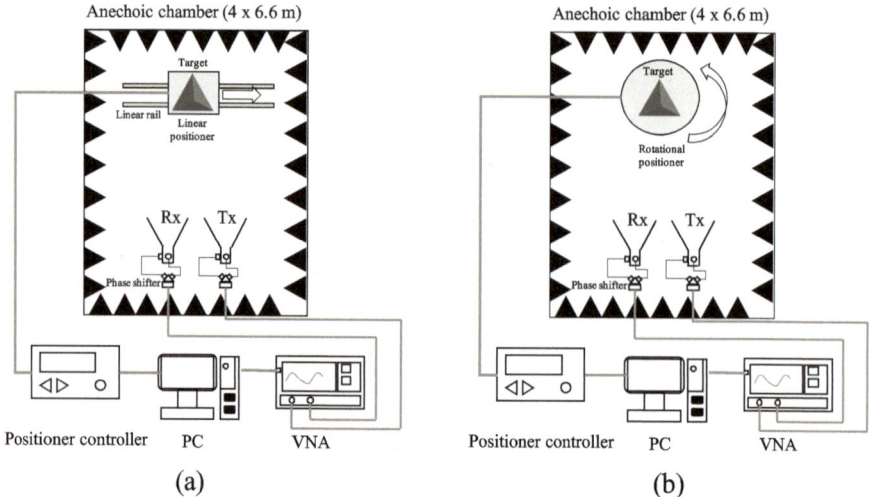

FIGURE 9.1   Circularly polarized SAR system configuration and geometry. (a) Linear ISAR geometry. (b) Circular ISAR geometry.

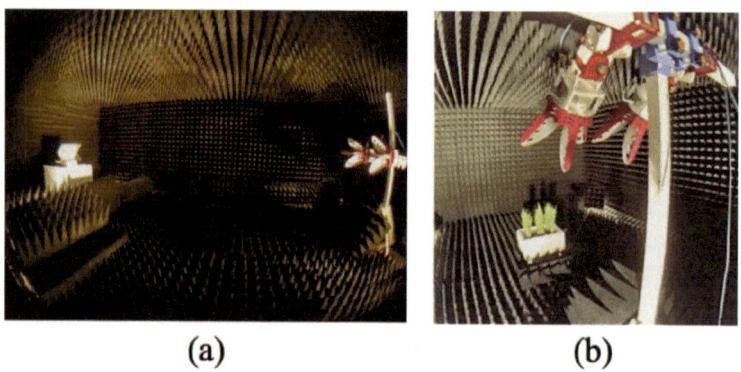

FIGURE 9.2   Sample photographs from the actual experiment. (a) Linear ISAR geometry. (b) Circular ISAR geometry.

where the target is rotated around its center. We illustrate both imaging geometries in Figure 9.1 and show sample photographs from the actual experiment in Figure 9.2.

Ground-based circularly polarized SAR system consists of a vector network analyzer (VNA), dual ridged horn antennas, an automated positioner controller, and a control PC, as illustrated in Figure 9.1. We introduce a 90° phase shifter on dual ridged horn antennas to generate the circular polarization signal. Applying a 90° phase shift into the V-port generates an LHCP signal while shifting the H-port generates an RHCP signal according to polarization theory in Chapter 2. The PC triggers and controls both the VNA and positioning mechanism. All parameters such as frequency, output power, frequency sampling points, azimuth sampling points, and the azimuth interval can be set via PC. An automatic positioning device moves the target linearly along a 2.2m rail for linear ISAR geometry and rotates the target for circular ISAR geometry. The operational frequency in this experiment is C-band spanning from 4.5 GHz to 7.5 GHz.

A VNA-based ground-based radar system transmits and receives the series of single frequency short continuous sub-waves, termed stepped frequency modulated wave (SFCW) [1]. The back projection algorithm [2] and $\omega - k$ algorithm [3] are employed as the SAR image reconstruction algorithm.

To check the deviation from circular polarization, the axial ratio parameter of the antennas is measured. Theoretically, 0dB of AR shows ideal circular polarization while an infinite value corresponds to ideal linear polarization, as demonstrated in Section 2.1.7. An AR value of less than 3dB is generally considered acceptable for most applications [4], and we follow this definition throughout this book. The AR of our circular polarization antennas was measured inside the anechoic chamber and was mostly under 2.5dB over the entire operational frequency bandwidth as shown in Figure 9.3, meaning that our antennas achieve good circular polarization purity within the investigated bandwidth.

All multipolarimetric radars should deal with challenging calibration problem which involves the removal of system-introduced polarimetric distortions such as polarimetric channel imbalances, crosstalk, and antenna gain. This contamination inevitably degrades the polarimetric decomposition results, and thus polarimetric calibration should be performed in all cases. Here, we introduce the polarimetric calibration method proposed by Wiesbeck et al. in [5] as a suitable and effective way of calibrating ground-based LP radar systems. To apply this method to circularly polarized radar, an error model on circular polarization basis is constructed [6] as

$$\begin{bmatrix} S_{LL}^{m} & S_{LR}^{m} \\ S_{RL}^{m} & S_{RR}^{m} \end{bmatrix} = \begin{bmatrix} R_{LL} & R_{LR} \\ R_{RL} & R_{RR} \end{bmatrix} \cdot \begin{bmatrix} S_{LL}^{c} & S_{LR}^{c} \\ S_{RL}^{c} & S_{RR}^{c} \end{bmatrix} \cdot \begin{bmatrix} T_{LL} & T_{LR} \\ T_{RL} & T_{RR} \end{bmatrix} + \begin{bmatrix} I_{LL} & I_{LR} \\ I_{RL} & I_{RR} \end{bmatrix}. \qquad \text{(Equation 9.1)}$$

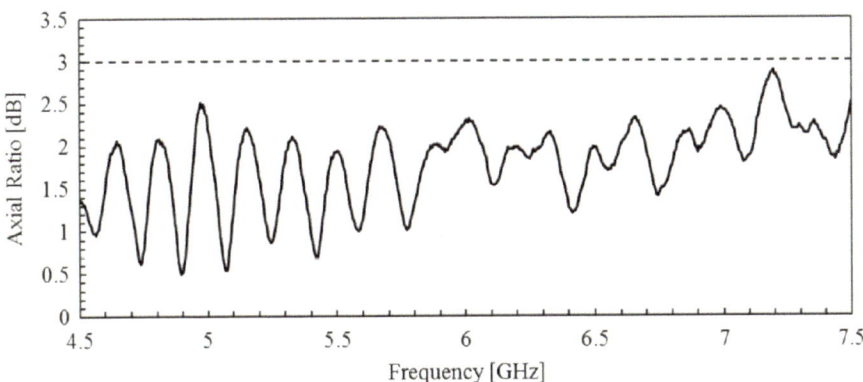

FIGURE 9.3   Measured AR plot in the range from 4.5GHz to 7.5GHz.

The model includes receiving $[\mathbf{R}]$ and transmitting $[\mathbf{T}]$ distortion matrices, an isolation distortion matrix $[\mathbf{I}]$, and correct $[\mathbf{S}^c]$ and measured $[\mathbf{S}^m]$ scattering matrices. Using three types of canonical reflectors: circular plate, dihedral, and 45° tilted dihedral, error coefficients in distortion matrices $[\mathbf{R}]$ and $[\mathbf{T}]$ can be estimated, as explained in detail in [6]. Empty room calibration is also needed within the steps to extract the isolation matrix $[\mathbf{I}]$, which can be performed easily by measuring a target-free scene.

9.2.2 SCATTERING EXPERIMENT ON CANONICAL TARGETS AND ITS RESULTS

We show the investigation results of the polarimetric response of several canonical targets with a known theoretical scattering matrix [7]. We first obtain focused SAR images for each target, followed by applying the polarimetric target decomposition technique to the target image. The steps of this experiment are in the following,

(i) Collect radar data of targets
(ii) Apply full polarimetric calibration to the collected data
(iii) Reconstruct SAR images
(iv) Derive coherency matrices from the focused image data
(v) Obtain H and $\bar{\alpha}$ through eigenvector-based decomposition based on the coherency matrices

The circularly polarized SAR images were reconstructed by applying the $\omega - k$ algorithm to the collected frequency domain data [3]. The chosen canonical test targets were a circular plate, a horizontal dihedral, a vertical wire, a horizontal wire, and a 45° tilted wire. The photographs of the measured targets are shown in Figure 9.4. Reconstructed circular polarized SAR images of the targets are shown in Figure 9.5 which are displayed on a normalized 40dB dynamic range. Since S_{RL} images are assumed to be equal to S_{LR} images under the reciprocity theorem, only the S_{LR} images are included in Figure 9.5. The wire targets in three orientation angles, that are vertical, horizontal, and 45° exhibit almost the same intensity levels for all polarizations,

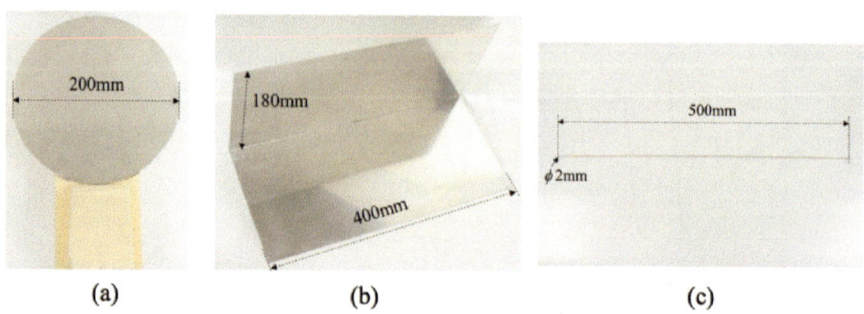

(a)   (b)   (c)

FIGURE 9.4 Photographs of the targets used in the experiment. (a) Circular plate. (b) Dihedral. (c) Wire.

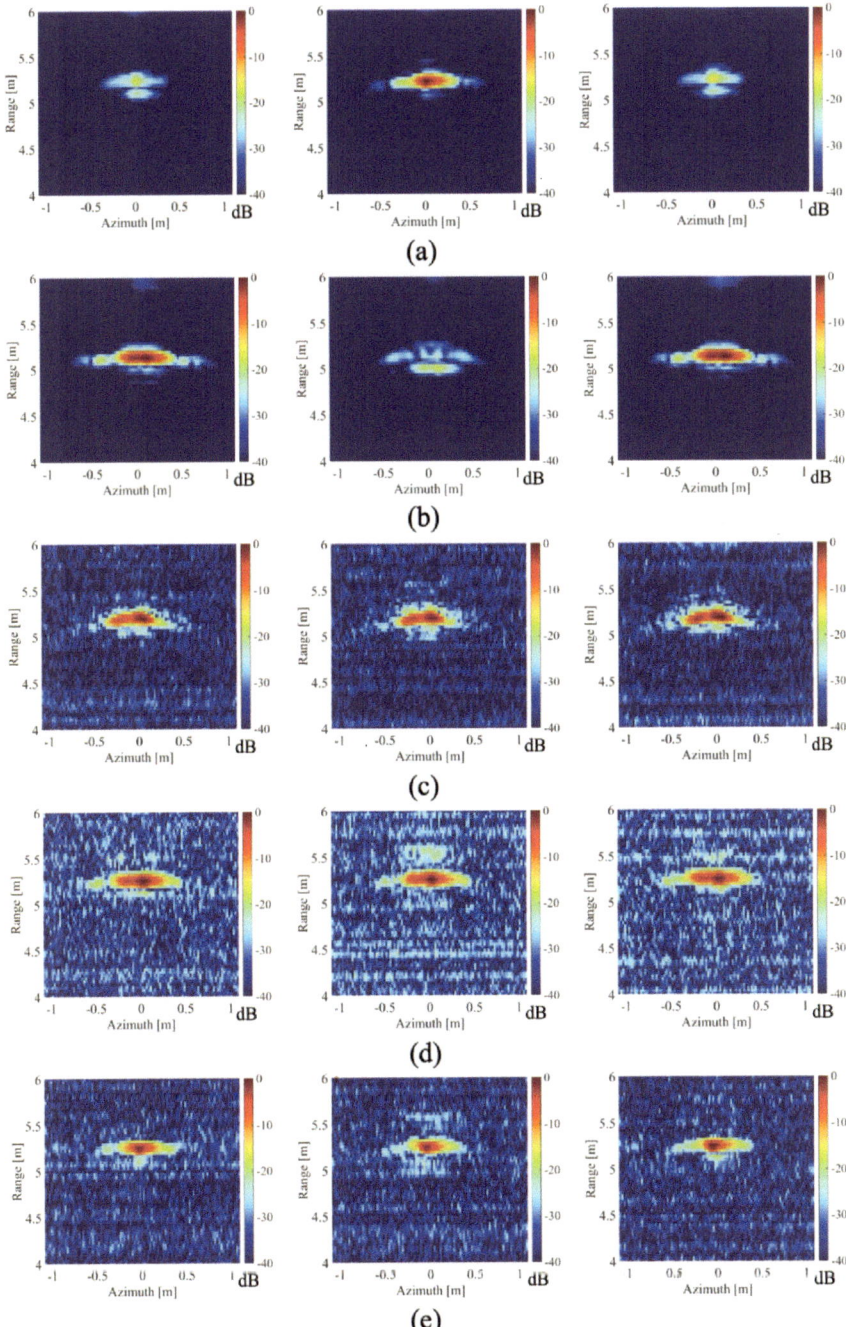

FIGURE 9.5 Circularly polarized SAR images reconstructed by an $\omega - k$ algorithm (left: LL, middle: LR, right: RR). (a) Circular plate. (b) Horizontal dihedral. (c) Vertical wire. (d) Horizontal wire. (e) 45° tilted wire.

TABLE 9.1

Scattering Matrix in (H-V) and (L-R) Basis of Some Canonical Targets

Target	Scattering matrix in (L-R) basis	Measured scattering matrix
Circular plate	$\begin{bmatrix} 0 & j \\ j & 0 \end{bmatrix}$	$\begin{bmatrix} 0.1451e^{-j170.4987°} & 0.9493e^{-j17.1594°} \\ 1.0000e^{j0°} & 0.1866e^{-j16.2642°} \end{bmatrix}$
Horizontal dihedral	$\begin{bmatrix} 1 & 0 \\ 0 & -1 \end{bmatrix}$	$\begin{bmatrix} 1.0000e^{j0°} & 0.1293e^{-j40.6913°} \\ 0.1244e^{-j53.5900°} & 0.9954e^{j176.1457°} \end{bmatrix}$
Vertical wire	$\dfrac{1}{2}\begin{bmatrix} 1 & j \\ j & -1 \end{bmatrix}$	$\begin{bmatrix} 0.9665e^{j172.3562°} & 0.9225e^{j93.2553°} \\ 0.9762e^{j97.1914°} & 1.0000e^{j0°} \end{bmatrix}$
Horizontal wire	$\dfrac{1}{2}\begin{bmatrix} -1 & j \\ j & 1 \end{bmatrix}$	$\begin{bmatrix} 1.0000e^{j0°} & 0.9995e^{j86.6877°} \\ 0.8612e^{j93.2656°} & 0.9659e^{j170.2806°} \end{bmatrix}$
45°tilted wire	$\dfrac{1}{2}\begin{bmatrix} j & j \\ j & j \end{bmatrix}$	$\begin{bmatrix} 0.9125e^{j3.7878°} & 0.8350e^{j22.9336°} \\ 0.8206e^{j7.1116°} & 1.0000e^{j0°} \end{bmatrix}$

which validates the invariance of the circular polarization signal's intensity to LOS rotation of targets.

A circularly polarized SAR image provides the knowledge of odd-bounce or even-bounce scattering mechanisms by examining the cross-polarization or co-polarization components, respectively. This can be confirmed from the images of the circular plate and the dihedral.

To investigate the phase and amplitude relations between polarimetric channels, the scattering matrices of targets are extracted from the target location (maximum intensity pixel). The results are shown in Table 9.1 together with theoretical scattering matrices of each target. The relative phases and amplitudes are in good agreement with their correct values.

The eigenvector-based decomposition is applied to the LL, LR, and RR SAR image datasets. These polarization data are converted to a coherency matrix with spatial averaging (3 x 3 multilooking). The H and $\bar{\alpha}$ results for the deployed canonical targets are displayed on the $H - \bar{\alpha}$, the 2D plane, as shown in Figure 9.6 (only target pixels are considered).

Since our canonical targets are regarded as point-target circular plate, dihedral, and wire, they should lie in the regions of low entropy surface scattering (Z9), multiple (even-bounce) scattering (Z7), and wire-like scattering (Z8), respectively [8]. The results from Figure 9.6 demonstrate that all canonical targets are within the expected zones. The $\bar{\alpha}$ of the circular plate and the dihedral show the values in consistent with theory as demonstrated in Section 2.4.2. The wire targets in three orientations exhibit similar $H - \bar{\alpha}$ distributions for all orientations, and $\bar{\alpha}$ of all wire orientations are consistent with theory (45°) either.

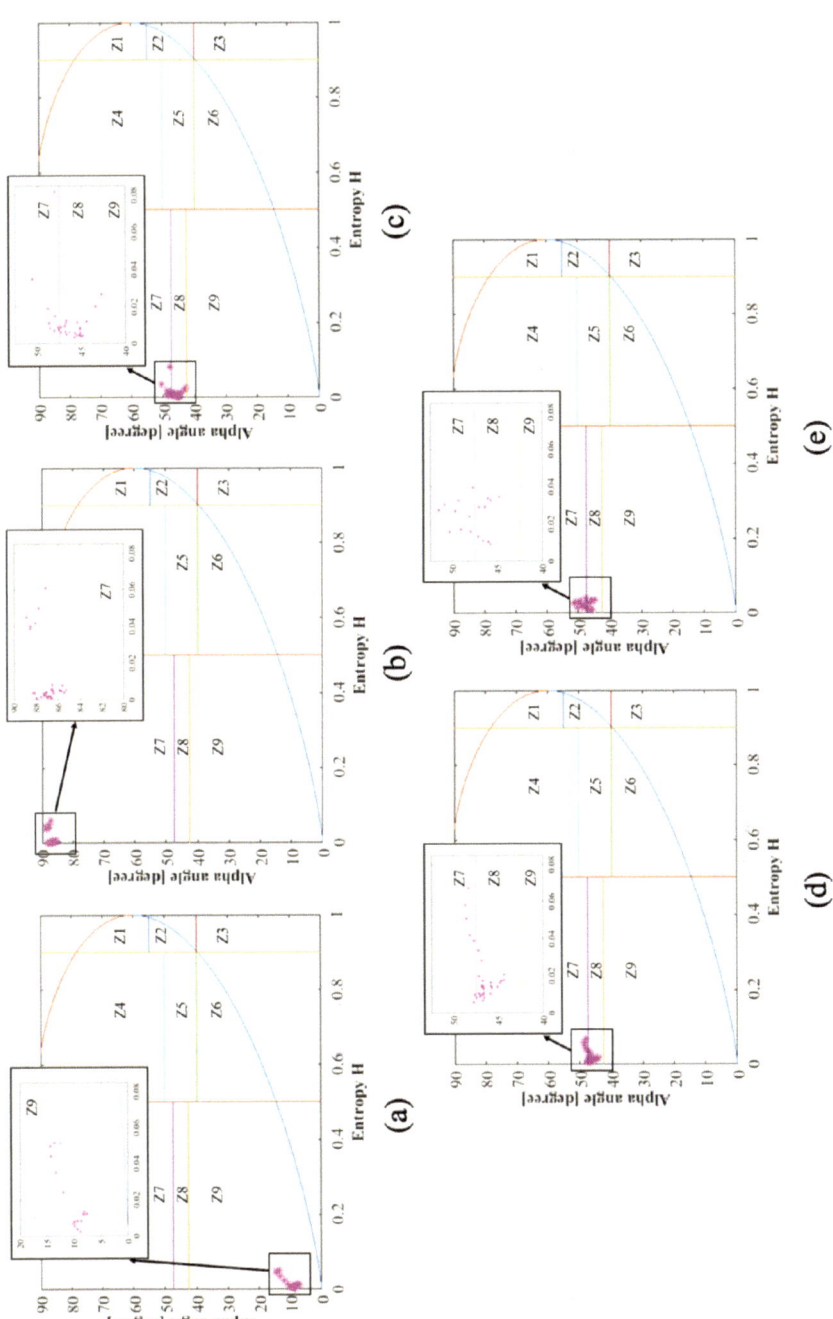

FIGURE 9.6 $H - \bar{\alpha}$ distribution for target location pixels. (a) Circular plate. (b) Horizontal dihedral. (c) Vertical wire. (d) Horizontal wire. (e) 45° tilted wire.

9.2.3 Scattering Experiment on Aircraft Model (N219)

We investigated the scattering of linear polarization and circular polarization by using the N219 aircraft model as shown in Figure 9.7 (a) by employing the circular ISAR technique in our anechoic chamber [9].

Figure 9.7 (b) shows that linear polarization depicts co-polarization (HH and VV modes) as strong scattering, but cross-polarization (HV and VH modes) has weak

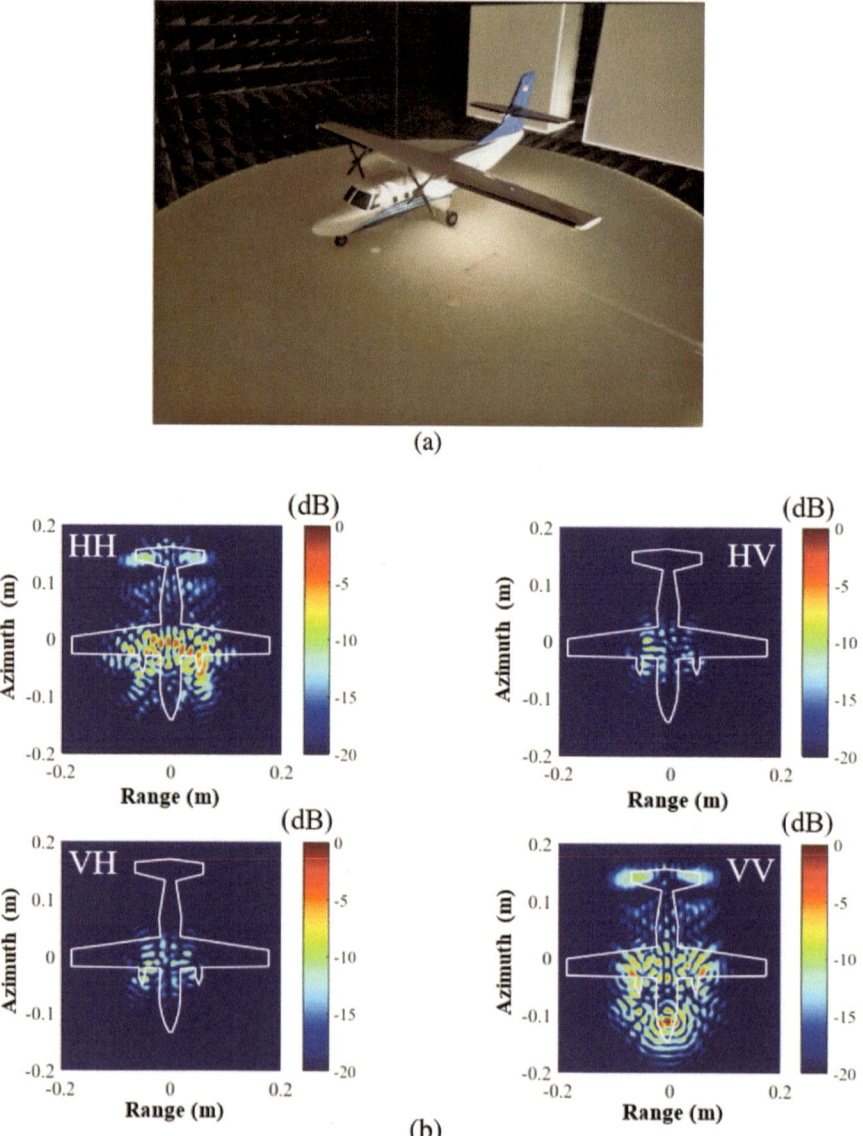

FIGURE 9.7   Circularly polarized scattering experiment. (a) Scattering target using N219 aircraft mode, (b) full polarimetric scattering in linear polarization, (c) full polarimetric scattering in circular polarization.

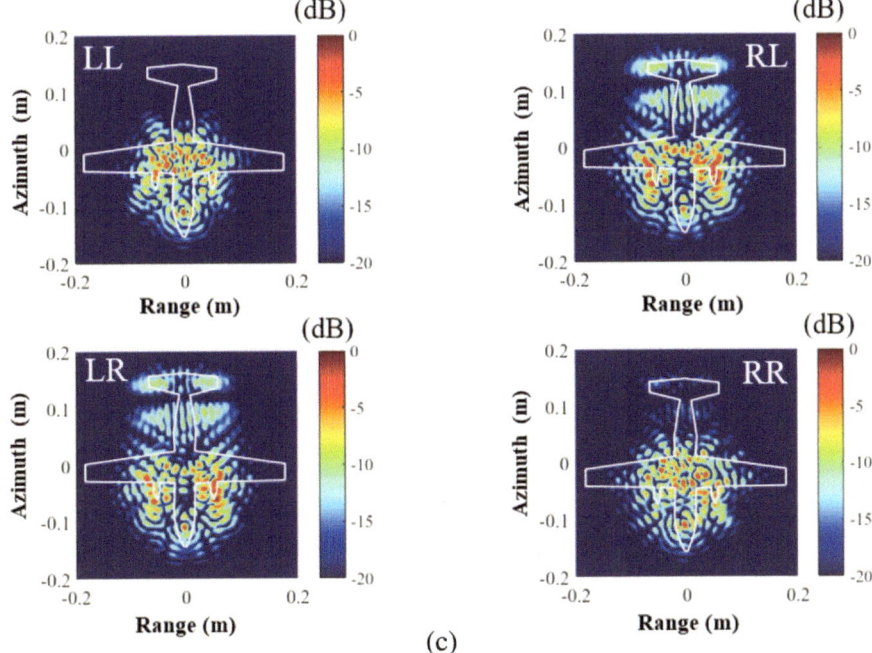

<figure>(c)</figure>

FIGURE 9.7   (Continued)

scattering. Cross-polarization shows volume scattering, and co-polarization shows surface scattering and double bounce [10]. It matches well with the full polarization theory demonstrated in Chapter 2. This figure shows that linear polarization has weak scattering by cross-polarization or volume scattering, especially scattering by the main body of the object. The co-polarization or surface scattering and double bounce show the strong intensity of scattering from the main body, main wing, nose, tail, and propeller.

Figure 9.7 (c) shows that circular polarization has strong scattering from the main body of the model on both co-polarization (LL and RR) and cross-polarization (LR and RL), but different scattering from the tail. Co-polarization shows even-scattering (double bounce) and cross polarization shows odd-scattering (surface scattering).

Figure 9.8 shows images of ellipticity angle χ and the *AR* of circularly polarized scattering of the N219 aircraft model from Figure 9.7 (c). The ellipticity angle χ (unit: radian) in Figure 9.8 (a) is calculated by using the intensity of LHCP and RHCP of scattering wave (E_L and E_R) using (Equation 2.31). Zero radian of χ means linear polarization (LP), positive value ($0 \sim 0.25\pi$) means LHCP, and negative value ($-0.25\pi \sim 0$) means RHCP. Figure 9.8 (a) shows that LHCP scattering wave scattered from tail wing and propellers. The main body scattered the RHCP wave. We can use this information to classify the structure of the targeted object as shown in this result.

Figure 9.8 (b) shows an AR image calculated using (Equation 2.32) with the input of ellipticity χ shown in Figure 9.8 (a). Zero value of *AR* means perfectly circular polarization. We define elliptical polarization for ellipticity from 0dB to 20dB, and

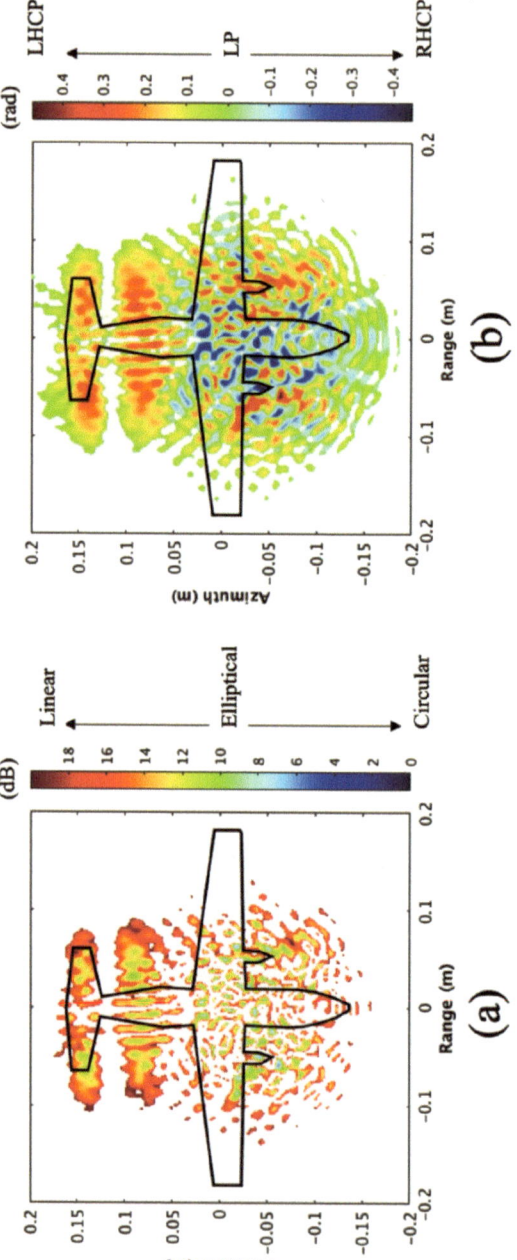

FIGURE 9.8   Axial ratio and ellipticity images of circular polarized scattering of N219 aircraft model, (a) ellipticity, (b) axial ratio.

linear polarization for ellipticity with a value of more than 20dB. This figure shows the various value of elliptical polarization of scattering waves from some parts of the model, where the main body has a low value and the tail has high scattering close to linear polarization.

The result shown in Figure 9.8 reveals that the AR and χ could be employed to classify the object characteristics, that are shape, roughness, thickness, height, and so forth, by using circularly polarized SAR. Those parameters are further investigated in Section 9.4 using airborne circularly polarized SAR.

9.3 APPLICATION: ANALYSIS OF POLARIMETRIC SAR MODES FOR RICE PHENOLOGY MONITORING

Among polarimetric SAR practices, rice monitoring has been one of the important applications as rice is a staple food for almost half the world's population. The rice phenology retrieval by remote sensing is of great importance in planning cultivation practices, yielding estimation, and water management. Various analyses of rice monitoring from polarimetric SAR data have been addressed [11]–[15]. Although the classical linearly polarized SAR is the most commonly utilized tool for this task, current studies have focused on CP SAR modes because of their advantages [11], [13].

Due to the growing interest in the use of circular polarization, we investigate, in this section, the feasibility of the FP and DCP modes (dual-pol mode with circular polarizations in Tx and Rx) of circularly polarized SAR and their performance on rice phenology monitoring. For this purpose, a ground-based circularly polarized radar system was adopted, and the time-series C-band backscattering data of Japanese rice paddy (*Oryza sativa* L.) were analyzed. The rice samples were repeatedly observed within an anechoic chamber from germination to ripening stages. To analyze the radar backscatter as a function of growth stages, different polarimetric signatures and target decomposition techniques are exploited for both FP and DCP data. In addition, $H / \overline{\alpha}$ decomposition for the DLP mode (dual-pol mode with linear polarizations in Tx and Rx) is also examined for comparison purposes.

Since CP modes have been gaining increasing attention, the next section aims to assess the performance of the DCP mode by comparing its information content with that of the FP mode in the case of rice backscattering [16]. The relevant scattering decomposition methods for both the FP and DCP modes are discussed in Section 2.5.

9.3.1 EXPERIMENTAL SCHEME

The VNA and rotational positioner controlled by the positioner controller were utilized to construct the whole circularly polarized radar system. The experimental geometry illustrated in Figure 9.9 was adopted where the circular ISAR geometry is adopted, and the rice measurements were conducted within an anechoic chamber to achieve a fully controlled environment. The incidence angle was set to 70°, and the operational frequency was adjusted to 4.5–7.5GHz (3GHz bandwidth) in C-band. An angular span of 0° to 359° was used to investigate the rice backscattering from all azimuth angles. Therefore, quasi-monostatic antennas always face the target direction.

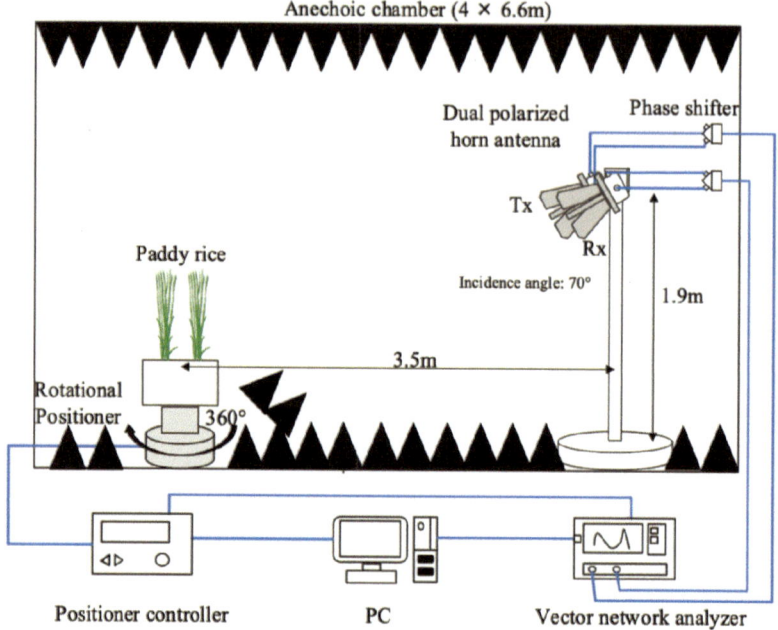

FIGURE 9.9 Experimental geometry inside an anechoic chamber.

9.3.2 PHENOLOGY DESCRIPTION OF CULTIVATED RICE

Figure 9.10 shows the photographs and the layout of the rice target used in experimental validation. A total of eight rice samples were uniformly planted in a rectangular box of dimensions 0.4×0.25×0.2 m (width×depth×height). The box was made of polystyrene to reduce undesired echoes and filled with 0.115m depth of soil and 0.125m depth of water, as shown in Figure 9.10 (b). Thus, the soil was flooded to realize the actual rice field condition. This condition was kept constant throughout the observation period from June to September to collect data that is sensitive only to the rice growth. The non-rice condition was also observed for investigation of the germination stage, where only flooded soil exists inside the box.

To describe the different rice phenological stages, we adopted *Biologische Bundesanstalt, Bundessortenamt und Chemische Industrie* (BBCH) decimal scale [17]. The BBCH scale describes the actual characteristics of an individual plant, such as the development rate of leaf, tiller, and panicle [12]. Therefore, this scale is helpful to express rice phenology but not for detailed morphological expression. BBCH values for all observed rice conditions are given in Table 9.2 together with day-of-year (DoY) and mean height as morphological values. Moreover, based on given BBCH codes, five principal stages were designated, namely, germination (stage 1), tillering (stage 2), stem elongation (stage 3), booting (stage 4), and ripening (stage 5) stages [12].

Employed rice plants were discovered to have an increasing number of tillers until stage 2. Finally, an average of 17 tillers in each stock was observed by 6 July. The increase in the number of tillers was stopped at the end of stage 2, and samples began

TABLE 9.2
BBCH Code and Phenology Stage of Each Observed Dataset

Date	DoY (Day of Year)	Mean Height (cm)	BBCH code	Phenological Stage
Soil and water	NA	NA	0	1: Germination
7 June 2016	159	19	21-29	2: Tillering
22 June 2016	174	27	21-29	2: Tillering
6 July 2016	188	34	30-39	3: Stem elongation
21 July 2016	203	42	30-39	3: Stem elongation
3 August 2016	216	49	41-49	4: Booting
22 August 2016	235	52	83-85	5: Ripening
30 August 2016	243	52	87-89	5: Ripening
14 September 2016	258	45	93	5: Ripening

to initiate panicles inside the stem during stage 3. Just before heading to stage 4, panicles went up and started to come out of the stems. Finally, we found head emergence on 6 August for this type of rice sample.

9.3.3 DATA ANALYSIS METHODOLOGY

We adopted a 2D circular ISAR data collection geometry with a fixed incidence angle. Thus, the frequency domain backscatter data of rice samples were acquired for the complete azimuth angles from 0° to 359°. Figure 9.11 shows the reconstructed images of one sample of data observed on 30 August 2016 for LL, RL, and RR polarizations. A spherical back-projection algorithm is used to process this wide-angle data [18]. From the reflectivity images in Figure 9.11, it can be clarified that all rice stocks are identified for each polarization thanks to the high-resolution capability. In practical rice monitoring applications, however, the resulting SAR images cannot usually maintain such a high-resolution feature, and a single resolution cell consists of the superposition of different scattering contributions from a few rice plants. Thus, for our situation, all eight rice plants should be confined inside a single resolution cell for reasonable analysis of rice scattering mechanisms, as Sagues et al. mentioned [19]. For this reason, our analyses are not performed on image data but on the scatterometric data obtained by ensemble averaging of the frequency domain data along with all the azimuth angles (0–359°) and frequencies (4.5–7.5GHz). As a result, the whole scattering behavior is combined into a single coherency matrix as adopted in [19]–[21], and the decomposition theories explained in Chapter 2 are applied to this matrix.

9.3.4 RICE MONITORING RESULTS AND DISCUSSION

In the following subsection, the backscattering coefficients and the polarimetric decomposition results deduced by the averaged single coherency matrix will be presented to extract the physical scattering mechanisms of rice growth.

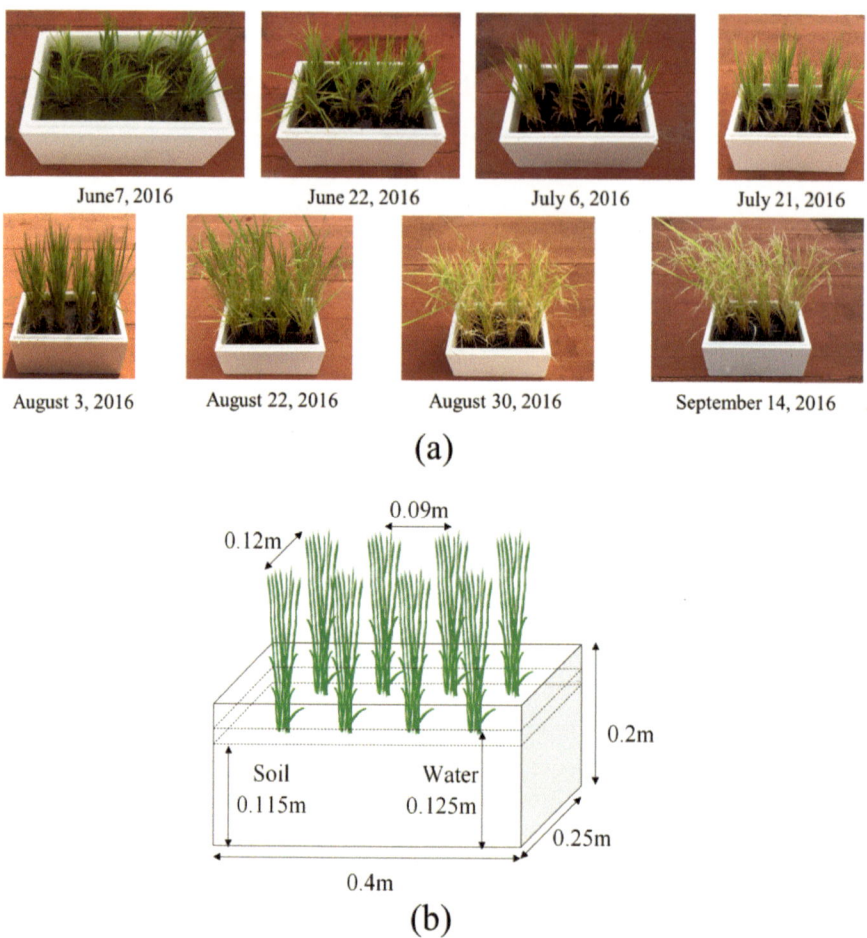

FIGURE 9.10 Photographs and layout of the rice used in experimental validation. (a) Photographs taken on each measurement date from 7 June 2016 until 14 September 2016. (b) The layout of the eight rice samples uniformly planted within a container box with 0.115m depth of soil and 0.125m depth of water.

9.3.4.1 Backscattering Coefficient

Three scattering matrix data; S_{LL}, S_{RL}, and S_{RR} were collected by assuming $S_{LR} = S_{RL}$ since the reciprocity theorem almost holds for our quasi-monostatic setup. Figure 9.12 shows the intensity of backscattering coefficients (LL, RL, and RR) as a function of DoY. The backscattering coefficients are normalized to a maximum value of the whole observation period. From Figure 9.12, we can see that the backscattering coefficients for all polarizations increase as rice plants grow until DoY 235 because these are related to the leaf area index, rice freshness, and rice height [22], where our rice plants stop to increase those heights from DoY235, as shown in Table 9.2. Before

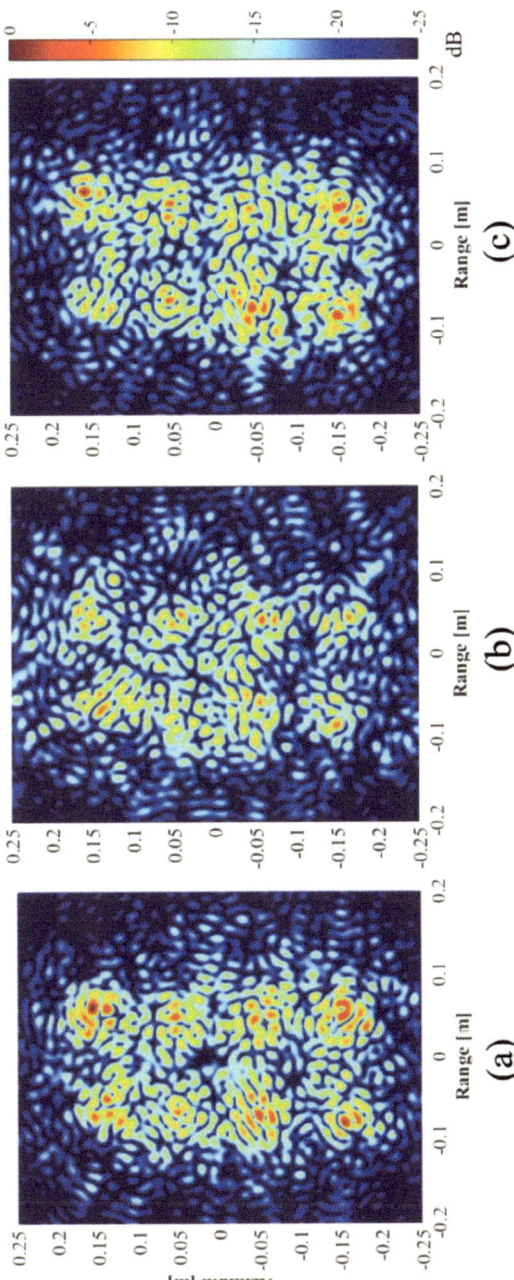

FIGURE 9.11 Reconstructed circularly polarized SAR images for the rice observed on 30 August 2016. The images are normalized to the maximum value of three images. (a) LL polarization. (b) RL polarization. (c) RR polarization.

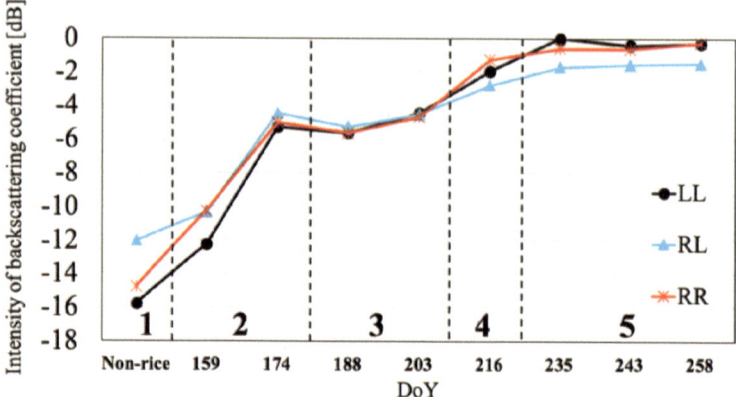

FIGURE 9.12   Backscattering coefficients of LL, RL, and RR polarization.

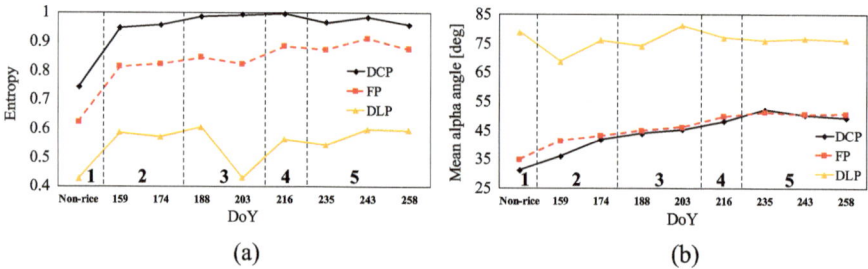

FIGURE 9.13 $H / \bar{\alpha}$ decomposition results of the FP, DCP, and DLP modes. (a) Entropy. (b) Mean alpha.

rice plants emerged from the soil (non-rice observation), the backscattering intensity exhibited very low values due to the specular reflection from the flooded ground.

Figure 9.12 shows that the surface (cross-polarization) scattering produces relatively higher intensities than the double-bounce (co-polarization) scattering until the end of stage 3 and the opposite is true for later stages. Nevertheless, the maximum difference between polarimetric channels is around 2 dB which is not significant.

9.3.4.2 $H/\bar{\alpha}$ Decomposition

The H and $\bar{\alpha}$ values for the FP, DCP, and DLP modes are plotted in Figure 9.13. Note that the reader can refer to $H/\bar{\alpha}$ decomposition methods for all modes described in Section 2.4.2, Section 2.4.3, and Section 2.5.4, respectively. In this section, $\bar{\alpha}$ for the DCP mode was treated with the condition $\bar{\alpha}' = 90° - \bar{\alpha}$ for comparison purposes as discussed in Section 2.5.4. It is important to note that the interpretation of α derived from DLP data differs from those of FP and DCP, where α of DLP accounts for a geometrical parameters of polarization ellipse shown in (Equation 2.14) and does not represent the scattering mechanism.

The entropy values shown in Figure 9.13(a) exhibit relatively lower values at stage 1, because of the non-planted soil condition, which produces highly deterministic scattering. Right after vegetation starts to grow, other scattering mechanisms such as double-bounce and volume scattering make entropy higher. Except for stage 1, entropy values fluctuate within the ranges 0.95–1 for FP, 0.8–0.9 for DCP, and 0.4–0.6 for the DLP mode. It can also be noted that the entropy values for FP and DCP represent a similar progressive trend and around 0.12 discrepancy between them. On the other hand, entropy for DLP ranges between 0.4 and 0.6, which is comparably lower than FP and DCP. Furthermore, an abrupt decrease at DoY 203 is observed in the DLP mode, which is not the case in the FP and DCP modes.

The $\bar{\alpha}$ characteristic is shown in Figure 9.13(b). FP and DCP patterns are approximately identical except for non-rice and DoY 159 observations, which differ by about ~5°. Despite this, each pattern shows an increasing trend, indicating a progression from surface scattering to dipole-like scattering. On the contrary, $\bar{\alpha}$ of DLP reveals a notably different result since it does not represent scattering mechanism information as demonstrated.

To gain a deeper understanding of the scattering mechanisms in FP and DCP data, the independent components of the $H/\bar{\alpha}$ decomposition is investigated separately. First, the appearance probabilities P_i for each scattering type defined by the associated eigenvector are calculated to interpret the entropy plots in Figure 9.13(a). Note that the decomposition of FP and DCP data produces three and two eigenvectors/probabilities, respectively. Also, the probability values are constrained by the expression, $P_1 + P_2 + P_3 = 1$ and $P_1 > P_2 > P_3$ for FP, and similarly for DCP. The variation of these probabilities as a function of DoY is displayed in Figure 9.14. The probability P_1 for DCP data shows a similar trend to FP over the whole observation period but experiences a slightly higher value (~10%) for stages 3 and 4. Figure 9.14(a) shows an almost constant and relatively large P_3 (recessive scattering) (~10%) is observed for FP data. This is the case when more than three scattering mechanism components contribute to the receiving signals. Usually, complex targets such as forested areas give rise to multiscattering. Suppose the third eigenvalue corresponds to P_3 strongly affects the receiving signal, the difference between the entropy values of FP and DCP becomes higher because entropy is formulated by summation of each independent scattering contribution which corresponds to scattering probability. As Cloude et al. pointed out in [23], CP systems typically produce higher entropy than the FP system.

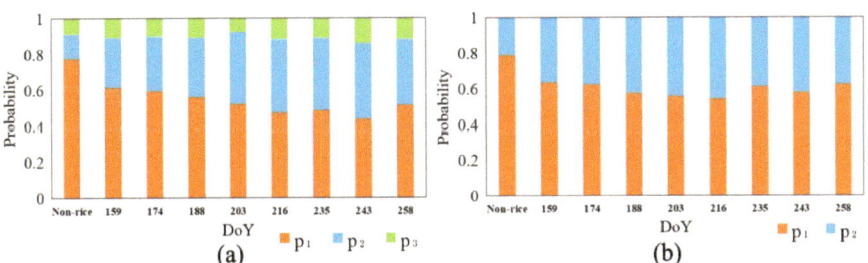

FIGURE 9.14 Scattering probabilities for the FP and DCP modes. (a) FP mode. (b) DCP mode.

Second, $\bar{\alpha}$ results given in Figure 9.13(b) are investigated in terms of their components. For each eigenvector u_i, the corresponding α_i is extracted by using $\alpha_i = \arccos\left(\left|u_{1i}\right|\right)$ where $\left|u_{1i}\right|$ is the absolute value of the first element of the eigenvector. From the results given in Figure 9.15, we notice that α_1 and α_2 for DCP indicate different trends and values from FP values. It is observed that α_1 for FP is close to 45° at stage 4 and fluctuates within the range of 40–65° at stage 5. Thus, these results confirm the necessity of statistical interpretation in terms of $\bar{\alpha}$ values and/or $H/\bar{\alpha}$ space for the discrimination of physical scattering mechanisms.

Figure 9.16 displays the $H/\bar{\alpha}$ 2D plane plots obtained from the entropy and $\bar{\alpha}$ values for FP, DLP, and DCP measurements. In the FP mode, the distributions mainly lie in zones 6, 5, and 4, indicating a vegetation-type scattering event. Roughly, four different clusters can be identified: germination stage, tillering stage, stem elongation stage, and booting plus ripening stage, and thus the booting and ripening stages cannot be separated effectively from FP data. The results for the DCP mode in Figure 9.16(b) also reveal a broadened $H/\bar{\alpha}$ patterns. Although the $H/\bar{\alpha}$ plane plots for the FP and DCP modes exhibit some different features, each mode yields satisfactory discrimination of the phenological intervals. In contrast, the $H/\bar{\alpha}$ plane plots for the DLP mode cannot achieve adequate discrimination capability for our rice targets, since almost all intervals are not resolved successfully, as seen from Figure 9.16(c).

It is also worth noting that entropy and $\bar{\alpha}$ values for crop backscattering usually increase with increasing incidence angle. For lower incidence angles, surface reflection from the soil is dominant whereas, for high incidence angles (e.g., greater than 20°), backscattering from vegetation stems and leaves makes dipole-like scattering dominant [20]. Therefore, because of our high incidence angle (70°) features, relatively higher H and $\bar{\alpha}$ values from our experiment than the usual spaceborne and airborne SAR incidence angle (common operational mode:<70°) can be expected.

9.3.4.3 Four- and Three-Component Decomposition

The four-component decomposition technique yields four elementary scattering mechanisms, that is, surface, double-bounce, volume, and helix scattering. This methodology was first proposed by Yamaguchi et al. [24], and we adopt herein the improved four-component decomposition proposed by Singh et al. [25] which fully accounts for coherency matrix coefficients for FP data. On the other hand, the DCP mode is capable of decomposing three scattering mechanisms (surface, double-bounce, and volume) with the method proposed by Cloude et al. [23]. We have demonstrated the compact decomposition theory in Section 2.5.3.

The four- and three-component decomposition results for FP and DCP modes are shown in Figure 9.17. P_s, P_d, P_v, and P_c show surface, double-bounce, volume, and helix scattering, respectively. The results are normalized to the maximum value of both modes. In Figure 9.17(a), the double-bounce and surface scattering contributions of four-component decomposition demonstrate a similar evolutionary trend as the backscattering coefficients of co-and cross-polarization in Figure 9.12. This similarity proves that co- and cross-polarization indicate even and odd bounce scattering mechanisms, respectively. Figure 9.17(a) also reveals a stronger volume scattering component than other scattering mechanisms at stages 4 and 5 and its evolutionary

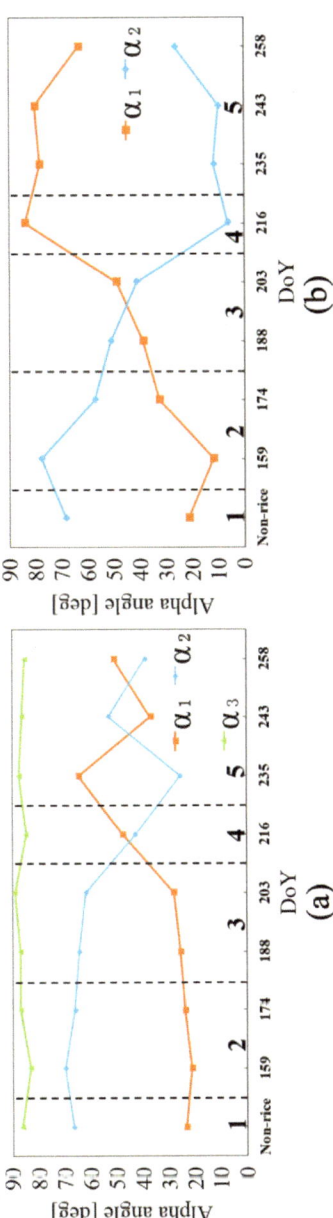

FIGURE 9.15 Alpha angle corresponding to each eigenvector. (a) FP mode. (b) DCP mode.

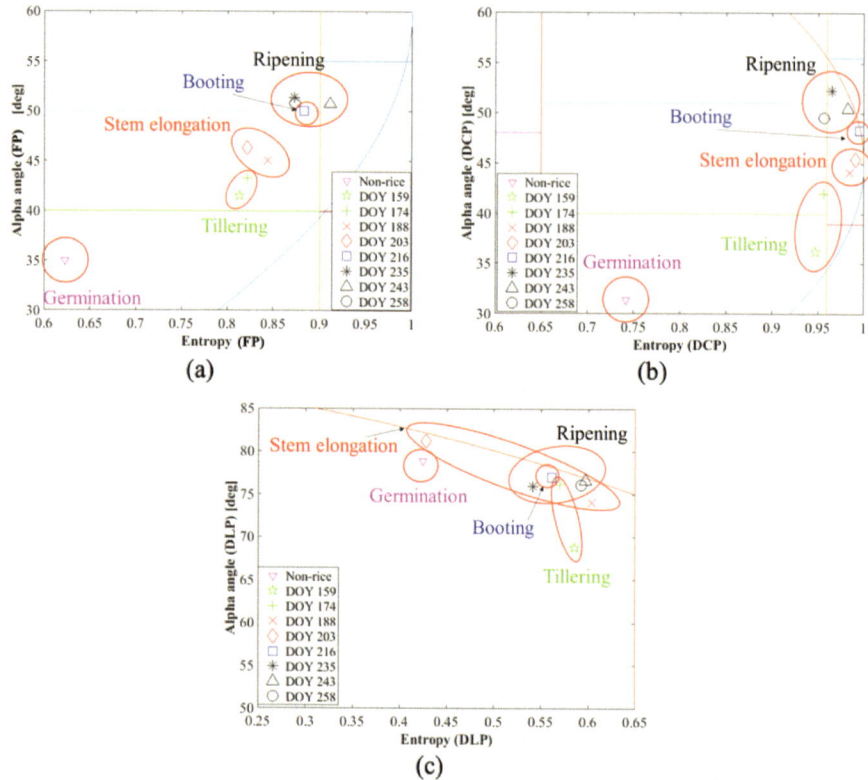

FIGURE 9.16 The $H / \bar{\alpha}$ 2D space. (a) FP mode. (b) DCP mode (c) DLP mode.

trend during rice growth. Relatively low helix scattering is also shown from rice samples.

When we compare the results of both modes, we see that both volume scattering components result in a similar evolutionary trend and value. In contrast, the surface and double-bounce scattering of the DCP mode show lower power than the FP mode.

To investigate the relative contribution of the scattering component, we present the rate of each scattering mechanism on a triangle plot, shown in Figure 9.18 for both FP and DCP modes. Note that we exclude the helix scattering contribution from the FP mode for comparison purposes between both modes. Figure 9.18(a) shows similar results to $H / \bar{\alpha}$ display, where the triangle plot for the FP mode loosely falls into four groups: germination, tillering, stem elongation, and a mix of booting and ripening stages. Therefore, the booting and ripening stages still cannot be separated. Moreover, we notice that double-bounce scattering gradually increases as rice grows (from 7% to 36%), while vice versa, the surface scattering decreases in the FP mode. This situation can also be seen in the DCP mode, where double-bounce scattering increases from 7.8% to 20%, as depicted in Figure 9.18(b). However, the DCP mode

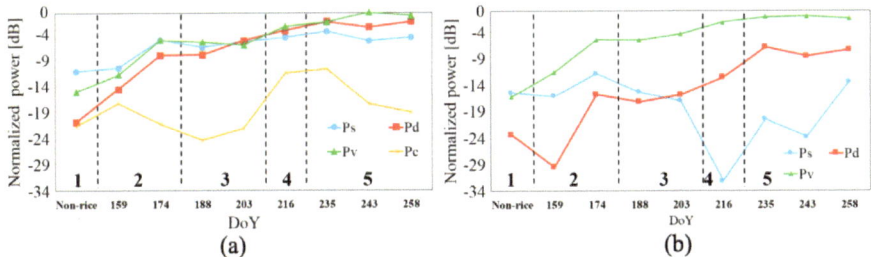

FIGURE 9.17 Four-/three-component decomposition results. (a) Four-component decomposition results for the FP mode. (b) Three-component decomposition results for the DCP mode.

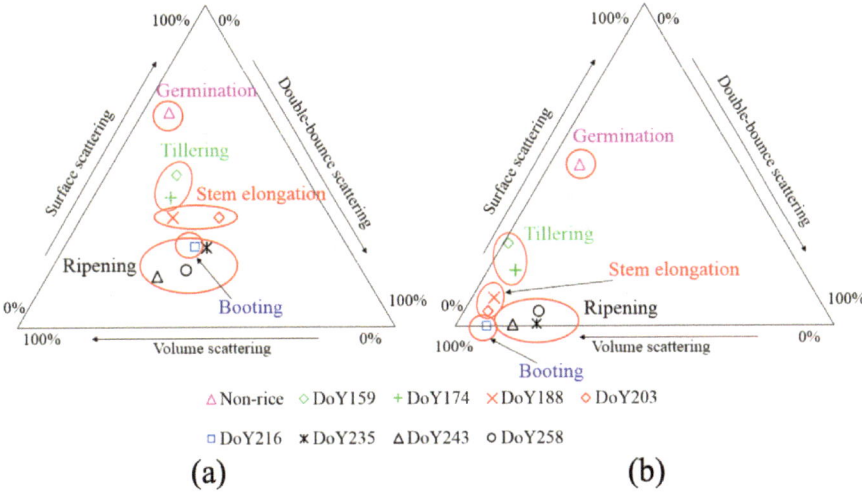

FIGURE 9.18 Relative contribution of surface, double-bounce, and volume scattering components on the triangle plot. (a) FP mode. (b) DCP mode.

shows a relatively stronger volume scattering component (lower surface and double-bounce scattering contribution) in Figure 9.18(b).

9.3.5 CONCLUSION IN RICE PHENOLOGY MONITORING STUDY

Overall, results for both the FP and DCP modes of circularly polarized SAR demonstrated adequate rice phenology classification capability. Differences in some polarimetric decomposition parameters between FP and DCP modes were noted. Moreover, DCP mode yielded a similar performance to the FP mode about $H / \bar{\alpha}$ decomposition parameters and volume scattering, and better classification performance of rice phenology compared to the DLP mode.

9.4 APPLICATION: AIRBORNE CIRCULARLY POLARIZED SAR IMAGE ANALYSIS

The flight mission in Chapter 7 has an agreement result with the theory and previous study [26], [27], hence the airborne circularly polarized SAR could be implemented for the further experiment and applications of circularly polarized SAR. Exploring the image classification method using circularly polarized SAR images, as an example, we processed images using conventional techniques of axial ratio AR, ellipticity χ, and polarization ratio ρ. The χ and AR can be derived by equations in Chapter 2 (Equation 2.31) and (Equation 2.32), respectively while polarization ratio ρ for circular polarization is given by

$$\rho = \frac{|E_R|}{|E_L|}.$$

(Equation 9.2)

For example, Figure 9.19 shows an image of axial ratio AR, ellipticity χ, and polarization ratio ρ, which the study area is shown in Figure 9.20. Here we defined the value of AR as 0 dB as circularly polarized scattering and 20dB as LP scattering. The value of AR between 0dB and 20dB shows elliptically polarized (EP) scattering. AR in Figure 9.19 shows runway and paddy field or flat area has 0dB or circularly polarized scattering, which is the same as the results illustrated in Figure 2.6, especially scattering on the low incident angle. Scattering on forest and steel roof buildings

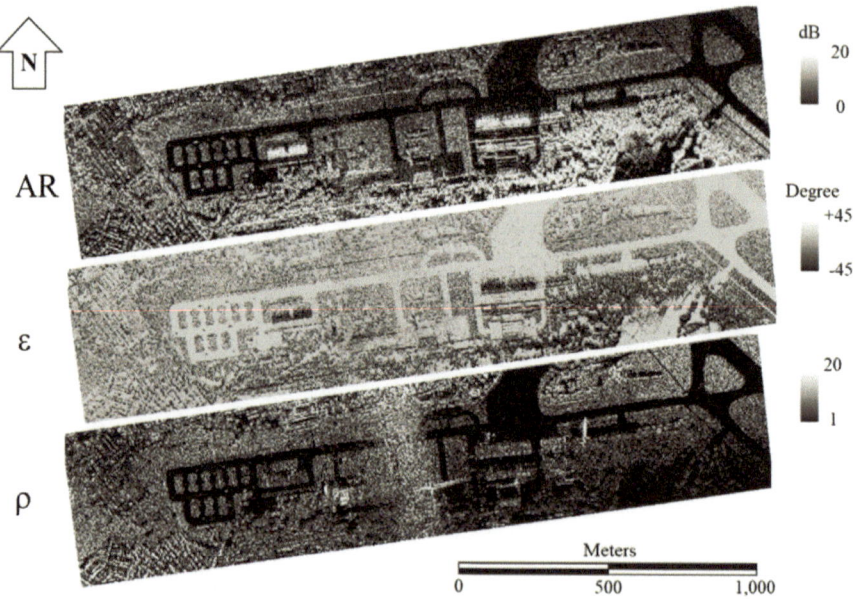

FIGURE 9.19 Circularly polarized images with various mode and its analysis (AR, χ, ρ): Hasanuddin International Airport.

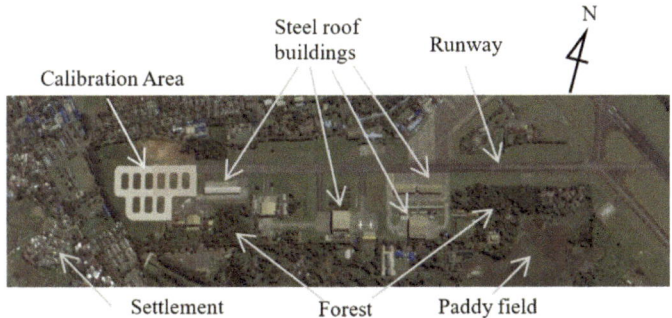

FIGURE 9.20   Study area (Source: Google Earth).

shows high *AR*, which means linear polarized scattering is dominant. The settlement area has various values of *AR*, which means elliptically polarized scattering as dominant scattering. .

The polarization ratio ρ image shows the scattering on runway and paddy field has a low value close to 1, and settlement or manmade object has a high value close to 20. Forest area has a value between 1 to 20. This result shows polarization ratio ρ image could be assessed for landcover mapping.

The ellipticity image χ shows a value between -45° and +45°. This result shows runway, paddy field, and settlement have a positive value. Then the forest or vegetated area has a negative value. The steel roof has a negative value on the surface facing circularly polarized SAR, and a positive value on the opposite surface. This image is good for urban mapping, especially to extract artificial objects.

9.5   SUMMARY

This chapter introduces several applications of airborne circularly polarized SAR by indoor circularly polarized SAR scattering experiment including the scattering experiment on canonical targets and aircraft model (N219), analysis of polarimetric SAR modes for rice phenology monitoring, and image analysis of airborne circularly polarized SAR flight mission, including the axial ratio images, ellipticity, polarization ratio, and polarimetric SAR (PolSAR) techniques. Several issues are remaining to explore, that are calibration and validation technique, circularly polarized SAR image processing, deep learning for classification using circularly polarized SAR images, and so forth.

REFERENCES

[1] C. Özdemir, *Inverse Synthetic Aperture Radar Imaging with MATLAB Algorithms.* Hoboken, NJ: John Wiley & Sons, 2012.

[2] W. Feng, L. Yi, and M. Sato, "Near range radar imaging based on block sparsity and cross-correlation fusion algorithm," *IEEE J. Sel. Top. Appl. Earth Obs. Remote Sens.*, vol. 11, no. 6, pp. 2079–2089, 2018.

[3] M. Soumekh, *Synthetic Aperture Radar Signal Processing*, vol. 7. New York: Wiley, 1999.

[4] S. S. Gao, Q. Luo, and F. Zhu, *Circularly Polarized Antennas*. West Sussex, UK: John Wiley & Sons, Ltd, 2013.

[5] W. Wiesbeck and D. Kahny, "Single reference, three target calibration and error correction for monostatic, polarimetric free space measurements," *Proc. IEEE*, vol. 79, no. 10, pp. 1551–1558, 1991.

[6] Y. Izumi, S. Demirci, M. Z. Baharuddin, M. M. Waqar, and J. T. S. Sumantyo, "The development and comparison of two polarimetric calibration techniques for ground-based circularly polarized radar system," *Prog. Electromagn. Res. B*, vol. 73, pp. 79–93, 2017.

[7] Y. Izumi, S. Demirci, M. Z. Baharuddin, J. T. S. Sumantyo, and H. Yang, "Analysis of circular polarization backscattering and target decomposition using GB-SAR," *Prog. Electromagn. Res. B*, vol. 73, pp. 17–29, 2017.

[8] S. R. Cloude and E. Pottier, "An entropy based classification scheme for land applications of polarimetric SAR," *IEEE Trans. Geosci. Remote Sens.*, vol. 35, no. 1, 1997.

[9] J. Tetuko S. S. *et al.*, "Development of circularly polarized synthetic aperture radar on-board UAV JX-1," *Int. J. Remote Sens.*, vol. 38, no. 8–10, pp. 2745–2756, May 2017.

[10] M. Migliaccio, F. Nunziata, A. Montuori, and C. E. Brown, "Marine added-value products using RADARSAT-2 fine quad-polarization," *Can. J. Remote Sens.*, vol. 37, no. 5, pp. 443–451, Oct. 2011.

[11] J. M. Lopez-Sanchez, F. Vicente-Guijalba, J. D. Ballester-Berman, and S. R. Cloude, "Polarimetric response of rice fields at c-band: analysis and phenology retrieval," *IEEE Trans. Geosci. Remote Sens.*, vol. 52, no. 5, pp. 2977–2993, 2014.

[12] J. M. Lopez-Sanchez, S. R. Cloude, and J. D. Ballester-Berman, "Rice phenology Mmonitoring by means of SAR polarimetry at x-band," *IEEE Trans. Geosci. Remote Sens.*, vol. 50, no. 7, pp. 2695–2709, 2012.

[13] Z. Yang, K. Li, L. Liu, Y. Shao, B. Brisco, and W. Li, "Rice growth monitoring using simulated compact polarimetric C band SAR," *Radio Sci.*, vol. 49, no. 12, pp. 1300–1315, 2014.

[14] N. Hayashi and M. Sato, "Measurement and analysis of paddy field by polarimetric GB-SAR," in *2009 IEEE Int. Geosci. Remote Sens. Symp.*, 2009, pp. IV-358-IV–361.

[15] K. Li, B. Brisco, S. Yun, and R. Touzi, "Polarimetric decomposition with RADARSAT-2 for rice mapping and monitoring," *Can. J. Remote Sens.*, vol. 38, no. 2, pp. 169–179, Jan. 2012.

[16] Y. Izumi, S. Demirci, M. Z. bin Baharuddin, T. Watanabe, and J. T. S. Sumantyo, "Analysis of dual- and full-circular polarimetric SAR modes for rice phenology monitoring: An experimental investigation through Gground-based measurements," *Appl. Sci.*, vol. 7, no. 4, p. 368, 2017.

[17] P. D. LANCASHIRE *et al.*, "A uniform decimal code for growth stages of crops and weeds," *Ann. Appl. Biol.*, vol. 119, no. 3, pp. 561–601, Dec. 1991.

[18] S. Demirci, E. Yigit, and C. Ozdemir, "Wide-field circular SAR imaging: An empirical assessment of layover effects," *Microw. Opt. Technol. Lett.*, vol. 57, no. 2, pp. 489–497, Feb. 2015.

[19] L. Sagues, J. M. Lopez-Sanchez, J. Fortuny, X. Fabregas, A. Broquetas, and A. J. Sieber, "Indoor experiments on polarimetric SAR interferometry," *IEEE Trans. Geosci. Remote Sens.*, vol. 38, no. 2, pp. 671–684, 2000.

[20] J. M. Lopez-Sanchez, J. Fortuny, S. R. Cloude, and A. J. Sieber, "Indoor polarimetric radar measurements on vegetation samples at l, s, c and x band," *J. Electromagn. Waves Appl.*, vol. 14, no. 2, pp. 205–231, 2000.

[21] Z. Zhou and S. Cloude, "Structural parameter estimation of Australian flora with a ground-based polarimetric radar interferometer," *2006 IEEE Int. Symp. Geosci. Remote Sens.*, 2006, pp. 71–74.

[22] Y. Inoue, E. Sakaiya, and C. Wang, "Capability of C-band backscattering coefficients from high-resolution satellite SAR sensors to assess biophysical variables in paddy rice," *Remote Sens. Environ.*, vol. 140, pp. 257–266, 2014.

[23] S. R. Cloude, D. G. Goodenough, and H. Chen, "Compact decomposition theory," *IEEE Geosci. Remote Sens. Lett.*, vol. 9, no. 1, pp. 28–32, 2012.

[24] Y. Yamaguchi, T. Moriyama, M. Ishido, and H. Yamada, "Four-component scattering model for polarimetric SAR image decomposition," *IEEE Trans. Geosci. Remote Sens.*, vol. 43, no. 8, pp. 1699–1706, 2005.

[25] G. Singh, Y. Yamaguchi, and S. Park, "General four-component scattering power decomposition with unitary transformation of coherency matrix," *IEEE Trans. Geosci. Remote Sens.*, vol. 51, no. 5, pp. 3014–3022, 2013.

[26] J. T. S. Sumantyo *et al.*, "Airborne circularly polarized synthetic aperture radar," *IEEE J. Sel. Top. Appl. Earth Obs. Remote Sens.*, vol. 14, pp. 1676–1692, 2021.

[27] M. Y. Chua *et al.*, "The maiden flight of Hinotori-C: The first C band full polarimetric circularly polarized synthetic aperture radar in the world," *IEEE Aerosp. Electron. Syst. Mag.*, vol. 34, no. 2, pp. 24–35, 2019.

10 Summary

Josaphat Tetuko Sri Sumantyo

10.1 SUMMARY

Since the synthetic aperture radar (SAR) was invented in 1951 by Carl Atwood Wiley, various segments of the SAR have continuously improved the performance for various polarization, frequency, platforms, and applications for civilian and military purposes. Many countries and institutions already have developed airborne SAR sensors with multi-polarization for various purposes. Almost of conventional systems were designed and operated in linear and compact SAR and no pure circularly polarized SAR was introduced before. This book discussed and focused on the airborne circularly polarized SAR which introduces concise knowledge on the development of airborne circularly polarized SAR, which is composed of the theory of circularly polarized waves, system design, hardware development, image signal processing, calibration, validation, ground test, and flight mission, and applications, based on the experience authors since 2002 to develop the airborne circularly polarized SAR system.

This book discussed in detail related to the history of airborne SAR in each country in the world, reviewing the basic knowledge of SAR Polarimetry by explaining the theory of the generation, propagation, and scattering of the circularly polarized wave and polarimetry related to the system design, scattering analysis, polarimetric target decomposition, and several polarimetric observation modes. The system design of the airborne circularly polarized SAR related to the determination of parameters on airborne circularly polarized SAR (i.e., power, altitude, off-nadir angle, beamwidth, resolution, etc.) is discussed to determine the parameters of the radiofrequency (RF) system related to the transmitter, receiver, and controller. The baseband and control unit is explained in the building of the chirp generator, modulator and demodulator, solid-state power amplifier (SSPA), low noise amplifier, ADC/DAC, and control unit to generate local oscillator frequency and timing controller including the installation of the circularly polarized SAR system in the CN235MPA aircraft to fulfill the requirement of system design, improvement of signal-to-noise ratio (SNR), collecting of full polarimetric raw data, GPS, and inertial measurement unit (IMU).

The other important part is the design and manufacturing of the circularly polarized antenna including the target, objective, specification, and characteristics to satisfy the requirement of system design. Then the image processing of SAR is explained by

DOI: 10.1201/9781003282693-10

explaining the code of circularly polarized synthetic aperture radar with a Range-Doppler algorithm (RDA) to process the signal (raw data) to obtain single-look complex (SLC) images for further application. The reader could use the MATLAB codes provided in this book to learn image processing and study further circularly polarized waves using several acquired images from in-flight missions in South Celebes, Indonesia, in March 2021. Finally, this book also introduces some applications with several examples of indoor polarimetric SAR experiments and airborne circularly polarized SAR analysis using axial ratio images, ellipticity, polarization ratio, and polarimetric SAR (PolSAR) techniques. This book also explains the experimental result of the scattering of circularly polarized waves on CN219 aircraft models and a rice paddy to compare to conventional linear polarized SAR images.

10.2   FUTURE RESEARCH

Based on the experience of the authors [1]–[5], the study and development of circularly polarized SAR remained several issues in theory, hardware, software, validation, calibration, observation mode, image processing, and applications. First, the theory issue needs further study on polarimetric SAR (PolSAR) based on circularly polarized waves, including circularly polarized wave generation, propagation, scattering, probability, statistics, and so forth. Hardware development needs more efficient, low power, lightweight, thermal management or control, broadband antenna, RF component, and so forth. Software needs the development of fast or near real-time image processing, including a more efficient control unit for the transmitter and receiver to combine with the image processor. Calibration and validation need the development of a specific corner reflector (CR) for broadband circularly polarized scattering waves, including the passive and active CR development. Observation mode using circularly polarized SAR needs to explore further including circular observation mode, geostationary SAR, compact SAR, and so forth. Image processing needs further exploring using the specific characteristics of circularly polarized SAR, including compact SAR. Finally, the authors hope that the circularly polarized SAR applications will be increasingly explored and utilized by many students, researchers, institutions, and companies by utilizing another trending method, that is, deep learning. The platform issue for the circularly polarized SAR is the integration technique to install this sensor to the drone or unmanned aerial vehicle, high altitude platform system or stratosphere drone, aircraft, microsatellite, and satellite.

REFERENCES

[1] Josaphat Tetuko Sri Sumantyo, Koo Voon Chet, Lim Tien Sze, Takafumi Kawai, Takuji Ebinuma, Yuta Izumi, Mohd Zafri Baharuddin, Steven Gao and Koichi Ito, "Development of Circularly Polarized Synthetic Aperture Radar Onboard UAV JX-1," *International Journal of Remote Sensing, Special Issue Papers on Drones, UAVs, RPASs for Environmental Research*, Vol.38, No.8–10, pp.2745–2756, Online: December 2016, Printed: July 2017, DOI:10.1080/01431161.2016.1275057.

[2] Josaphat Tetuko Sri Sumantyo, Chua Ming Yam, Cahya Edi Santosa, Good Fried Panggabean, Tomoro Watanabe, Bambang Setiadi, Franciscus Dwi Sri Sumantyo,

Kengo Tsushima, Karna Sasmita, Agus Mardiyanto, Edi Supartono, Eko Tjipto Rahardjo, Gunawan Wibisono, Muhammad Aris Marfai, Retnadi Heru Jatmiko, Sudaryatno, Taufik Hery Purwanto, Barandi Sapta Widartono, Muhammad Kamal, Daniel Perissin, Steven Gao, and Koichi Ito, "Airborne Circularly Polarized Synthetic Aperture Radar," *IEEE Selected Topics in Applied Earth Observations and Remote Sensing (JSTARS)*, Vol.14, pp.1676–1692, January 2021, DOI:10.1109/JSTARS.2020.3045032.

[3] Ming Yam Chua, Josaphat Tetuko Sri Sumantyo, Cahya Edi Santosa, Good Fried Panggabean, Franciskus D. Sri Sumantyo, Tomoro Watanabe, Ya Qi Ji, Peberlin Parulian Sitompul, Mohammad Nasucha, Farohaji Kurniawan, Babag Purbantoro, Asif Awaludin, Karna Sasmita, Eko Tjipto Rahardjo, Gunawan Wibisono, Retnadi H. Jatmiko, Sudaryatno, Taufik H. Purwanto, Barandi S. Widartono, and Muhammad Kamal, "The Maiden Flight of Hinotori-C: The First C Band Full Polarimetric Circularly Polarized Synthetic Aperture Radar in the World," *IEEE Aerospace and Electronic Systems Magazine*, Vol.34, No.2, pp.24–35, February 2019, DOI:10.1109/MAES.2019.180120.

[4] Katia Nagamine Urata, Josaphat Tetuko Sri Sumantyo, Cahya Edi Santosa, and Tor Viscor, "Development of an L-Band SAR Microsatellite Antenna for Earth Observation," *MDPI Aerospace*, Vol.5, No.4, 128, December 2018, DOI:10.3390/aerospace5040128

[5] Katia Nagamine Urata, Josaphat Tetuko Sri Sumantyo, Cahya Edi Santosa, and Tor Viscor, "A Compact C-band SAR Microsatellite Antenna for Earth Observation," *Acta Astronautica*, Vol.159, pp.517–526, June 2019 (Elsevier), DOI:10.1016/j.actaastro.2019.01.030

Index

Note: Page numbers in **bold** refer to tables; those in *italics* refer to figures.